Industrial Chocolate Manufacture and Use

Industrial Chocolate Manufacture and Use

edited by

S.T. BECKETT
Chocolate Research Manager
Group Products Research and Development
Rowntree plc, UK

Blackie

Glasgow and London

Published in the USA by
AVI, an imprint of
Van Nostrand Reinhold Company
New York

Blackie & Son Limited,
Bishopbriggs, Glasgow G64 2NZ
7 Leicester Place
London WC2H 7BP

Published in the USA and Canada by
AVI, an imprint of
Van Nostrand Reinhold Company Inc.
115 Fifth Avenue
New York, New York 10003

Distributed in Canada by
Macmillan of Canada
Division of Canada Publishing Corporation
164 Commander Boulevard
Agincourt, Ontario M1S 3C7

British Library Cataloguing in Publication Data

Industrial Chocolate Manufacture and use.
1. Chocolate industry
I. Beckett, S.T.
664'.153   TP640

ISBN 0-216-92258-5

Library of Congress Cataloging-in-Publication Data

Industrial chocolate manufacture and use.

Includes bibliographies and index.
1. Chocolate.   2. Cocoa.   I. Beckett, S.T.
TP640.I53   1987   664'.153   87-20865
ISBN 0-442-20494-9 (AVI)

Phototypeset at Thomson Press (India) Limited, New Delhi
Printed in Great Britain by Bell & Bain (Glasgow) Ltd

Preface

The manufacture of chocolate until recently has tended to be a craft industry, relying heavily on the skills of its workforce. With the development of large, complex machinery, higher manufacturing throughput and the need for increased automation, it has become necessary to understand more fully the principles behind the processes involved.

There is no such thing as the best chocolate, whatever individual manufacturing companies may say. There are, however, preferred chocolates for different products and locations. It is strange, but true, that a chocolate bar with high sales, say, in North America can be regarded as unpalatable in Europe.

This book provides an up-to-date scientific and technical approach to the principles of chocolate manufacture, from the growing of cocoa beans to the packaging and marketing of the end product. Individual topics are covered in depth by internationally acknowledged experts from both the manufacturing industry and academic research. In addition to traditional chocolate manufacturing processes, many new and unconventional techniques are included. Separate marketing and packaging chapters reflect the importance of these subjects to the industry. The book is written especially for food technologists, food scientists and process equipment manufacturers involved in the confectionery and baking industry.

I would like to thank the authors who have contributed to the book. I appreciate the care taken and the time spent over what it is hoped the reader will find to be a varied and instructive group of chapters. I also wish to thank my company, Rowntree plc, for their help and support, my family for putting up with me talking about and working on little else for the past year, and Blackie for giving me the opportunity of carrying out an interesting, if somewhat demanding, job.

The following organizations are thanked for permission to reproduce illustrative material and for the use of registered trade names: Aasted L. International; Alfa Laval Inc.; BSS Group plc; Baker Perkins BCS Ltd.; Maschinenfabrik G.W. Barth GmbH & Co.; Gebrüder Bauermeister & Co.; Brabender Messtechnik KG; Buhler Brothers Ltd.; Buss-Luwa AB; Cadbury-Schweppes plc; Carle & Montanari SpA; Ferranti plc; Richard Frisse GmbH; Gainsborough Engineering Co.; J.W. Greer Co.; Otto Hansel GmbH; Kreuter GmbH; F.B. Lehmann Maschinenfabrik GmbH; Lesme Ltd.; Lindt Sprüngli Ltd.; Low & Duff (Developments) Ltd.; Mars

Confectionery Ltd.; Micronizing Co. (UK) Ltd.; Nagema VEB Kombinat; Maschinenfabrik Petzholdt GmbH; Wilhelm Rasch & Co. GmbH; Rowntree plc; Sollich GmbH & Co. K.G; Thouet KG Maschinenbau; Werner & Pfleiderer Maschinenfabrik; George D. Woody Associates.

S T B

Contents

Contributors

Stephen T. Beckett, BSc (Durham), D.Phil. (York, UK) in physics, worked for 8 years on asbestosis. Author of over 20 publications on airborne asbestos monitoring, he is co-originator of the 'Walton–Beckett' graticule. The original interest in particle size distribution monitoring and manipulation gained here has continued into the confectionery industry, where for the past 8 years he has been involved in research, particularly into chocolate processing. Presented paper on cocoa mass at the Solingen (West Germany) Interpack Seminar, 1984.

B.L. Hancock qualified in 1941 with a first class honours degree in Botany. He was involved in research into cocoa growing in Trinidad and Ghana before joining Rowntree's Research Department in 1945. Here he was Divisional Research Manager concerned with all aspects of usage of cocoa beans when he retired in 1981.

Christof Krüger studied chemistry and sugar technology at the Braunschweig Institute of Technology. After his graduation as 'Diplom-Chemiker' (M.Sc.), he worked in German sugar companies and was the first applications manager in the German sugar industry. Concurrently with this function he was also senior manager of a company producing caramel colours and sugar syrups and was involved in the commercial and technical planning of a new liquid sugar plant. For seven years, he was chief chemist at Rowntree Mackintosh, Hamburg, where his responsibilities included management of the laboratories, quality control, product development and the sensorics department. He also worked actively on the committee for food law and on the scientific committee of the Association of the German Confectionery and Chocolate Industry, by which association he was commissioned to serve as research representative on the confectionery section.

Since 1986, Christof Krüger has been technical applications manager at Finnsugar Xyrofin, Hamburg. In this capacity, he advises customers in the food industry, and especially in the confectionery and chocolate industry, in the use of various bulk sweeteners. He frequently presents papers and acts as moderator at international symposia at the Central College of the German Confectionery Trade in Solingen.

Professor E.H. Reimerdes has a Ph.D. in Pharmaceutic Food Sciences and is currently the head of the Department of Food Chemistry and Technology at the Bergische University, Wuppertal, and Dean of the Faculty of Chemistry and Biology. He is also a corresponding member of the Académie Nationale de Pharmacie and a member of many national and international organizations. His main research interests are protein chemistry, functional properties of food proteins, protein modifications, enzymology related to stereospecific organic synthesis and immobilized enzyme systems.

Dr Hans-A. Mehrens has a Ph.D. in human nutrition and food science and is a member of the faculty in the Department of Food Chemistry and Technology at the Bergische University, Wuppertal. His main research interests are functional properties of food proteins, protein modification and nutritional aspects of protein utilization.

Dr J Kleinert has for many years been one of the foremost figures in European chocolate making. Following military service he obtained a degree in food engineering followed by a Dr. sc. techn. at the Swiss Federal University, ETHZ, in Zurich. He started to work for Lindt & Sprüngli Ltd in 1951, and was their laboratory manager for R&D until his retirement in February 1987. In 1970 he was appointed a Vice-Director of the company. A prolific writer, he has over 66 published papers on chocolate-related subjects as well as five published patents. He has been a frequent lecturer at the Fraunhofer-Institut in Munich, the German Confectionery School in Solingen and the ETHZ in Zurich. In North America he gave presentations at the PMCA Conferences in 1961, 1966 and 1971 and at the University of Guelph in 1974. As well as acting as a consultant on numerous occasions, he was a member of the Commission of Experts of Office International da Cacao et du Chocolat, President of the Commission of Chemists of Chocosuisse, President of the Research Group of the Food Branch Association, and a member of the Swiss Federal Commission to Promote the Scientific and Applied Research (Swiss Nationalfonds) and the commission to revise and update the Swiss *Food and Drug Handbook*.

Dr E.-A. Niediek is one of Europe's best-known authorities on chocolate manufacture. After studying machine building and chemical engineering he worked under Professor Rumpf at the University of Karlsruhe. His thesis in 1968 was concerned with the improvement of chocolate processing. He was a member of the Food Technology Department of the University of Karlsruhe from 1964 to 1985. Since 1983 he has been a lecturer in Food Technology at the Technological University of Hamburg-Harburg and in 1985 was appointed Director of the Deutsches Institut für Lebensmitteltechnik, a new institute carrying out research for the food and confectionery industries.

Professor Paul Dimick has had over 140 technical papers published, produced over 24 years of research. His areas of interest include cocoa butter crystallization, chocolate flavour chemistry and chocolate processing technology. He has advised 12 students in the area of chocolate and confectionery that have been supported by industry. Since 1975 he has had 12 papers presented at PMCA.

Dr Jonathan Hoskin has co-authored with Professor Dimick a review paper on non-enzymatic browning during the processing of chocolate. He is currently an assistant professor at Clemson University where his main interests include chocolate flavour chemistry, flavour chemistry of dairy products, dairy processing technology and food and light interaction. He has over 15 technical publications and has been doing research for four years since obtaining his Ph.D. degree.

Dieter Ley, Dipl.-Ing., worked for the chocolate and confectionery firm of Messrs Sprengel, Hannover, from 1964–79, initially as manager of technical planning of the chocolate factory and eventually as General Technical Manager. Since 1979 he has been Technical Director of Messrs Frisse, Herford. He represented them at the Pennsylvania Manufacturing Confectioners' Association and at the study group for chocolate at the Fraunhofer Institute for Food Technology and Packing in Munich. A frequent lecturer, his venues have included Solingen (West Germany), PMCA, Japan and China. In addition he is a member of the European Candy Kettle Club, and on the working committees for machines and plants in the confectionery industry and the VDI Commission Air Purity Preservation and Emission Reduction in the chocolate industry.

Dr J. Chevalley studied at Lille where she obtained a degree in chemical engineering. Having obtained a doctorate in physical chemistry in Paris, she joined the Nestlé Company in 1963. Her work has been primarily concerned with the rheology of chocolate, and she published a review of this topic in *Journal of Textruder Studies* in 1975.

Dr Lars Hernqvist is from the Food Technology Section of the Chemical Centre of the University of Lund, where he has worked with Professor K. Larsson. In 1984 he published a thesis at the University of Lund on the polymorphism of fats.

Roy Nelson had had 35 years in the confectionery industry, starting as a confectionery machinery designer, then with Baker Perkins for nine years as a design engineer. He presented a paper to the 21st PNCA Production Conference in 1967 on Tempering and Enrobing. He joined John Mackintosh & Son Ltd in 1968 as Research Manager, and after six years moved to Tourell for $1\frac{1}{2}$ years developing raw ideas in gas cooking and systems designs. He has been with RMCL for 12 years as chief designer and CAD manager, using computer and state-of-the-art developments.

Dr G.G. Jewell's earlier work on chocolate resulted from his time at the Leatherhead Food Research Institute where he used techniques such as electron microscopy and X-ray diffraction to study the structure of chocolate with particular emphasis on bloom. He then spent 10 years working in the corporate research laboratory of Cadbury-Schweppes, and during this period contributed invited papers to technical meetings on the role of fat in chocolate in arenas such as the American Oil Chemists and the PMCA. He has also lectured extensively on wider aspects of food structure chemistry, including chocolate and vegetable fats in these topics. He is currently Director of Research & Development/Quality Assurance for Quaker Oats Ltd.

Ken Jackson has worked for Rowntree plc for over 30 years. The first eight years were spent in the Product Development Department at York (UK), and during this period he obtained his City and Guilds qualification in Chocolate and Sugar Confectionery. He transferred to Rowntree Mackintosh Canada in 1963, and has worked in product development, research, production, technical services, trouble shooting and packaging development. He is currently Technical Services Manager located at the Laura Secord plant. Whilst in Canada he helped organize, and then taught for 17 years in, the only night school course in industrial candy-making in North America. Half of the 20–26 week course was devoted to chocolate manufacture, and involved visits to most major confectionery companies in the Toronto area, as well as much practical work.

Ian McFarlane (MA, Ph.D. in applied physics) has twenty years' experience of process instrumentation. From 1972 to 1982 he worked for United Biscuits, where he was Manager of Process Control Development. He is at present Director of the Beaconsfield Instrument Company, which he founded on leaving United Biscuits. He is author of *Automatic Control of Food Manufacturing Processes*, published by Applied Science Publishers (1983).

Norman Ferguson trained as a chemist and **Vince Martin** as an economist and their chapter draws upon over 50 years of packaging experience from two contrasted yet complementary viewpoints. Both authors have been involved in the development of award-winning packs and have been members of various industry committees, judging panels, etc.

Colin Nuttall, by profession a statistician, worked for Mars Confectionery from 1945 to 1982, retiring as Corporate Planning Manager and Secretary to the UK Board of Management. He was Chairman of the Statistics Committee of the Cocoa, Chocolate and Confectionery Alliance, and for ten years the first President of the Joint International Statistics Committee of the ISCMA and the International Office of Cocoa and Chocolate, which awarded him their Gold Medal in 1982. He was a founder member of the Henley Centre for Forecasting, and remains on the Management Council.

1 Traditional chocolate making

S.T. BECKETT

1.1 History

Cacao trees were cultivated by the Aztecs of Mexico long before the arrival of the Europeans. The beans were prized both for their use as a currency and for the production of a spiced drink called 'chocolatl'. The Aztec Emperor Montezeuma is said to have drunk 50 jars or pitchers per day of this beverage, which was considered to have aphrodisiac properties, a belief still held as late as 1712, when *The Spectator* advised its readers to be careful how they meddled with 'romances, chocolate, novels and the like inflamers ···'. The chocolate was prepared by roasting the cocoa beans in earthenware pots, before grinding them between stones. The mixture was added to cold water, often with other ingredients such as spice or honey, and whipped to a frothy consistency (1).

The first cocoa beans were brought to Europe by Columbus as a curiosity, but were later exploited commercially by Don Cortes as a new drink (2). The Spaniards preferred their drink sweetened, and in this form its popularity spread to Central and Northern Europe. In 1664 it was mentioned in England in Pepys' *Diary*, but was essentially still restricted to the wealthy. The introduction of milk into this chocolate drink was first recorded in the UK in 1727, by Nicholas Sanders (3), although his reasons for doing so are uncertain.

A mixture of the ground cocoa beans and sugar would not by itself produce the solid chocolate so familiar to the modern consumer. Instead it would give a very hard substance which would not be pleasant in the mouth. In order to enable it to melt easily it is necessary to add extra fat. This can be obtained by pressing the cocoa beans and removing some of the fat content, known as cocoa butter. The ability to extract this fat was developed in 1828 by Van Houten of Holland, and had a double advantage: the expressed fat was used to make the solid chocolate bars, while the remaining lower-fat cocoa powder could still be incorporated into a drink. This 'drinking chocolate' was in fact usually preferred as it was less rich than the original high fat mixture.

The solid form of milk chocolate is normally attributed to Daniel Peters of Vevey, Geneva, in 1876. In Switzerland, water-powered machines were able to operate for long periods at an economic rate. This enabled the extra water from the milk to be driven out of the chocolate without incurring a large extra cost. Chocolates with moisture contents of above about 2% are normally unacceptable as they have poor keeping qualities, as well as a poor texture.

1

Over the years many different flavours of both milk and plain (dark) chocolate have been developed. Sometimes there has been a definite policy to develop a 'house' flavour within a company, e.g. in Cadbury's Dairy Milk, or the Hershey Bar. At other times the flavour is adjusted to complement the centre of the sweet which is to be coated with chocolate. A very sweet centre such as a fondant may be best complemented by a relatively bitter chocolate and vice versa. For milk chocolate, one of the biggest flavour differences is between the chocolates made from milk powder which are predominantly found in Continental Europe, and the 'milk crumb' ones of the UK and parts of America. Milk crumb (see Chapter 4) is obtained by dehydrating milk sugar and cocoa mass. This was developed where milk production was very seasonal. As cocoa is a natural antioxidant, it was possible to improve the keeping properties of the dehydrated form of milk over extended periods without refrigeration. The drying process also produced a distinct cooked flavour, not normally present when the milk is dried separately.

1.2 Outline of process

Chocolate has two major distinguishing characteristics: its flavour and its texture. Although many different flavours of chocolate exist, all must be free from objectionable tastes, and yet incorporate at least some of the pleasant ones, which the consumer will associate with the product. A primary feature of the texture is that it must be solid at a normal room temperature of 20–25 °C (70–75 °F) and yet melt rapidly in the mouth at 37 °C (98.5 °F) giving a liquid which appears smooth to the tongue. The processing of chocolate is related to obtaining these two criteria, and is therefore devoted either to developing the flavour of the product—using a raw bean would produce a very unpleasant taste—or treating it so that it will flow properly and be free from large gritty material.

Although many different methods of chocolate-making exist, most traditional ones are based on the process outlined in Figure 1.1 and briefly described below. Further details are given in the relevant chapters of the book.

1.2.1 *Preparation of cocoa nib—flavour development*

The cocoa tree produces pods containing a pulp and the raw beans. The outer pod is removed together with some of the pulp and the beans are fermented. This enables chemical compounds to develop inside the beans, which are the precursors of the flavour in the final chocolate. Failure to carry out this stage properly cannot be rectified by processing at a later date. This is also true of the subsequent stage when the fermented beans are dried. Poor control at this stage can give rise to moulds, which give a very unpleasant-flavoured product even if the fermentation has been correct. In addition, correct transport conditions are required when the beans are moved from the country of growing to that of chocolate manufacture.

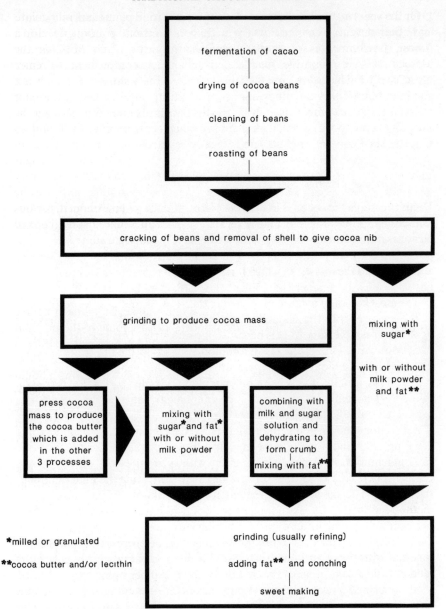

Figure 1.1 Schematic diagram of traditional chocolate-making process.

On arrival it is then necessary to clean the beans to remove metal and stones and other extraneous material which might contaminate the product. Further flavour development is subsequently obtained by roasting the beans. This also loosens the shell round the outside of the bean, and enables them to break more easily.* The beans are then broken, and the relatively lighter shell particles removed by a winnowing action. The presence of shell in the final chocolate is undesirable as it will impair the flavour, as well as causing excessive wear to the subsequent grinding machine. It should also be noted that the shell content of chocolate is legally restricted in some markets.

1.2.2 Grinding—particle size reduction

Up to this stage the cocoa is in discrete pieces, several millimetres in diameter. Subsequent processing may take several forms, but all require the solid cocoa particles, sugar and any milk solids to be broken so that they are small enough not to be detected on the tongue. The actual size depends upon the type of chocolate and the market in which it is sold, but in general the vast majority of particles must be smaller than $40 \mu m$ (1.5×10^{-3} inch).

The most common method of achieving this is by the use of a five-roll refiner. In order to enable the chocolate ingredients to pass through the refiner, however, it is necessary to get them into a paste form. This may be done in a variety of ways. One of the most common is to grind the nib to form cocoa mass, which is a liquid at temperatures above the melting point of cocoa butter. This usually involves hammer mills, disc mills, ball mills, three-roll refiners or a combination of the four. The sugar can then be added in a granulated or milled form, and the two mixed with extra fat (and milk powder if milk chocolate is being manufactured). The mixing may include some grinding, and traditionally a melangeur pan was employed for the purpose. This machine has a rotating pan, often with a granite bed, on which two granite rollers rotate. Scrapers ensure mixing by directing the material under the rollers. The modern requirement for a continuous higher throughput method has lead to the development of devices like the two-roll refiner which perform an equivalent amount of crushing. An alternative approach, which avoids producing mass, is to use the melangeur type process to grind the nib, together with the other ingredients (2). This method of making chocolate has tended to lose favour since the demise of the melangeur pan.

As described previously, the cocoa mass can also be incorporated into chocolate crumb. This also must be crushed and mixed with fat in order to provide a suitable paste for processing in a refiner.

---

* Some chocolate manufacturers prefer to heat the surface of the beans, to facilitate shell removal, and to carry out the full roasting of the cocoa at either the nib or the cocoa mass stage. This is described more fully in Chapter 5.

1.2.3 *Conching—flavour and texture development*

Although the fermentation, drying and roasting are able to develop the precursors of chocolate flavour, there are also many undesirable chemical compounds present. These give rise to acidic and astringent tastes in the mouth. The object of conching is to remove the undesirable flavours, while developing the pleasant ones. In addition, the previous grinding process will have created many new surfaces, particularly of sugar, which are not coated with fat. These prevent the chocolate flowing properly when the fat is in a liquid state. Because of this the chocolate cannot yet be used to make sweets, and does not have the normal chocolate texture in the mouth. The conching process, therefore, coats these new surfaces with fat and develops the flow properties as well as the flavour. This is normally carried out by agitating the chocolate over an extended period in a large tank. Some manufacturers prefer to limit the conching time by restricting the conching process to primarily one of liquefying the chocolate. This is made possible by treating the cocoa mass at an earlier stage in order to remove some of the less desirable volatile chemicals.

1.3 Concept of the book

Chocolate making was for over one hundred years a traditional industry governed by craftsmen who developed individual methods of working as well as 'house' flavours for products. With increasing economic demands for higher throughputs and less labour, the industrial manufacture of chocolate has become more and more mechanized. There has also been an increased application of science and technology to control production plants and enable them to operate efficiently. In this situation the equipment manufacturers are continually producing new machinery, whilst the literature abounds with new methods of manufacture and patents for 'improved' techniques. Certain basic principles of chocolate making exist, however, and the aim of this book is to show what these are, and how they can be related to the processes used in its manufacture. It has been intended to avoid making the book a catalogue of a selected number of machines and products. In order to try and achieve this, and give the book as wide a coverage as possible, authors have been chosen from a range of industries and research institutions in Europe and North America. Chapters have deliberately been kept relatively short, and to a certain extent follow the order of processing described in the previous section. Certain topics have been divided into two, for example the chemical changes involved during conching have been presented separately from the physical and engineering aspects, as most authorities tend to concentrate predominantly on one or other of these aspects of conching. Where other industries are directly involved with chocolate manufacture, such as sugar, instrumentation and cooling tunnel/temperer manufacture, authors have been selected who have worked for periods in both industries.

The manufacture of chocolate goods would not exist but for the consumer. What is seen on the market shelves is seldom the chocolate itself, but usually the container. For this reason the packaging and marketing of the product is of considerable importance and chapters on these two topics are included in the book.

Every author has contributed to the book as an individual. Each chapter, therefore, is the author's responsibility, and may or may not be in agreement with the theories or principles adopted by the company by whom he is employed, or by the editor. As the chapters were written concurrently with little contact between the authors, several topics were duplicated. This has been minimized where possible, but retained where authors have given additional or even contradictory information. The latter is bound to occur owing to the present incomplete understanding of the processes involved. Minor differences in machinery or ingredients can produce major changes in the product. Each author, therefore, is merely reflecting his own experience within the wide range of combinations possible in chocolate making.

The multinational authorship of the book highlighted the differences in terminology and units found throughout the industry. For example, the term 'refinement' means flavour development in some countries and grinding in others. For this reason, and to aid people unfamiliar with the industry, a glossary of terms has been included (p. 373). The units given are those with which the author is most familiar, but frequently the most widely used alternative is also quoted.

References

1. Whymper, R. *Cocoa and Chocolate, Their Chemistry and Manufacture*. Churchill, London (1912).
2. Minifie, B.W. *Chocolate, Cocoa and Confectionery*. 2nd edn., Avi Publishing Co. Inc., Westport, Connecticut (1980).
3. Cook, L.R. (revised by E.H. Meursing) *Chocolate Production and Use*. Harcourt Brace Jovanovitch, New York (1984).

2 Cocoa bean production and transport

B.L. HANCOCK

2.1 Introduction

Cocoa beans, the essential ingredient of chocolate, are the seeds of a small tree known botanically as *Theobroma cacao*, the second word of which is the common name applied to the tree by agriculturalists. However, in English it is usually called 'cocoa'. The tree was already being cultivated in its native continent, South America, where it is still part of the natural flora, when the Spaniards first went there in the sixteenth century. It is now grown in all the wet tropical forest regions, mostly within 17 degrees of latitude of the equator (Figure 2.1).

Cocoa trees are small, growing up to about 20 feet (6m) in height in the shade of the big trees of the wetter areas of the tropical forests. The leaves are evergreen, resembling a laurel leaf in shape, and up to about 8 inches (20 cm) long. Cocoa is an unusual tree in several ways. It has two kinds of branches, the chupons which grow vertically upwards for about 5 feet (1.5 m) and bear leaves arranged in a spiral, and the fan branches, up to five of which grow out horizontally like the spokes of a wheel from the top of each chupon where its vertical growth stops. This is called a jorquette. The leaves on the fan branches are arranged in two rows, one on either side of the branch. This dimorphic growth results in a clear distinction between the fan branches and the vertical growth of the main stem, particularly in the young tree. The main stem usually reaches higher than the 5 feet (1.5 m) or so to the first jorquette as a result of a bud appearing just below the fan branches; this grows up vertically as another chupon from near the top of the previous one (1). The comparatively horizontal fan branches with their alternate arrangement of leaves grow and branch to form the leafy head of the tree which, in a cocoa plantation, joins with those of its neighbours to give a dense canopy (Figure 2.2). The heavy continuous shade of a well-grown planting of cocoa trees prevents to a large extent the growth of weeds beneath. Cocoa flowers are small, a little more than half-an-inch across, with petals varying from white to pink in different varieties (Figure 2.3). The minute fertilized ovary, however, grows over a period of nearly six months into a huge, waxy looking, oval cocoa pod which can be as much as 8 inches (20 cm) long. A further unusual feature is that many of the flowers, and hence the pods, are borne on the trunk or main branches of the tree below the leafy branches. Flowers continue to be borne often in clusters on the same spot on the stem, originally the site of a leaf stalk. This gradually

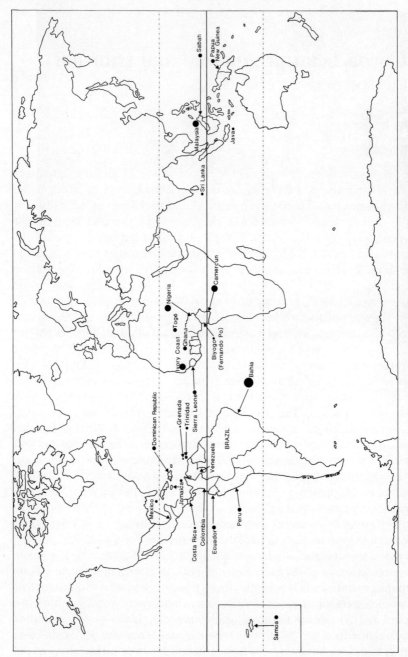

Figure 2.1 The named countries produce most of the world's cocoa in the areas marked. The circles indicate relative areas under cultivation.

Figure 2.2 A West African cocoa farm.

becomes thickened and is known as a flower cushion. A mature tree can be an arresting sight with its raised flower cushions spaced on the bare trunk, some only a foot (30 cm) above ground level, bearing tiny pink flowers and pods at various stages of their growth.

Ripe pods (Figure 2.4) have a waxy coating over the wall of dense tissue which is lignified to a varying extent and sometimes more than 0.5 inch (1 cm) thick. It has proved impossible to find an economically worthwhile use for this material. Inside the pod, there are some 30 to 40 seeds embedded in a mildly sweet mucilaginous pulp. Each seed, or bean as it is called because of its resemblance to the typical leguminous bean seed, consists of two convoluted and swollen seed leaves or cotyledons and a small germ or embryo plant, all enclosed in a skin or testa. The cotyledons serve both as the storage organs containing the food for the development of the seedling and as the first two leaves of the plant when the seed germinates. Much of the food stored in the cotyledons consists of a fat known as cocoa butter which amounts to about half the weight of the dry seed. Cocoa butter is an unusual fat in being quite hard at temperatures below 90 °F (32 °C), softening over a comparatively small

Figure 2.3 (Left) Cocoa tree trunk with flowers and young pods.

Figure 2.4 (Right) Ripe cocoa pods.

range of temperature, and is quite liquid at our blood heat. It is these properties which allow chocolate to be quite hard and breakable at cooler temperatures and to melt completely in the mouth.

2.2 Botanical types

There are two appreciably different types of cocoa, the Forastero which has purple cotyledons, and the Criollo which has white cotyledons. The colour is given by anthocyanins, the group of chemicals which give most blue and red flowers their colours. These are present in the cotyledons, confined in special pigment cells comprising about 10% of the storage cells. Their function in the seed is not known though it has been suggested that they have a protective role. In Criollo cocoas, the coloured anthocyanins are replaced by leuco forms. Nowadays, a very large proportion of the world's cocoa crop is Forastero and it is difficult to obtain pure Criollo cocoa. There is a third type, generally described as Trinitario, which has, often within the same pod, beans with cotyledons ranging from almost white to a full purple colour. Trinitario cocoa appears to have originated by hybridization between Forastero and Criollo cocoas. This hybrid mixture was extensively planted in Trinidad after the island's Criollo plantations had been devastated in the eighteenth century. In the following century, this small island was becoming a major producer of cocoa and the hybrid mixture which was found to be successful was quite widely planted in various countries. Consequently, it has become known as a cocoa type by the name of the island. As would be expected of a hybrid mixture, there is quite a wide variation between individual trees.

The colour of the cotyledon is the main difference between Criollo and Forastero cocoas. It is also a very important one because the anthocyanins are involved in the production of the unique flavour of cocoa. The purple anthocyanins are associated with the stronger, more astringent and robust flavours. It is not surprising that Criollo, without these anthocyanins, is an altogether milder cocoa. Chocolate made from it is of a light brown colour quite like a milk chocolate, and has a pleasant flavour of a nutty type with only a gentle indication of what is now considered a typical chocolate flavour. Nevertheless, Criollo cocoa used to be regarded as the superior quality and was much in demand. However, it is a less vigorous plant than the Forastero and is more vulnerable to some diseases. Developments in processing and of new products both enabled the less desirable features of the Forastero to be minimized and use made of their stronger chocolate flavour, so that the tendency to replace Criollo by higher-yielding Forastero has been encouraged. The third type, the Trinitarios, while including pale beans within their pods, produce a larger proportion of beans with the purple colour. Certainly, they can have sufficient of the purple anthocyanins to show the strong chocolate flavour of the Forastero but it is refined and displays in addition the so-called ancillary flavours. The Trinitarios, derived from both the main types, thus include the most fully-flavoured cocoas which some now regard as the superior cocoas of the highest quality. They continue to be grown in some areas. However, the popularity with the growers of the Forasteros which are more robust and higher-yielding than even the Trinitarios, combined with their suitability for some popular products which are now being consumed in great quantity, has resulted in their becoming by far the larger proportion of the cocoa crop.

2.3 The preparation of cocoa beans

Cocoa pods are harvested when ripe by cutting the woody stalk. This is easily done with the pods borne low on the trunk, but less so with the pods on the upper branches where it is necessary to use a special knife fixed on a long pole. The thick-walled pods are opened (Figure 2.5) to release the seeds, either by cutting with a knife or cracking with a simple wooden club. The former requires care in practice to avoid piercing the pod wall too deeply and cutting into the shells of some of the beans. The pods generally change colour as they ripen, green pods become a golden yellow, while the red immature pods of other varieties become orange-yellow. The crop does not all ripen at the same

Figure 2.5 Opening pods to collect the beans for fermentation.

time so that harvesting has to be carried out over a period usually of several months.

There are some 30 to 40 seeds, cocoa beans, in a cocoa pod. Unlike seeds of most temperate plants they are not long-lived and do not dry out naturally. Indeed, if these large seeds are not planted within a week or two, they fail to germinate. In the ripe pod, the moisture content of the beans is in the region of 65% and they are embedded in a sugary, mucilaginous pulp. Such materials are of course, very liable to decay, especially at tropical temperatures, and the sticky pulp is both difficult to remove mechanically and difficult to dry.

It is fortunate both for cocoa growers and, as has subsequently been realized, for chocolate manufacturers and consumers that there is a simple solution to this difficulty. When the beans in their surrounding pulp are removed from the pod and left for a few days, yeasts and bacteria develop, causing fermentation and the breakdown of the sugars and mucilages in the pulp which can then drain away as liquid. Subsequently, the beans can be spread out in the sun or dried in mechanical driers down to a moisture content of about 7%. These are the processes of fermentation and drying by which the vulnerable moist seeds are prepared for transport and storage, the first stages of their journey into chocolate.

These processes of preparation are essential for another reason. If Forastero beans are carefully cleaned of pulp and dried without any fermentation of the surrounding pulp, the dry cotyledons within the bean are not the brown or purple-brown colour of fermented cocoa beans but an unattractive dark slaty grey. When such beans are made into chocolate this same colour persists and the taste is not like that of chocolate at all. Chocolate made from slaty, unfermented beans tastes extremely unpleasant, being very bitter and astringent without any apparent chocolate flavour. The grower of cocoa therefore, as well as producing the beans, carries out the first and most vital stage of flavour development.

Criollo cocoa, even though it is milder, also requires fermentation to develop its flavour. However, much less fermentation can be sufficient and this seems to be possible with little more than heaping the beans for a few nights before much drying has taken place. It, therefore, seems to have been no accident that Criollo was the variety first in cultivation. Criollo beans are not unpleasant to the taste even without undergoing fermentation, whereas severely under-fermented Forastero cocoa is most unpleasant. The milder Criollo cocoa, therefore, was the early standard and the Forastero was then regarded as a coarse cocoa of poor quality. However, with proper fermentation, the Forastero flavour can be very attractive in many products and, with the added advantage of better yields, this chocolate flavour and the more complex Trinitario chocolate flavours have become predominant. Now, the very limited amount of pure Criollo still grown would not be regarded by most consumers as making the best chocolate.

2.4 Fermentation

Fermentation is carried out in a variety of ways, but all depend on heaping a quantity of fresh beans with their pulp sufficiently for the micro-organisms to produce heat, which raises the temperature, while limited access of air is allowed to the beans. It can, as is usual on small farms of a few acres in West Africa, be done in heaps enclosed by banana leaves or, as is normal on plantations with their larger production, in boxes with the beans covered with banana leaves or sacking. The boxes must have provision for the liquefied pulp, the sweatings, to drain away and for some entry of air, either by means of small holes in the bottom of the box or preferably through a floor of slats each separated by about 0.25 inch (6 mm). Heaps can be used to ferment anything from around 200 to about 3000 lb (90–1100 kg) of wet cocoa beans, although intermediate amounts are desirable. Boxes should be at least 2 feet 6 inches (0.75 m) across and should not be filled to much more than this depth. The required quantities of cocoa beans can take a considerable time to collect, but it is essential to use ripe pods and desirable to open all the pods and fill the box or heap on a single day. This can necessitate holding some of the harvested pods for a few days. There is some evidence, notably from comparing practices in some areas where Trinitario is grown, that holding unopened pods for a few days results in better flavour development. Recent work on behalf of the Cocoa, Chocolate and Confectionery Alliance has confirmed this conclusion in West Africa and Malaysia. Some drying of the beans in the unopened pods results in more air reaching among the beans so that the fermentation starts more actively.

The fermentation begins with yeasts converting the sugars in the pulp to ethyl alcohol. This produces the initial anaerobic conditions, but then bacteria starts to oxidize the alcohol to acetic acid and further to carbon dioxide and water, producing more heat and raising the temperature by more than 18 °F (10 °C) during the first 24 hours to over 104 °F (40 °C) in a good active fermentation. As the pulp starts to break down and drain away during the second day, bacteria further increase, lactic acid is produced and the acetic acid bacteria are, under the slightly more aerobic conditions, more actively oxidizing alcohol to acetic acid. By this time the temperature should have reached almost 122 °F (50 °C). In the remaining few days of a normal Forastero fermentation of five or six days, bacterial activity continues under conditions of increasingly good aeration as the remains of the pulp drain away allowing air to diffuse between the beans. The high temperature is maintained by the bacterial activity. In box fermentations it is usual to turn the beans. In the older cocoa-growing regions, this is done after 1 or 2 days and again after 3 or 4 days, but in the Far East the practice of turning every day has developed. Turning is often done by shovelling the beans into another box set in front of and below the one used previously. The process of turning has the immediate effect of increasing the aeration and consequently the bacterial activity, which is reflected in a rapid rise in temperature that may exceed the cooling effect of turning. One of the purposes of turning is to secure an even degree of fermentation but it has found that there is considerable variation between different parts even where turning has been carried out. Turning is also recommended for heap fermentations where it would be expected that aeration would not be as good as in a suitably ventilated box. In practice, it is doubtful whether West African heaps are turned more than once, and many are not turned at all. Such evidence as there is suggests that the relatively smaller bulks fermented in heaps may receive quite similar aeration to the larger amounts in boxes. However, there is clearly a greater risk that a few beans in a heap which is not turned at all will receive virtually no fermentation. Certainly, a very small percentage of slaty and scarcely fermented beans is not unusual even in good West African cocoa.

Although five or six days can be regarded as the usual period for Forastero cocoa in the fermentation heap or box, there is appreciable variation in the practice. The West African heap is normally opened after about 5 days; in Trinidad the period used to depend on the judgement of the estate for each fermentation. It was usually within the range of five to seven days. Grenada differed in fermenting for a longer period, usually nine to eleven days. Criollo, on the other hand, needs much less fermentation and two days usually suffices. Surprisingly, although Trinitario cocoa requires the longer fermentation for its purple beans, the Criollo type beans do not appear to be over-fermented. There is evidence from taking samples at daily intervals from a Trinitario fermentation heap that the most important changes, especially the reduction in astringency and the increase in potential chocolate flavour, mainly occur during the first few days. Moreover, there is little doubt that some ferment-ation activity takes place during drying before the moisture content is reduced too much. Unless drying is very rapid, deficiencies in the fermentation are remedied to some extent during the first days of drying.

The chemical changes within the cocoa bean depend on the cells of the cotyledon dying so that the cellular membranes break down and allow the different constituents, which are kept separate in the living tissue, to come into contact. Death, which takes place during the second day, is caused primarily by the acetic acid which is then being produced in the pulp. The high temperature is also a contributory cause of the death of the cells. The anthocyanins and other polyphenolic compounds in the pigment cells can then diffuse out into the adjacent main storage cells where they meet various enzymes that bring about hydrolytic reactions while the conditions within the bean are anaerobic. These include the breakdown of the coloured anthocyanins of the Forastero beans so that there is some bleaching of the cotyledons at this stage. As more air reaches and enters the beans, oxidative or browning reactions start to predominate and the tissue darkens. This stage occurs in the latter part of a normal Forastero box fermentation of 6 or 7 days and can continue during drying, provided this is not carried out too rapidly. At the same time as these more visible reactions are taking place, various other chemical changes, which are not fully understood but certainly involve the polyphenols, take place and are essential for the development of chocolate flavour.

2.5 Drying

When fermentation is completed, the beans are removed from the box or heap for drying. They are then reasonably free of adherent pulp but still have a high moisture content and are somewhat soft. In areas where the weather is comparatively dry at harvest time, the beans are usually dried by being spread out during the day in layers a few inches thick on trays or mats which are exposed to the sun. The layers of beans are raked over at intervals, usually heaped up at night and protected when it rains. The trays can, as in the West Indies, be whole floors either with the roof on wheels to be pushed back when it is dry and closed over the floor at night or when it rains. Alternatively, the floors themselves are arranged on wheels so that several can be run under a roof, one above the other to save space. In some West African countries, the beans are spread on mats made of split bamboo which are placed on low platforms. These can be rolled up to protect the beans and can be carried into the nearby huts when it rains. Such mats have the advantage that raking about tends to dislodge any fragments left adhering to the shells of the bean, and these tend to fall away from the cocoa between the strips of bamboo. It usually takes about a week of sunny weather to dry down to the 7% moisture content needed to prevent mould growth during storage.

Where the weather is less dry and sunny at harvest time, drying is done artificially. In the simplest form, the beans are spread over a surface which is heated from below by the flue gases from a wood fire, wood being the only local fuel. More complex drying equipment involves the use of heat exchangers so that warm, and clean, air can be blown through the layer of beans spread on a perforated surface or in a rotating cylinder. Modern equipment can give complete combustion of oil, or of solid fuel including wood, so that the flue gases can be used to dry the beans after some admixture of air. Such equipment is widely used by the large plantations in the Far East. Artificial drying introduces two problems. The beans can be dried too quickly so that the enzymes within the bean are inactivated by lack of moisture before the various changes have been completed. To some extent this can be overcome by ensuring that the chemical processes have gone sufficiently far before the beans are removed from the fermentation heap or box, although it is normal with sun drying for some of these processes to take place during the first few days of drying. The second problem is that the smoke may find its way on to the beans. This is liable to produce an unpleasant smoky or hammy taste, which cannot be removed from the resulting chocolate by processing. While it is comparatively easy to design a drier in which the smoke is kept away from the cocoa, when it is operated for a number of years under the conditions on many cocoa farms and estates which lack skilled maintenance facilities, the risk of smoke reaching the cocoa beans too often becomes a reality. This is the reason why cocoas from some areas are in less demand and consequently command lower prices.

2.6 The cut test

Cocoa grading schemes in producing countries are based on a visual assessment of quality by means of a procedure known as the cut test. The procedure has been laid down as an international cocoa standard and its text is included in a pamphlet, *Cocoa Beans—Chocolate Manufacturers Quality Requirements* (2), available from the Cocoa Chocolate and Confectionery Alliance in London. The cut test involves cutting at least 100 beans lengthways to reveal in section the greater amount of the middle of the cotyledons. These are examined individually and the percentage of beans of the various categories determined. When a measure of the degree of fermentation is required, three additional categories are used. The first is fully fermented beans which are brown in colour with the convolutions of the cotyledons tending to separate when the bean is properly dry. If there is some blue or purple colour evident, then these beans, which are partially fermented, are classified as partly brown–partly purple. If all the cut surface is blue or purple without any brown patches, and the cotyledons are pressed tightly together, the bean is classified as fully purple. The least fermented beans, which are defined in the ordinance, are slaty in colour. This is because the pigment cells have not released their contents, and none of the chemical changes of fermentation have taken place.

In practice no fermentation is uniform, and since too much fermentation results in a loss of both the strength and quality of flavour, the ideal degree of fermentation shown in the cut test is some 70–80% of the beans fully fermented and 20–30% partly brown–partly purple. Slaty beans should be absent, their presence indicates a lack of sufficient turning, and any more than about 5%, depending on the degree of fermentation of the rest of the beans, will be reflected as some astringency apparent in the chocolate. Fully purple beans are those in which the anthocyanins have been liberated from the pigment cells by the acetic acid of fermentation, but very little further chemical change has taken place. These beans also should not be present, and they have an effect on flavour which is similar to although slightly less extreme than that of the slaty beans. While this cut test is useful in defining the degree of fermentation of a cocoa, it has been found to be extremely difficult to define these categories of beans in such a way that examiners working in different places can separate the categories in the same way, to produce similar numerical results on the same cocoa. As an alternative, a chemical measure of the amount of anthocyanin was tried, since fermentation involves its breakdown. The quantity of anthocyanin, however, in the slaty and fully purple beans is so much greater than that in the partly purple beans that the results simply reflect the few unfermented beans present. One fully purple bean contains more anthocyanin than about 30 partly purple beans. Consequently the cut test, which also reveals defects other than the degree of fermentation, remains the accepted test is spite of its limitations.

The other categories defined in the model ordinance are mouldy beans, where mould is visible on the cotyledons. These are especially undesirable— even as few as 3% can give an unpleasant musty or mouldy flavour to chocolate. Insect-damaged beans are those where the bean has been penetrated by an insect, which feeds on the cotyledon. These should not be present. Any number will involve loss of material and a risk of contamination with fragments of the insect. Also, certain beetles, when present in large numbers, can result in an undesirable stale taste. Germinated beans are those where the seed has started to grow before being killed by the fermentation, and the shell has been pierced by the growth of the first root. In the dry germinated bean, the embryo plant usually drops out, leaving a hole, which makes the bean more easily attacked by insects and moulds. Flat beans are those which have begun to form, but have not developed or filled out. There is no useful cotyledon in them so they simply add to the shell, which is waste. They should be excluded from the fermentation.

2.7 Sources of cocoa bean supplies

2.7.1 *Bulk cocoas*

Much of the world's supply of cocoa is grown in West Africa. Just before the beginning of this century cocoa was brought to the Gold Coast, now called Ghana, from the island of Fernando Po (now called Biyogo). It was a purple-seeded variety of Forastero known as Amelonado. This proved to be well suited to the conditions where the indigenous farmers established small farms of a few acres on land which they cleared from the forest, leaving a few big trees to give some shade to the cocoa trees. At first the quality of this cocoa was poor because of the lack of experience and care in preparation, but an effective grading system was introduced and applied when the farmers brought their cocoa for sale to the buying firms and later the Government buying centres. This was based on the cut test, and undue numbers of defective beans were penalized. As a result, a remarkably uniform quality of cocoa was exported which was free of major defects such as mouldy flavours. Also, in spite of the absence of any requirement in the grading regulations for limiting the partly brown–partly purple beans, it was found in practice that fermenting sufficiently well to keep the slaty beans down to below 3% resulted in the average level of fermentation, as shown by the numbers of brown and purple beans, usually being sufficiently good.

Satisfactorily fermented Amelonado cocoa, as grown in West Africa, has a full almost heavy chocolate flavour of a relatively simple kind, without undue acidity or bitterness or the various ancillary flavour effects, which are characteristic of Trinitario cocoa. Cocoa from Ghana and Nigeria was almost all from the Amelonado variety which is, at least in the trees grown there, unusually uniform genetically and produces a uniform type of flavour. This West African cocoa has also developed an enviable reputation for freedom from mould and other taints. This has been assisted by the relatively dry climate. The rainfall is actually less than had previously been considered the minimum needed for successful cocoa growing. Consequently, it is comparatively easy to dry cocoa in the sun sufficiently to inhibit mould growth, and the absence of artificial drying obviates the risk of smoke contamination. In addition, the use of fresh banana leaves for the fermentation heap eliminates the risk that an inadequately clean box will introduce taints. The uniformity of the flavour of the cocoa, with its freedom from undesirable taints combined with the suitability of its comparatively simple but full chocolate flavour for the manufacture of milk chocolate, has resulted in its becoming most acceptable to manufacturers for general use. At the same time the West African crop has increased in quantity with the result that this cocoa has become the world's bulk cocoa against which other cocoas are judged. Although it is not generally regarded as the highest quality of cocoa, good West African with its uniform and full chocolate flavour is not merely a bulk filler, to which other more flavourful cocoas must be added, but a type of cocoa which is well suited for making certain products and particularly milk chocolate, which is now consumed in large quantity.

During the years around 1960, Ghana and Nigeria were producing half the world's crop, but since then other areas have greatly increased production and Ghana's crop is now much reduced. The Ivory Coast, the next country west of Ghana, has produced a similar cocoa from the same Amelonado variety, although more recently there has been a great deal of planting of a more variable but more vigorous and earlier-bearing hybrid Forastero cocoa, which is now contributing substantially to the Ivory Coast crop. This is based on selection of cocoas collected in the region of the Upper Amazon rivers in South America. This hybrid cocoa, when prepared in a similar way to the Amelonado, is similar in flavour although it does tend towards somewhat sharper overtones of flavour and is not as uniform as is the Amelonado. The same cocoa is also now planted to some extent in Ghana and Nigeria. However, the Ivory Coast has not organized such an effective grading scheme as these two countries have applied for many years at the stage when the farmer first brings his beans for sale. The result is that a great deal of Ivory Coast cocoa is badly prepared, being inadequately fermented and dried. It often includes mouldy beans and is not free from taints which can be described as dirty foreign flavours. In consequence, Ivory Coast cocoa is less desirable for chocolate manufacture, but the Ivory Coast is now West Africa's biggest producer and one of the two biggest in the world.

A moderately big West African cocoa producing area is Cameroun. Here again the cocoa is mainly Amelonado, although there has been some planting of Trinitarios, but the area suffers from a much heavier rainfall which necessitates artificial drying. This leads to occasional smoke contamination of the beans. Unfortunately, smoky beans are not easy to detect, except by tasting the cocoa after roasting, and, since the contamination may only be present in a proportion of bags of a consignment, even after careful checking of samples, the smoky taint may not be found until the consignment is being used in manufacture.

The world's other big producer of bulk cocoa is Brazil in South America, whose produce is often referred to as Bahia cocoa. This name is based in the state in which most of the cocoa is grown. Here there is extensive growing, both in small farms and big plantations, of a similar Amelonado cocoa. In fact, it is believed that West African Amelonado came originally from the Brazilian cultivation. More recently hybrid Forasteros, including the Upper Amazon hybrids, have been planted. Brazilian cocoa, however, has tended to be less fermented with distinctly fewer brown beans than West African and is quite distinctly more acidic. In addition, artificial drying is the practice and as in

Cameroun, it is not always done satisfactorily. There is a similar, ever-present risk of finding smoky flavours. This risk can be reduced for the manufacturer by a recent development, the initial processing of the bean within the producing country, which is, however, in some other respects less desirable for the manufacturer. This involves the producing country roasting the cocoa beans, removing the shells, and grinding the beans into the liquid stage known as cocoa puree or cocoa liquor. Hence the industry term 'origin liquor' for cocoa bought in this form. Since the grinding inevitably involves mixing, it is easier to check effectively for smoke by testing samples of the origin liquor. The procedure of selling cocoa as origin liquor has advantages in avoiding the transport of the waste shell and saving some space, but it prevents the chocolate manufacturer exercising a control over a quality of the beans or carrying out the roasting process, the precise operation of which does affect the development of the flavour.

The Dominican Republic is another producer, big for the West Indies, of a bulk cocoa, but here there is a tradition of minimal fermentation. The proportion of slaty and fully purple beans is quite high and the cocoa is, as would be expected, distinctly astringent, dark-coloured, and poor in chocolate flavour. In the Far East, New Guinea and Malaysia have more recently become substantial producers. New Guinea contains a good deal of Trinitario cocoa as well as hybrid Forastero from the Upper Amazon. Without a long tradition of fermenting cocoa, this source produces very variable results, including a good deal of unsatisfactory cocoa caused by mould, smoke and other dirty unpleasant flavours. It all tends to be weak in chocolate flavour and somewhat acidic, but some New Guineas are quite strong in some of the desirable ancillary flavours of the Trinitario. Malaysia, which includes Sabah, is the most recent additional producer of bulk cocoa, and is rapidly becoming one of the bigger producers. The cocoa is virtually all the selected hybrid Forastero from the Upper Amazon and most is grown on large estates. Fermentation is based on the box practice of the West Indies but does involve more aeration with frequent turning of the beans and usually very rapid artificial drying. Smoke contamination is unusual but the cocoa is mostly altogether too sharp and acidic and lacks sufficient chocolate flavour.

2.7.2 Flavour cocoas

The other main type of cocoa grown is the Trinitario, itself a hybrid mixture of Criollo and Forastero, which, together with the small amounts remaining of Criollo cocoa, produces the 'fine' or flavour cocoas. These have flavours ranging widely from the pale nutty Criollos to the full but titillating flavours of some of the Trinidad estates. The latter can produce cocoas which have a strong but not overpowering chocolate flavour showing depth and fullness together with various attractive ancillary flavours reminiscent of raisin and of wines.

The development of these flavours is variable and appears to depend on a number of factors which are not well understood. Even cocoa from a single small estate usually develops its characteristic flavour to a variable extent. It is apparent that the kind of cocoa grown is important and, where the flavour potential of particular selections of cocoa is being examined, it should be remembered that the cocoa bean is a seed and the pollen parent may be expected to make as big a contribution to the flavour potential of the cotyledons as the tree which bears the pod. There is some reason to believe that a mixture of genetic types like that of the Trinitario populations is essential for this type of flavour development. The right fermentation is certainly necessary to develop these flavours properly and the same objective as for bulk cocoa of mostly, but not all, brown beans is probably right. However, assessment of the degree of fermentation by the cut test does not enable the better flavours to be selected. Certainly, defective beans should be absent and normally are in cocoa from the better estates. But some cocoas with an apparently ideal degree of fermentation are dull in flavour while others with what appears to be a distinctly different degree of fermentation have been found to be well developed in flavour. There is, in fact, no substitute for tasting after roasting as a method for determining the flavour quality of Trinitario cocoas.

An attempt was made in the past by one manufacturer to overcome this problem of variation in the flavour development of Trinitario cocoa by having unfermented cocoa shipped to his factory and carrying out a fermentation there. It did not prove satisfactory and was discontinued after a few years.

Prior to the beginning of this century, most of the cocoa used by chocolate manufacturers was either Criollo or Trinitario. The main use was for plain chocolate, where blends of different proportions of a number of origins would be used to produce a variety of flavour effects. In milk chocolate these interesting flavour effects would be much less apparent and it was found that the simple full chocolate flavour of the Amelonado bulk cocoa produced a satisfactory result and avoided the need for the careful tasting and selection, which is involved in the use of fine cocoas. Some plain chocolate also is made from bulk cocoa, and has a distinctly full and heavy chocolate flavour, although some manufacturers reduce the strength by modifications to the processing. Sometimes, a proportion of flavour cocoa is added to modify and brighten the flavour, while in other plain chocolates, where the heavier effect of the full Amelonado chocolate flavour is not wanted, the chocolate is made from a blend of fine cocoas. In such a chocolate, a small proportion of a good bulk cocoa may usefully be added if the chosen flavour cocoas are lacking in sufficient simple chocolate flavour to carry the ancillary flavours. In this way, a suitable bulk cocoa can serve as a flavour cocoa.

Trinidad has for a long time been a producer of good Trinitario flavour cocoa showing, with the right processing to develop it, a bright full chocolate flavour made more interesting by fruity and winey overtones. It also produces cocoa which cannot be described as good; indeed it is found that the flavour-

cocoa-producing areas produce a much wider range of flavour quality than most bulk cocoa areas. Grenada produces an essentially similar cocoa to Trinidad, although the ancillary flavours tend not be as strong. Jamaica is a small producer of a similar type of cocoa which is rather more acidic. Here the cocoa is collected from the farmers and fermented in a few central fermentaries but there is some variation in the flavour. Most of the south and central American countries which grow cocoa, apart from Brazil, are also growers of Trinitario. Venezuela produces various cocoas of the Trinidad type. Recognizable differences occur between the production of the various river valleys in the mountains along its northern coast. It is interesting that geographically Trinidad is the eastern end of this mountain range. Ecuador is now mainly a bulk cocoa producer, but used to produce a very characteristic cocoa known as Arriba which appeared to require less fermentation, to such an extent that the best Arriba with its unique strong floral flavour usually had at least 5% of slaty beans.

Sri Lanka (Ceylon), now only producing a small quantity, was unusual in that the beans were washed after fermentation. This resulted in the shells being both lighter and much more easily broken, facilitating the entry of insects and making uniform roasting difficult. Java, now part of Indonesia, was one of the larger producers of white-seeded Criollo cocoas, and Western Samoa, two small islands in the Pacific Ocean, still produces some Criollo. New Guinea, where the earlier plantings were of Trinitario, produces substantial quantities of a flavour cocoa, as well as a great deal of poor-quality cocoa. This flavour cocoa is rather weak in chocolate flavour with an interesting acidity, and occasional shipments show strongly developed ancillary flavours of various types.

Good flavour cocoas command higher prices than the bulk cocoas, but the proportional difference has declined with the increasing importance of bulk cocoa to manufacturers. Moreover, the quantity of flavour cocoas required is relatively small and most manufacturers, certainly for milk chocolate use, would not willingly replace their bulk with flavour cocoa. There is, consequently, no possibility that a major bulk cocoa producing country would receive a relatively higher price by altering its bulk cocoa to show flavour cocoa characteristics. The premium for flavour cocoas over bulk cocoas does depend on there being a sufficient demand for the particular flavour cocoa. Within the field of bulk cocoas, relative prices reflect both the useful yield expected by the manufacturers and experience of the quality as it affects the flavour of cocoa from the area. The shell has to be removed and so the slight differences in its average weight will affect the yield of dry cotyledon or nib, as it is called at this stage. A further important character is the amount of fat in the cocoa bean. This is because chocolate, which needs sugar to balance the flavour of the cocoa, requires more fat than that which is sufficient to liquefy cocoa nib simply by grinding to release its fat (see Chapter 1). So additional fat is required in chocolate manufacture and the higher the fat content of the cocoa beans used the less extra fat will suffice. However, the differences in the fat contents between cocoas from different areas are relatively small.

Experience of the quality of supplies of bulk cocoa also affect acceptability. Ghana, the generally preferred standard for bulk cocoa, produces a relatively uniform cocoa with a strong chocolate flavour which is free from serious defects. Nigeria has a similar reputation, but its cocoa has marginally more shell and less fat. A small reduction in price, compared with the price of good Ghana cocoa, is based on this difference in yield. Ivory Coast cocoa is slightly lower again in yield, but, more seriously, it suffers from the presence of mould and other undesirable flavours. There is, therefore, a larger discount in the Ivory Coast price, and bulk New Guinea is cheaper still because, as well as giving a lower yield as a result of a high shell content, the presence of undesirable flavours including mould and smoke is relatively common.

2.8 Shipment of cocoa beans

Cocoa beans are transported and stored in hessian sacks which are strong, can easily be stacked and can be sampled by means of a sampling stick which can be pushed between the woven threads. The hessian allows water vapour to pass through, so that the cocoa beans gradually come into equilibrium with the humidity in which they are stored. This can have the advantage of allowing an odd damper bag to dry a little, but is part of the reason for the need to dry cocoa beans down to between 6 and 7% of moisture to allow for some slight uptake while stored in the producing country, where relative humidities even within stores tend to be fairly high.

Difficulties, however, can still be caused by this moisture in commercially dry cocoa when it is transported to manufacturers in colder countries. The West African crop is mainly shipped during the northern winter months. Cocoa at 8% moisture is, for example, in equilibrium with a relative humidity of 70%. It may be loaded, usually filling the holds of the ship, at a temperature of about 85 °F (30 °C). After a few days, the temperature both of the air and the sea will start to fall and, within a few more days, on reaching the North Atlantic it may be down close to freezing point. These are the conditions under which dew will condense on to the cold metal of the ship. But the bulk of cocoa will take quite a long time to cool appreciably so that it will keep the air around it warm and at its equilibrium relative humidity. This air will diffuse outwards and be cooled by the cold metal of the ship till its relative humidity reaches 100% and water starts condensing on the cold metal. A drop in temperature of only 12 °F (7 °C) is sufficient for this air to become fully saturated. A stack of cocoa allows some movement of air in the spaces between the rounded bags and even between the beans. There is, therefore, in practice a distinct convection circulation of the warmed air up through the middle of the stack and down the outside, where it is cooled by the metal of the ship's side in contact with, and cooled by, the cold sea. This results in a steady movement of water evaporated from the more central part of the stack of cocoa beans on to the cold metal of the ship where it is condensed, running off as liquid water, and giving up its latent heat to the metal, which remains at a low temperature because the heat is conducted away to the cold sea and air outside.

Although the cocoa beans are dry, even at 7% moisture the water in a hold containing 1000 tons of cocoa would as liquid water amount to 15 000 gallons. If the moisture content of the cocoa was reduced only a quarter percent, there would be 560 gallons available for condensation. Such amounts are liable to involve water literally raining down from the deck above and running down the walls of the hold without apparently reducing the average moisture content of the dry cocoa. There can easily be a sufficient quantity to wet patches of the dry cocoa in bags. Liquid water is taken up very quickly by cocoa beans, which can in a few minutes go from their dry condition to 20 or 30% moisture. At these moisture contents and temperatures around 75 °F (24 °C), moulds can develop in a few days. As a result, a number of shipments have arrived in Europe and North America with patches of beans which have been wetted and spoilt by the growth of mould.

Precautions to prevent such damage must be taken in ships. The best precaution is to operate mechanical ventilation. This continually removes the warm air, carrying more moisture than it can hold when it cools, before the water can condense on to the cold surfaces. In addition, the bags are stacked so that they are not in contact with the cold metal and are not placed where any condensation water is likely to run down and wet the bags, and absorbent mats are spread over the tops of the stacks to catch drips from the deck head. But there can be problems in ventilating during bad weather, so that difficulties do arise at times. It is usual to leave channels for ventilation in large stacks. This helps the ventilation to cool the cocoa by removing the heat which, with the rapid change which is possible in the external temperature, is the cause of the problem.

It has been suggested that mould growth can be a result of moisture migrating actually through the cocoa from the warm central area of the stack to the cool outer layers. There is, of course, a tendency for such a movement to take place, because the vapour pressure of water in air at a relative humidity of, for example 70%, which is in equilibrium with the dry cocoa at 8% moisture, is relatively lower at the lower temperatures near the outside of the stack. The cooler cocoa will therefore gradually absorb moisture vapour, which diffuses out from the air around the warmer beans. This is because the vapour pressure of the water in the warm air around the warm cocoa beans will be higher than that in the cooler air in equilibrium with the cooler cocoa. Water vapour will therefore diffuse out to the area of lower vapour pressure, raising this pressure above the vapour pressure in equilibrium with the cocoa there. Water will then be absorbed by the cocoa to bring it into equilibrium with its surrounding air and its moisture content will be raised while the vapour pressure of the air will be reduced. This process is caused by the difference in temperature between different parts of the stack due to the cooling of the outer layers and will continue as long as there is a temperature difference, unless the cooler cocoa has a moisture content sufficiently raised that its equilibrium vapour pressure is as high as that of the warmer cocoa.

However, this movement of water vapour by diffusion is a slow process and, even though it may be assisted by some movement of air, the uptake of water in the vapour state by cocoa beans takes days rather than the minutes involved in the uptake of liquid water. Instead of liquid water wetting relatively few beans, the vapour will be absorbed by any beans of moisture content low enough to have a vapour pressure below that of the air. In practice, the increase in moisture content by moisture migration is not more than 1 or 2%, even under the most favourable circumstances. This may be enough to allow moulds to grow, but their growth at this sort of moisture content is very slow indeed. Moreover, it would only be the cool cocoa that would be increased in moisture by this means. The low temperature would also delay growth. For example, beans at temperatures up to 60 °F (16 °C) fully exposed to 90% relative humidity took several weeks to rise to 10% moisture content, even though their equilibrium moisture content in 90% relative humidity is as high as 13%, and these beans did not show any mould growth until they had been exposed for 7 weeks. This kind of moisture movement could not, therefore, cause mould damage to cocoa during even a slow voyage. The rapid uptake of substantial amounts of water into relatively few cocoa beans as a result of condensation on to the cold inner surfaces of the hold is, however, a constant risk when moving from tropical to colder regions. The necessary precautions must be taken against this risk.

A more recent development has been the introduction of containers. An early attempt to ship cocoa in bulk in a specially designed container proved unsatisfactory due to condensation. More recent trials have involved the normal hessian bags of cocoa being stacked into containers. It has been found possible to ship cocoa in the standard closed containers between West Africa and Europe during the warmer summer months, but during the winter excessive condensation can be expected on the inside of the container in the same manner as in a ship's hold. Special containers have therefore been designed with grilles to allow ventilation air movement involving circulation to remove the warm air. Provided there is sufficient air movement in the ship's hold, such ventilated containers have proved satisfactory, again overcoming the difficulty of the temperature change by removing the warm moist air and gradually cooling the cocoa to remove the cause of the trouble.

The cocoa trade accepts that there is a loss in weight when cocoa is transported from the producing to the consuming countries. A loss of 0.5% of the weight is allowed in the London contract, and statistics of cocoa supply and usage include this 0.5% loss as a difference between production and consumption figures which is known as shrinkage. In fact, it appears to be an over-estimate. Careful measurements made a few years ago showed no average change when a number of bags in different parts of a stow were tested in West Africa and the same bags sampled and checked for moisture on arrival in England as part of an ordinary shipment. Much earlier, it was believed that cocoa as used in factories in Europe had a substantially lower moisture content in the region of 4 or 5%, presumably partly as a result of drying while in store. It was not then realized that the established method of sampling, which involved putting the sample of beans into a cloth bag, allowed appreciable drying before the sample was examined and tested in the laboratory. However, such inaccuracies, although no doubt explaining some perplexing differences between the results of a reliable method of moisture determination when used in different laboratories, did not affect the actual yield of dry roast nib from each bag of beans. This is determined by the accuracy of the weight of beans filled into the bag and their average moisture content at that time. It is unimportant as to how much moisture is lost subsequently, since a greater loss will merely reduce the amount of moisture lost during the roasting process, the product of which is dry roast nib, the characteristic ingredient of chocolate.

2.9 Infestation of cocoa

Once they have been dried, cocoa beans are not subject to deterioration, provided they are kept under good conditions. These should include an absence of the various insect pests which feed on cocoa beans. Cocoa is particularly vulnerable to two small tropical moths belonging to the genus *Ephestia*. The larva stage, the caterpillar, enters a bean, usually where the shell is damaged, and feeds and grows there until it is ready to pupate. It then leaves the bean and usually finds its way out of the bag and up the wall of the store. Insect damage to cocoa beans is easily seen in the cut test where the droppings (frass) and woolly filaments produced by the caterpillar are obvious in the cut beans.

Various beetles and their larvae also feed on cocoa beans and can be recognised by the powdery frass they leave. Some only rarely enter the beans, feeding mostly on the residue of pulp adherent to the outside. Their presence can result in a consignment being described as severely infested, even though the cut test on a sample of beans does not reveal any insect damage. Control of light infestations is usually effected by spraying with an insecticide such as a pyrethrum type, which is virtually non-toxic, suspended in a non-tainting oil or in water. More severe infestations are better controlled by fumigation, either in suitable chambers or under gas-proof sheeting. The *Ephestia* moth, often known as the cocoa moth, is a particularly successful pest. Its habit of pupating away from its food enables the moths to emerge in considerable numbers to lay their eggs to continue infesting foodstuffs even after an infested consignment has been moved and treated to kill the insects. Control measures need therefore to be applied to the structure of the store as well as the actual consignment. One of the two species of *Ephestia, E. elutella* is more tolerent of the cooler temperatures of most consumer countries, and is particularly troublesome to manufacturers. Both species can feed on chocolate products as well as cocoa beans. A most distasteful appearance is produced when a larva finds its way into a product and feeds there.

2.10 Microbiology of cocoa

Cocoa beans are free from micro-organisms when they are inside the cocoa pod, but the pulp is an excellent medium for their growth and this is encouraged in the fermentation process. Yeasts and various kinds of bacteria proliferate and it is to be expected that even after drying there will be a great many micro-organisms, including moulds, adhering to the shell. In fact, cocoa shell is coated with very large numbers of harmless bacilli remaining from the fermentation and, depending on the hygiene of the drying process, there may also be some of the less resistant but potentially harmful bacteria. The bacteria are dormant while the bean is dry. The roasting process normally involves sufficient heat to kill all but the more heat-resistant of the bacilli. However, it is important to separate the intake of beans, which necessarily involves considerable evolution of bacteria-laden dust, from the handling of the beans after they have been roasted and from subsequent processes. Most of the bacteria on the beans are quite harmless and merely contribute to the bacterial count. However, it is possible for some contamination to occur with harmful bacteria like *Salmonella*. Fortunately, these are much more sensitive to heat and are normally killed in the roasting process. However, they are more resistant under dry conditions and there have been a very few instances where chocolate has been found to contain sufficient of these bacteria, believed to have been brought into the factory with cocoa beans, to cause some illness. Chocolate and most chocolate products are too dry and sugary to allow multiplication of bacteria, but chocolate does appear to allow bacteria in it to remain in a dormant, but potentially active, state for long periods and may exert a protective effect on bacteria in the stomach. If use is made of origin liquor, freedom from *Salmonella* depends on its manufacture and routine checks should be made of its bacterial condition. Some factories actually heat-treat the cocoa liquor when producing origin liquor to ensure that the bacteria are controlled. Unfortunately, the high temperature needed to kill bacteria under the conditions within the liquefied cocoa has an adverse effect on the flavour.

Moulds attack cocoa beans and, as has been pointed out, are most objectionable since as few as 3% of mouldy beans can taint chocolate in a most unpleasant fashion, and this taint cannot be removed by processing. When it was found that some moulds could produce the highly toxic substance, aflatoxin, the question arose as to whether it was produced in cocoa beans. Numerous tests have been carried out but the presence of aflatoxin in cocoa beans has never been confirmed, although tests have been carried out actually on the mouldy beans on which *Aspergillus glaucus*, a species of mould capable of producing aflatoxin, has grown actively. There is evidence that the mould must have a particular medium on which to grow to produce the aflatoxin and it appears that the necessary medium is not provided by cocoa beans. Nevertheless, the objectionable effect of moulds upon the flavour of chocolate does necessitate constant attention to the prevention of moulds from growing on cocoa beans. This is achieved by drying reasonably rapidly after fermentation and avoiding subsequent uptake of water.

References

1. Wood, G.A.R. and Lass, R.A. *Cocoa*, 4th edn., Longman (1985).
2. *Cocoa Beans, Chocolate Manufacturers Quality Requirements*, 3rd edn:, Cocoa, Chocolate and Confectionery Alliance, London (1984).

3 Sugar

Ch. KRÜGER

Sugar is the sweet-tasting crystallized saccharide extracted from sugar cane or sugar beet. Both the beet and the cane produce an absolutely identical natural substance, which is chemically termed 'sucrose' or 'saccharose'. Sugar is a disaccharide, composed of the chemically linked monosaccharides glucose and fructose. This linkage, however, may be cleaved hydrolytically by acids or by the enzyme invertase (β-D-fructo-furanosidase). The resulting mixture consisting of equal parts of glucose and fructose is called invert sugar.

A great many other types of sugar are also in existence, such as the monosaccharides, glucose (dextrose) and fructose, the disaccharide lactose, as well as sugar alcohols, for instance sorbitol and xylitol. For the production of chocolate, however, sucrose is by far the most important type of sugar.

3.1 The production of sugar

For the production of beet sugar, the sugar beet, which contain about 14–17% of sucrose, are cleaned and cut into beet slices. These are then extracted in hot water by means of a counter-flow process known as a diffuser. Together with the sugar, however, mineral as well as organic substances from the beet find their way into the raw solution produced. Since these non-sugar substances strongly inhibit the crystallization of the sugar, the solution must be purified. This is carried out by adding slaked lime in order to flocculate or precipitate the majority of the contaminants and even decompose a small proportion of them. When carbon dioxide is subsequently bubbled through the solution it precipitates the excess calcium hydroxide from the slaked lime in the form of calcium carbonate. This is then filtered off together with the precipated non-sugar substances.

The clarified weak sugar solution thus produced has a solids content of approximately 15% and this is subsequently evaporated to about 65–70% dry solids. This is then further concentrated under vacuum until crystallization sets in. The cooling down of this syrup-crystal mix leads to further crystallization of the sugar. The separation of the sugar from the first molasses or mother syrup is frequently performed in a centrifuging process. The crystallization always requires several processing stages, since it is not possible to recover all of the sugar in a single step. Thus white sugar has to be crystallized in three or four different steps. Raw sugar factories produce brown raw sugar as an interim product by a simplified crystallization process. This type of sugar contains impurities which are removed by means of further recrystallization in sugar refineries. A by-product, called molasses, is the syrup obtained during the final crystallization step, from which sugar can be no longer crystallized due to the high concentration in non-sugar substances. Beet molasses with 80–85% dry solids contains about 50% of sucrose, but also up to 11% of inorganic substances, 4–5% of organic acids and about 13% of nitrogenated organic compounds. Furthermore, about 1 to 2% of the trisaccharide raffinose and traces of invert sugar may be found (1).

Sugar cane has a sucrose content of 11–17%. The raw juice in this case is manufactured by squeezing the crushed stalks on roller mills or by means of extraction equipment in combination with roller mills. What remains is the so-called bagasse, which serve either as fuel or as raw material for the production of paper, cardboard, hard particle board etc. Since raw juice from sugar cane contains more invert sugar than the equivalent beet solution, a more gentle treatment is required to clarify the juice. A lime treatment as used for the beet would degrade the invert sugar, leading to the formation of undesirable brown colours. The chemical treatment required may be carried out by one of the following three methods: (i) a gentle lime treatment; (ii) a lime treatment plus 'sulphitation' with sulphur dioxide; (iii) a lime treatment plus 'carbonation' with carbon dioxide. This resembles juice clarification in the beet sugar industry, without, however, using the high temperatures normally employed in the latter. The chemical juice clarification is preceded by a mechanical separation of suspended plant particles, e.g. by using hydrocyclones and bow-shaped sieves. Evaporation and crystallization closely resemble the same procedures used for the production of beet sugar. However, special purification as well as the above clarification steps are necessary in the refining of raw cane sugar due to the different composition of the non-sugar contaminants. These include carbonation with slaked lime and carbon dioxide or phosphatation, that is a slaked lime phosphoric acid treatment as well as a special colour removal with activated charcoal.

Cane molasses of 75–83% dry solids contains about 30–40% of sucrose as well as approximately 10–25% of invert sugar, 2–5% of other carbohydrates, 7–15% of inorganic substances, 2–8% of organic acids and 2–5% of nitrogenated organic compounds (2).

3.2 Sugar qualities

The sugar industry supplies a wide range of crystallized and liquid sugars. Crystallized sugar is graded according to its purity and crystal size. The purity of all types of white crystallized sugars is extremely good; the sucrose content is generally more than 99.9% and only rarely falls below 99.7%. Any differences in quality result from the minute quantities of non-sugar substances, which are mostly present in the syrup layer surrounding the sugar crystals.

In Europe, quality criteria for sugar are determined by the European

Table 3.1 Quality criteria of white crystallized sugar according to European Community market regulations

Category	1 Refined sugar	2 White sugar, standard quality	3	4
Optical rotation, °S, min.	—	99.7	99.7	99.5
Water, %, max.	0.06	0.06	0.06	—
Invert sugar, %, max.	0.04	0.04	0.04	—
Braunschweig colour type colour type unit, max. points (0.5 units = 1 point)	(2) 4	(4.5) 9	(6) 12	—
Ash by conductivity, %, max. points (0.0018% = 1 point)	0.0108 6	0.0270 15	—	—
Colour in solution ICUMSA units, max. points (7.5 units = 1 point)	(22.5) 3	(45) 6	—	—
Total score according to EEC point score, max.	8	22	—	—

Community (EEC) market regulations and the National Legislation on sugar types. The EEC sugar market regulations break crystallized sugars down into four categories, the quality criteria of which are summarized in Table 3.1.

The degree of optical rotation is a yardstick for the sugar purity. Refined sugar, as a rule, gives values of 99.9 °S.*The water content must not exceed 0.06% and in general does not even exceed 0.03% in good quality crystallized sugar. The amount of invert sugar present should not be more than 0.04%. Once again the values obtained in practice are frequently much lower. Colour in solution is measured in a 50% solution with a pH value of 7.0 at a wavelength of 420 nm. The coefficient of extinction established in this way and multiplied by 1 000 represents one ICUMSA unit. The Braunschweig colour type is determined by visual comparison with calibrated colour standards which may be obtained from the Institute for Agricultural Technology and Sugar Industry (Institut für landwirtschaftliche Technologie und Zuckerindustrie), D-3300 Braunschweig, West Germany. The ash content is determined in a 28% solution by conductometric measurements as the so-called 'conductivity ash'. According to the method employed, 1 μS/cm in a 28% solution represents 5.76×10^{-4}% of ash. The values thus determined are converted into points for a score; the points add up to the total score of the sugar. The lower the total score, the better the quality of the sugar. As a rule, category II sugar is used for the manufacture of chocolate. Category III sugar is slightly cheaper and of a quality which in most cases quite suffices for the manufacture of chocolate; however, it is not available in many countries.

*This is a purity measurement, not degrees of rotation.

White crystallized sugar should be free-flowing and have crystals of uniform particle size. There are no legal stipulations regarding grain or particle size. Nevertheless, the following grading may be carried out by sieving and is more or less generally accepted (3).

Coarse sugar	1.0–2.5 mm grain size (0.04–0.1 inch)
Medium fine sugar	0.6–1.0 mm grain size (0.02–0.04 inch)
Fine sugar	0.1–0.6 mm grain size (0.004–0.02 inch)
Icing sugar	0.005–0.1 mm grain size (0.2×10^{-3}–0.004 inch)

The manufacture of chocolate masses is predominantly based on the use of medium fine sugar. On the other hand, some chocolate manufacturers insist on certain specifications concerning the particle spectrum. Some factories which refine masses in a two-step procedure (see Chapter 6) specify a spectrum of 0.5–1.25 mm (0.02–0.05 inch) with the unavoidable amount of fine grain, < 0.2 mm (0.008 inch), of up to but not exceeding 2%.

3.3 The storage of sugar

In the great majority of cases, sugar is delivered to the chocolate industry by means of tanker vehicles and not, as in former times, in bags or sacks. The sugar is pneumatically discharged from the vehicles into the silos, where it is stored until further processing. Four factors are important in the silo storage of crystal sugar (4): grain structure, moisture content, apparent or bulk density, and angle of repose.

The grain structure of crystallized sugar is determined by the grain size, shape of the grains and the grain size distribution. The sugar industry supplies sugar which has been sifted to well-defined particle sizes, shows a good fluidity and is thus suitable for silo storage. In European countries like the United Kingdom, France and the Federal Republic of Germany, for instance, the chocolate industry is mainly being supplied with sugar within the size range 0.5–1.5 mm (0.02–0.06 inch). The amount of dust (grain size $\leqslant 0.1$ mm, 0.007 inch) is generally below 1% at the time of delivery in a tanker vehicle.

The sugar should, if possible, consist of regular-shaped individual crystals, since irregular conglomerates have a detrimental effect on the bulk sugar's rheological properties. Problems may also arise due to the sugar dust. This may be formed during pneumatic handling due to friction against the inner surfaces of pipelines, especially at the site of manifolds. The amount of fine grain (< 0.2 mm, 0.008 inch) and dust (< 0.1 mm, 0.004 inch) present in the sugar correlates with its flowability and both should therefore be kept as low as possible. With 10% of dust, storage and discharge problems are likely to occur. At dust levels of 15% or above, proper silo storage becomes impractical (5). Thus special attention should be paid to ensuring an optimum layout of handling equipment such as pipeline design, manifold radius and feeding rate in order to keep the mechanical breakage of the sugar as low as possible.

A minimum feed rate of about 11 m/s (36 ft/s) is required for the sugar transport. However, the handling conditions should be designed in a way that the feeding rate does not exceed a speed of about 22 m/s (72 ft/s) (6).

The moisture content of sugar is extremely low. Its actual level, however, has a decisive influence on its storability. After drying and cooling, freshly produced sugar has a total water content of about 0.1%, which is further reduced to some 0.03–0.06% by conditioning in the silo of the sugar factory.

Freshly-produced sugar which has not been conditioned should be processed immediately without any intermediate storage, because it might otherwise form lumps and become hard. The surface moisture of sugar changes as a function of the relative humidity of the ambient air. Figure 3.1 shows this dependency by means of the sorption isotherms of sugars of differing purities at 20 °C (68 °F) (4). Each sorption isotherm is a function of the sugar's purity, but even more of its ash content and its traces of invert sugar and its crystal size. As may be easily seen, the curve runs almost parallel to the abscissa at 20–60% air relative humidities. This would imply that there is practically no water uptake even when the relative humidity of the air is increasing. Only when the relative humidity of the air exceeds 65% does water become increasingly absorbed on the crystal surface. This in turn implies that

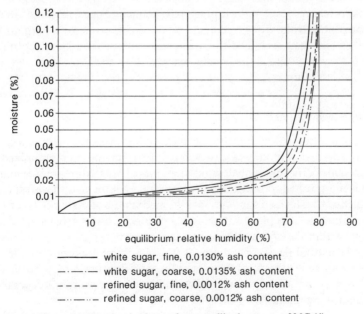

Figure 3.1 Sorption isotherms for crystallized sugars at 20 °C (4).

proper storage requires a temperature of 20 °C (68 °F) and a relative humidity ranging from 20% to 60%.

A relative humidity of more than 65% exponentially increases the water content in the sugar. Wet sugar must be avoided at all costs, as it can become chemically and microbiologically contaminated. In addition, if wet sugar is stored at a relative humidity below its equilibrium relative humidity (ERH), the sugar will cake during drying and become lumpy.

Purer sugars (category 1 with lower ash content) and sugars with coarser particle structure possess superior storage properties. On delivery, the surface water content of any sugar should not exceed 0.03% (7).

According to the 1972 EEC sugar regulations, the water content in crystallized sugar measured by loss during drying may reach a maximum of 0.06%; the 1976 Federal Republic of Germany regulation for sugar types even permits a maximum of 0.1%.

External sugar silos should be equipped with adequate insulation as well as with heating elements in the silo walls to avoid the problem of condensation. However, some outdoor silos are constructed without insulation but have the headspace in the silos continuously filled with dried air fed in through pipelines. Sometimes only the supporting case is heated.

The maximum bulk density for the sugar may be taken to be $850 \, kg/m^3$ (53 lb/ft^3). When designing silo systems, it is, however in practice, somewhat less and only reaches $750 \, kg/m^3$ (47 lb/ft^3) during filling. At a surface moisture of 0.02–0.04%, the angle of repose of the sugar varies between approximately 35–41°. The layout of sugar silos is often based on an angle of repose of 45–50°, with the discharge level in the tapered part being inclined at least 55° towards the horizontal. If the discharge angle is too small, proper discharge may be impeded. This may even lead to partial segregation, since larger crystals fall out more easily whereas the smaller ones stick to the slip plane where they build up into a layer.

3.4 Sugar grinding and the prevention of sugar dust explosions

In chocolate production there has been a trend towards the process where the granulated sugar is ground in a two-step process having been pre-mixed with the cocoa mass, the milk powder and other ingredients. More traditionally the sugar is pre-crushed to powdered or icing sugar before mixing with the other ingredients. Crystallized sugar is a brittle, medium-hard material. Its crushing during grinding takes place following fracturing processes, triggered by elastic tensions within the crystal. These fractures start propagating in the areas of minute structural flaws, which are always present in solid bodies. Since the frequency of such structural flaws is reduced with decreasing particle size, a higher energy input is required for the production of new interfaces when very fine sugar is required.

Three stress mechanisms are involved in the crushing of solid bodies by mechanical means (8).

Table 3.2 Classification of sugar mills according to their operating mechanisms

---

1. Mills with stress exerted by means of one solid surface or by the collision of two particles
   1.1   Mills with rotating grinding devices
       1.1.1   Mills with a grinding track
           —hammer mill
           —turbo mill
           —impact pulverizer
       1.1.2   Mills without a grinding track
           —pin mill
   1.2   Mills without moving mechanical parts
       —air jet mill
2. Mills with stress exerted between two
   solid surfaces
   —roll mill
   —ball mill

  (i) Compression crushing between two solid surfaces
 (ii) Impact crushing by a solid surface or by particle–particle collisions
(iii) Using the surrounding medium to shear the material i.e. does not involve
      solid surfaces.

Impact stressing is the most effective way of crushing sugar (9), and crushing by impacting with a single solid surface is the most common principle of present-day sugar mills. Table 3.2 shows a classification of sugar mills according to the type of breaking mechanism employed (10).

Obtaining a closely defined particle spectrum during the grinding process is a major objective in the crushing of granulated sugar, because it results in well-defined physical properties within the chocolate masse produced. However, each mill will give its own particle size distribution in practice. For organoleptic reasons the maximum particle size should not exceed about 30 μm (1.2 × 10^{-3} inch) (see Chapter 13) in chocolate. On the other hand, 6 μm (0.5 × 10^{-3} inch) is said to be the mimimum size if the optimum rheological properties are to be achieved in the chocolate masse (9) (see Chapter 9). Numerous attempts have been made to fine-grind crystallized sugar within these limits. However, practice has shown that these restrictions are extremely difficult to conform to, even when the grinding process is combined with a separation/classification stage. It is equally impractical to achieve a controlled crystallization of the sugar within this narrow particle range at the production stage in the sugar factory (10). In order to obtain the ultimate fineness in chocolate with a satisfactory texture, the masse containing relatively rough sugar is normally processed by means of roll refining.

Icing sugar shows a tendency to form lumps. This is because freshly ground crystallized sugar possesses amorphous surface layers which are able to take up moisture at higher rates and at lower relative humidities than is the case for the crystallized parts. Those amorphous surface layers take up water vapour until they recrystallize, expelling water in the process. This leads to the formation of sugar solution which bonds the particles together, and becomes even firmer as the solution solidifies on drying. It is these solid bonds between the particles that cause a hardening of the icing sugar (11). It is thus advisable to process any ground sugar as speedily as possible and to use a tank with an agitator and screw discharge.

The risk of sugar dust explosions calls for special attention. The explosion of a sugar dust–air mixture may be triggered when a minimum concentration of sugar dust has been reached simultaneously with a high ignition energy. This may be due to electrostatic charging, friction or impact sparks. Tests have shown that a concentration of $30\,g/m^3$ ($0.8\,oz/yd^3$) of sugar dust with particle sizes up to $250\,\mu m$ (0.01 inch) is sufficient to cause an explosion (12). These conditions can be roughly determined by the visibility. With a concentration of $20\,g/m^3$ ($0.5\,oz/yd^3$), the visibility is 1 m (3.3 ft) at the most. At a concentration of $50\,g/m^3$ ($1.3\,oz/yd^3$), a 25 W light bulb is barely visible from a distance of 3 m (9 ft) (13).

A great number of preventive measures should be taken to avoid explosions, and it is not possible to discuss them all here in detail. However, it is useful to briefly discuss some of the most important safety measures (12).

Powder mills should be situated in separate, enclosed rooms which should, if possible, have a high ceiling, be well ventilated and situated on the top floor. Massive, sufficiently pressure-proof walls and fire-resistant doors opening to the outside should separate these rooms from all the other production facilities. A lightweight exterior wall can serve as a pressure release system, or a 'louvred' ceiling which pressure vents to outside atmosphere.

The equipment in the silos, the mill casing and the drive shaft must be electrostatically earthed. The feed devices should be equipped with magnetic solenoids to detect and remove any metal impurities which might cause sparks, leading to an explosion inside the mill. These magnets should be interlocked with the milldrive in such a manner that the powder mill automatically shuts down under failure. Mill and elevator heads must be fitted with explosion ducts which vent a long distance away over the roof. Any dust accumulations must be removed from the point of formation.

Within the production localities, the concentration of the sugar dust in the air, when kept within normal operating conditions, is far below the level at which explosions can occur. However, dust deposits can give rise to an unexpected danger. This is because they may be carried up within any piece of production equipment by the pressure wave of a primary explosion within that machine and they may then ignite themselves. It is absolutely necessary to prevent any dust deposits building up in the production rooms. This is aided by reducing dust production to a technically feasible minimum and by frequent cleaning and a high standard of housekeeping.

It goes without saying that all electrical installations must be intrinsically safe. Furthermore, electrical installations should be kept out of rooms with an explosion risk if at all possible. For example, it is preferable to install switches for electric lighting outside the milling rooms.

3.5 Amorphous sugar

As a rule, sugar is present in the crystalline state. Even the so-called 'acucar amorfo' made in Brazil is not a truly amorphous sucrose but only refined sugar of extremely small crystal size made from crude cane sugar (14). Truly amorphous sugar may be produced by spraying a sucrose solution into a hot and very dry atmosphere or by drying a thin film of sucrose solution at an extremely low relative humidity. The material produced in this way is highly hygroscopic (15).

Whilst crystallized sugar produces a sharp x-ray diffraction pattern, freshly prepared freeze-dried concentrated sugar solution, for instance, as well as sugar, which has been ground over a longer period of time, does not produce such a diffraction pattern; that is to say it is amorphous. The heat of solution for crystallized sucrose is $+ 16.75 \, J/g$; however, for amorphous sucrose a negative heat of $- 41.45 \, J/g$ may be recorded (16). It is known furthermore that amorphous surface layers are formed when icing sugar is produced by grinding crystallized sugar. The estimated proportion of this amorphous phase comes to approximately 2% with an average layer thickness of $0.75 \times 10^{-2} \, \mu m$ $(0.3 \times 10^{-6}$ inch) (11).

Apart from the rapid moisture uptake from the ambient air, the other interesting facet of amorphous sugar from the chocolate manufacturer's point of view is its flavour absorption properties (17). It has been estimated that about 30–90% of the sugar becomes amorphous during the roll refining of chocolate masses. This has a decisive influence on the chocolate taste, since amorphous sugar can absorb large quantities of different flavour compounds.

Chocolate masses made experimentally with sugar already ground to its final fineness were unsatisfactory in taste. This was said to be because the amorphous sugar normally produced during the refining stage was missing and thus not available for flavour absorption.

However, the rate of recrystallization increases with increasing temperatures and relative humidities. At very high air humidities, the amorphous sugar which has just been produced may recrystallize immediately after passing through the rollers. The flavour components absorbed by the amorphous sugar during the grinding of the chocolate masse are then released again during re-crystallization. Under normal conditions, even after a storage period, there is still some amorphous sugar to be found in the finished chocolate. This has been found to happen, to a certain degree, in the moulded chocolate.

Amorphous sugar may also have a positive influence in improving the heat stability of chocolate (17). A heat-stable chocolate may be produced by adding 1–10% of finely ground amorphous sugar pastes to the conched chocolate masse. This is then subjected to several days' heat treatment at temperatures between 20–35 °C. A network of matted sugar particles is thus formed within the moulded chocolate which will not soften again even at higher temperatures. The amorphous sugar may be made, for example, from something like a high-boiled candy masse which includes sucrose and invert sugar or dextrose respectively.

The positive advantage of freshly manufactured crumb, as far as heat stability is concerned, compared with using the single components like cocoa mass, milk powder and sugar, is due to the fact that the sugar present in this crumb is at least partially in the amorphous state. This is because the other ingredients of crumb, such as lactoproteins, fat and non-fat cocoa components, are very likely to have a pronounced retarding effect on the recrystallization of the sugar.

3.6 Other bulk sweeteners

Apart from sucrose, there are numerous other types of sugar and sugar alcohols, some of which have become important in the production of confectionery articles or have received special attention in recent years. More often than not there are major differences as far as physical, chemical or physiological characteristics are concerned, and not all of them are suitable for the production of chocolate masses. The various degrees of sweetening power of these sweeteners compared with sucrose are discussed in section 3.7.

Unlike saccharides, sugar alcohols are not carbohydrates but polyalcohols or polyols. However, they are considered to belong to the so-called 'bulk sweeteners', given their sweetening power and their caloric value. Some are suitable for use as sugar substitutes for diabetics. Furthermore, some are regarded as 'tooth friendly' sweeteners due to their reduced rate of metabolism by oral micro-organisms. A certain laxative effect of all sugar alcohols must, however, be taken into account.

Table 3.3 Amounts of sugars and sugar alcohols which can be tolerated in the diet (19)

Average daily intakes	g/day (oz/day)	g/day (oz/day)	g/day (oz/day)
	Per 1 kg (2.2 lb) body weight	70 kg (154 lb) body weight (adult)	20 kg (44 lb) body weight (child)
Mannitol	0.14 (0.005)	10 (0.35)	2.8 (0.1)
Sorbitol	0.43 (0.02)	30 (1.1)	8.6 (0.3)
Xylitol	0.43–0.71 (0.02–0.025)	30–50 (1.1–1.8)	8.6–14.3 (0.3–0.5)
Fructose	0.71 (0.025)	50 (1.8)	14.3 (0.5)
Sucrose	1.00–1.28 (0.035–0.045)	70–90 (2.5–3.2)	20–25 (0.7–0.9)

Table 3.3 summarizes daily intake values as a function of body weight to illustrate levels at which they can be safely eaten (17).

3.6.1 *Invert sugar*

Invert sugar is a mixture consisting of equal parts of the monosaccharides fructose and glucose which is produced during hydrolytic cleavage of the disaccharide sucrose, by using either the activity of specific enzymes or that of acids. It is commercially available as a syrup or as partially crystallized paste with a dry solids content of 65–80%. Invert sugar is naturally present in many fruits as well as in honey. This product is not suitable for the manufacture of chocolate, because it is almost exclusively supplied as an aqueous syrup.

Publications from Japan, however, refer to the manufacture of a pseudo-chocolate using freeze-dried dates. In this case the natural sweetness is almost exclusively made up of invert sugar (20). The sweetening power of invert sugar corresponds closely to that of sucrose.

3.6.2 *Glucose*

The monosaccharide glucose, also known by its trade name dextrose, is present in nature where it is found, together with fructose, in many fruits and in honey ('invert sugar'). Since it was prepared in former times from grapes, glucose is also known as 'grape sugar'.

Today, glucose is normally produced industrially by an extensive hydrolysis of starch into high-conversion glucose syrup, from which it is crystallized in the form of dextrose monohydrate. About 9% of water is retained as a constituent of the glucose, which has a melting point of 83 °C (181 °F). Part of this water of crystallization is released at temperatures which are commonly reached during conching and produces an adverse effect on the rheological properties of the chocolate (Chapter 8). It is preferable, therefore, to use glucose with no water content (anhydrous dextrose) for the manufacture of chocolate (21).

Even though there have been many attempts to produce 'grape sugar' (or dextrose) chocolate, these chocolates sweetened with only glucose have never become very popular because their taste characteristics differed from the standard product. This, however, does not prevent glucose being used in small quantities together with sucrose for the sweetening of chocolate. The sweetening power of glucose is very much less than that of sucrose.

A variety of glucose syrups (corn syrups) with solids contents of about 70–80% may be produced by means of partial saccharification of starch. These syrups, together with glucose, may also contain maltose and other saccharides of higher molecular weight. Furthermore, syrups with various degrees of fructose are made by means of partial enzymatic isomerization of glucose. Glucose syrups are not used commercially to make chocolate on account of their water content, and nothing has yet been reported concerning the use of spray-dried glucose syrups. On the other hand, large amounts of glucose syrup are used for the manufacture of toffees, fondants, etc.

3.6.3 Fructose

Fructose is the monosaccharide, known also as fruit sugar, which is present, together with glucose, in almost all fruits and in honey ('invert sugar'). Fructose is currently being produced in most large-scale processes by isolation and subsequent crystallization from fructose containing glucose or invert sugar syrups. Fructose is naturally hygroscopic and has a melting point of 102–105 °C (215–221 °F). The sweetening power of fructose is usually considered to be higher than that of sucrose. This, however, depends on a number of different factors (see also section 3.7). Fructose is of special importance for the manufacture of chocolate and confectionery articles suitable for diabetics. This is because there is no acute blood sugar increase in the body after eating fructose, unlike sucrose or glucose. Fructose is metabolized more slowly than glucose and only a small proportion is converted into glucose during absorption. This slow absorption, as well as the speedy metabolism, of the fructose, reduces any rapid peaks in the blood glucose concentration. Furthermore, fructose is mainly metabolized in the liver, i.e. without insulin involvement, whereas the corresponding glucose metabolism reaction depends upon insulin. This makes fructose a valuable sugar substitute in diabetic diets. It should, however, be used only in moderate quantities in order to avoid overstressing the fructose metabolism. The recommended quantity per day is 50–100 g (1.8–3.5 oz) of fructose taken in two or four doses (27).

Certain factors need to be taken into account, however, when fructose is being used in the manufacture of chocolate. For instance, the water content of the other ingredients such as milk powder should be kept as low as possible, and the temperature during the conching processes should not exceed 40 °C (104 °F). Failure to do so may result in a 'sandy' mouthfeel and/or the formation of degradation or reaction products which give off-flavours. This risk is very high since fructose caramelizes and takes part in Maillard reactions fairly easily.

3.6.4 Lactose

Lactose, also called milk sugar, is a disaccharide consisting of the mono-saccharides glucose and galactose, and is an integral part of all types of milk. In cow's milk, it amounts to about 4.5%. The present day large-scale production of lactose is based on whey, from which it is isolated to a very high degree of purity, following several purification steps. Lactose crystallizes with one molecule of water as a monohydrate. It does not expel this water even when heated to 100 °C (212 °F). Lactose has been used traditionally in the production of milk chocolate as a constituent of full cream milk powder, skim milk powder or chocolate crumb. However, in recent years pure lactose is more and more frequently being added in small quantities to sucrose in the manufacture of chocolate (21).

Lactose monohydrate is non-hygroscopic and forms crystals harder than those of sucrose which melt only at temperatures between 202–252 °C (396–486 °F). Compared with sucrose, its sweetening power is very low.

3.6.5 *Isomaltulose*

Isomaltulose, which is also known by the trade name 'Palatinose', has been detected in very small quantities in honey and in cane sugar extract. This material is produced by enzymatic action from sucrose. Isomaltulose is a disaccharide made up of the monosaccharides glucose and fructose, which crystallizes with one molecule of constituent water. Its sweetening power is clearly inferior to that of sucrose. Isomaltulose has been recommended as a sugar substitute for candy and chocolate articles (23).

3.6.6 *Polydextrose*

Polydextrose is a synthetic polymer made up of glucose and small amounts of sorbitol, which, due to its manufacturing process, also contains minor residues of citric acid. It is an amorphous, rather hygroscopic and non-sweet tasting powder with a melting point above 130 °C (266 °F). Since it is only partly metabolized in the human body, it is a frequently used raw material for low-calorie food. The FDA has recognised polydextrose as containing not more than 4.2 J/g, while sucrose and other carbohydrates contain about 16.8 J/g (24). Since polydextrose is not sweet in itself, it is always used together with other sweetening agents. Owing to its hygroscopicity and the residual amounts of citric acid, the use of polydextrose is not recommended in the manufacture of chocolate (25). There are chocolates on the market, however, for which polydextrose and additional fructose as sweetening agent are used.

3.6.7 *Sorbitol*

Sorbitol is a monosaccharide alcohol which is present in small quantities in numerous fruits. The commercial production process is based on the catalytic hydrogenation of glucose. In its crystallized form, sorbitol is hygroscopic and has a melting point between 93–97 °C (199–207 °F). Its sweetening power is noticeably inferior compared with that of sucrose; the heat of solution of sorbitol generates a slightly cooling effect when dissolved in the mouth. Sorbitol is used in the production of chocolates for diabetics, with an artificial sweetener being generally used in order to enhance its sweetness (26).

3.6.8 *Mannitol*

Mannitol is a monosaccharide alcohol present in manna, the dried juice of the flowering or manna ash. The large-scale industrial production is a catalytic hydrogenation process based on pure invert sugar, which results in a mixture of sorbitol and mannitol, from which mannitol is separated in a multi-step process. Mannitol is non-hygroscopic with a melting point ranging from 165–169 °C (329–336 °F). Its sweetening power is comparable with that of sorbitol. In combination with sorbitol and enhanced in its sweetening power by artificial sweeteners, it is occasionally used for the production of chocolates for diabetics. However, of all the sugar alcohols, mannitol manifests the greatest laxative effect; the safe daily intake for adults is only about 10 g (0.35 oz) (19).

3.6.9 *Xylitol*

Xylitol is another monosaccharide alcohol which, however, unlike sorbitol and mannitol with six carbon atoms each, has only five carbon atoms. Xylitol exists in many natural forms and is present in numerous mushrooms, vegetables and fruits. For example, Jamaica plums contain about 1% (dry weight) of xylitol. Furthermore, xylitol is an inherent part of the normal human metabolism. Every human being generates between 5–15 g (0.2–0.5 oz) of xylitol every day as part of the special carbohydrate metabolism. The large-scale production of xylitol is based on birch wood, corn cobs, straw and other plant material containing a high amount of xylan. Xylan, a polymer of xylose, is hydrolysed to xylose by means of acids, once it has been isolated from the raw material. Following further isolation and purification of the xylose, it is hydrogenated into xylitol.

Xylitol is non-hygroscopic and melts into a clear liquid at temperatures between 92° and 96 °C (178–205 °F). Of all the known sweetening agents it has the greatest heat of solution (153.1 J/g) which results in a remarkable cooling effect when melting in the mouth. The sweetening power of xylitol is the highest of all the sugar alcohols and comparable to that of sucrose (27).

The physiological properties of xylitol are quite remarkable. Like any sugar alcohol, xylitol exhibits laxative effects when consumed in excess. However, the body can adapt to the continuous use of this material and eventually xylitol becomes more highly tolerated even than sorbitol (19). As with other mono- and disaccharide alcohols, xylitol causes only negligible increase in blood glucose concentration.

However, xylitol's exceptional characteristic consists in its prophylactic benefit with respect to dental caries. It cannot be fermented by the cariogenic bacteria of the oral cavity, especially *Streptococcus mutans*. As the highly publicized Turku studies proved, the progression of caries was arrested in people eating xylitol-containing sweets over a period of two years. Furthermore, two extensive studies initiated by the World Health Organisation, each encompassing more than 900 school pupils in Hungary and Polynesia over the period 1981–1984, showed that the caries progression may be significantly slowed down if 15 to 20 g (0.5–0.7 oz) of xylitol in the form of xylitol-containing chocolate and candy are eaten together with a normal, sucrose-containing diet (29, 30). Chocolates may be produced with xylitol without any special problems, and these generate a cooling mouthfeel.

3.6.10 Maltitol

The disaccharide alcohol maltitol is produced by hydrogenation of maltose. Maltitol is non-hygroscopic and melts between 130–135 °C (266–275 °F). It is slightly less sweet than sucrose, but sweeter than sorbitol and it is suitable for confectionery articles for diabetics. Even though oral lactobacilli may ferment maltitol, it is not fermentable by streptococci, and this is why maltitol is considered a 'tooth-sparing' bulk sweetener. At only 8.4 J/g, maltitol is claimed to be a low-calorie agent (31) though this is sometimes disputed.

Maltitol is, however, not yet available commercially in pure form but only as a product which contains sorbitol and hydrogenated higher polysaccharides together with 86–90% of maltitol. In order to compensate for a sweetening power which is slightly inferior to that of sucrose, other artificial sweeteners are sometimes added to chocolates sweetened with maltitol. The combination of maltitol and xylitol enables a very good-tasting sugar-free chocolate to be produced with a sweetness which can hardly be differentiated from chocolates made with sucrose (see also Chapter 13).

3.6.11 Isomalt

Isomalt is a mixture of equal parts of two disaccharide alcohols and is generally known by its trade name 'Palatinit'. It is produced by means of enzymatic conversion of sucrose to isomaltulose, which is then hydrogenated. One of these two sugar alcohols crystallizes with 2 molecules of water, the other exhibits an anhydrous crystallization. The total bound water content of the product reaches 5%. Isomalt is non-hygroscopic and is clearly inferior to sucrose in its sweetening power. The melting point is reached at about 145–150 °C (293–302 °F) (32). The laxative effect of isomalt is said to wear off as the body becomes adapted to it. The product may be used for the production of 'tooth-sparing' candy and confectionery articles for diabetics. The claim is that isomalt is utilized at only 8.4 J/g and thus with only 50% of its caloric potential. The low sweetening power of isomalt in chocolate may be enhanced by means of other artificial sweeteners.

3.6.12 Lycasin

Hydrogenated glucose syrups containing sorbitol, maltitol and hydrogenated higher polysaccharides are commercially available under the trade name of

'Lycasin'. The product is used as a crystallization inhibitor of, for instance, xylitol in 'tooth-sparing' confectionery articles. Lycasin is not used in chocolates due to its high moisture content, and is not considered suitable for diabetics.

3.7 The sweetening power of bulk sweeteners

In the sensory evaluation of the sweetening power, sucrose generally serves as a standard to which the sweetness of other bulk sweeteners is compared. 'Sweetening power' and 'degree of sweetness' are the terms employed. The sweetening power is the intensity of sweetness expressed in percent in comparison with sucrose, which is assumed to be 100%. The degree of sweetness expresses the intensity of the sweet taste as a fraction of the sucrose's sweetness, which is equal to 1.00. Table 3.4 lists the degree of sweetness of the most important types of sugar and sugar alcohols (33).

Table 3.4 Relative degree of sweetness of different sugars and sugar alcohols (33).

Sucrose	1.0
Fructose	1.1
Glucose	0.6
Xylitol	1.0
Maltitol	0.65
Sorbitol	0.6
Mannitol	0.5
Isomalt	0.45

As can be seen from Table 3.4, fructose is sweeter than sucrose and the sugar alcohol xylitol is just as sweet as sucrose. The table is typical and representative of many publications. However, it only provides guideline values for the confectioner because many comparisons were apparently based on aqueous solutions of pure sugars and more often than not statements regarding the solids content of the sampled solutions are missing. It is, moreover, important to remember that the sweetening power of the various types of sugar does not increase linearly with concentration, but is highly dependent upon the temperature and the other raw material ingredients in the foodstuffs, as well as the pH value.

Furthermore, there are synergistic effects. These phenomena cause sugars in mixtures to mutually enhance their sweetening power as well as their other sweetness characteristics. Fructose in milk chocolate has been found to taste only slightly sweeter than sucrose, yet mixtures of both sugars, on the other hand, were very much sweeter. Even though dextrose in a 10% aqueous solution shows a markedly lower sweetening power than sucrose, milk chocolates which were sweetened by anyhydrous dextrose tasted almost as sweet as those made with sucrose. The sweetness also increased when part of the sucrose was replaced by anhydrous dextrose. However, dextrose leaves a burning taste in the mouth which rather enhances the impression of sweetness instead of moderating it. Fructose also gives a 'pungent' sweetness in milk chocolate, but on the other hand, this product is felt to be pleasantly mild in chocolate drinks for instance (34).

If sucrose in cocoa-containing or any other foodstuff is to be partially or completely replaced by other types of sugar or by sugar substitutes, theoretical evaluations are at present unable to predict the sweetness of the product. These will only provide rough guidelines and it is always necessary to carry out sensory evaluations on actual samples of the product. In order to obtain the best results, the recipes will normally need adapting and a series of samples should be manufactured.

References

1. Hoffmann, H., Mauch, W. and Untze, W. *Zucker und Zuckerwaren*. Verlag Parcy, Berlin (1985).
2. Meade. G.P. and Chen, J.C.P. *Cane Sugar Handbook*. 11th edn., John Wiley, New York (1985).
3. Diefenthaler, T. *Zucker- und Süsswarenwirtschaft* **27** (1974) 238, 240, 255–257.
4. Kelm, W. 'Anforderungen an die Silolagerung von Zucker'. Paper given at the international seminar 'Schoko–Technik', Central College of the Germany Confectionery Trade, Solingen, 5 December 1983.
5. Gaupp, E. *Kakao und Zucker* **11** (1972) 456–458.
6. Tills, J.W. 'An introduction to pneumatic conveying with particular reference to granular sugar'. *Brit. Sugar Corp. 20th Techn. Conf.*, 1970.
7. Neumann, E. *Zucker- und Süsswarenwirtschaft* **27** (1974) 200, 202–203, 206.
8. Rumpf, H. *Chem–Ing. Techn.* **31** (1959) 323–337.
9. Niediek, E.A. *Zuckerindustrie* **21** (1971) 432–439, 492–498; **22** (1972) 21–31.
10. Heidenreich, E. and Huth, W. *Lebensmittel-Ind.* **23** (1976) 495–499.
11. Roth, D. *Zucker* **30** (1977) 464–470.
12. Schneider, G. *Zucker* **22** (1969) 166–171, 473–479, 573–577.
13. Dietl, H. *Zucker* **14** (1961) 594–599.
14. Anon. *Sugar y Acucar* **72** (5) (1977) 55–57.
15. Powers, H.E.C. *Int. Sugar H.* **82** (1980) 315.
16. Van Hook, A: *Sugar J.* **43** (12) (1981) 31–32.
17. Niediek, E.A. *Zucker- und Süsswarenwirtschaft* **34** (1981) 44–57.
18. Pirsch, L.A., Schubiger, G. -F. and Rostagno, W. Brit. Patent 12 199996 (1969).
19. Kammerer, F.X. *Kakao und Zucker* **11** (1972) 184–190.
20. Anon. 'Morinaga begins selling sugarless chocolate using date as sweetening'. *Conf. Prod.* **47** (1981) 62.
21. Hogenbirk, G. *Man. Conf.* **65** (10) (1985) 27–34.
22. Askar, A. and Treptow, H. *Ernährungs-Umschau* **32** (1985) 135–141.
23. Kaga, T. and Minzutani, T. *Proc. Res. Soc. Japan. Sugar Ref. Technol.* **34** (1985) 45–57.
24. Liebrand, J. and Smiles, R. *Manuf. Conf.* **61** (1981) (Nov.) 35–40.
25. Carpenter, J. *Manuf. Conf.* **64** (1984) (May) 63–67.
26. Caliari, R. *Manuf. Conf.* **63** (1983) (Nov.) 25–30.
27. Voirol, F. 'Fructose und Polyalkohole in Diat- und Spezialprodukten'. Paper given at International Confectionery Symposium, Central College of the German Confectionery Trade, Solingen, 30 October 1985.
28. Makinein, K.K. *Conf. Prod.* **3** (1977) (Nov.) 464–466.
29. Scheinin, A., Banoczy, J., Szöke, J. Esztari, I., Pienihäkinen, K. Scheinin, U., Tiekso, J. Zimmermann, P. and Hadas, E. *Acta Odont. Scand.* **43** (1985) 327–348.
30. Kandelmann, D., Hefti, A. and Bär, A. *Caries Res.*, submitted for publication.
31. Lichtel, R. 'Maltit–Eigenschaften und Anwendungs möglichkeiten'. Paper given at the

International Seminar 'Ernährungsphysiologisch angepasste Süsswaren', Central College of the German Confectionery Trade, Solingen, 9 October 1985.
32. Bollinger, H. 'Palatinit: Technologische Eigenschaften und Einsatzmöglichkeiten in Süsswaren'. Paper given at the International Confectionery Seminar, Central College of the German Confectionery Trade, Solingen, 9 October 1985.
33. von Rymon Lipinski, G.-W. *Alimenta* **25** (1986) 119.
34. Krüger, Ch., Sievers, B. and Vonhoff, U. *Zucker und Susswarenwirtschaft* **40** (1987) 7–13.

4 Milk

E.H. REIMERDES and H.-A. MEHRENS

4.1 Introduction

Milk and milk products are very important food ingredients due to their nutritional, organoleptic and processing properties. One of their main applications is, however, in the manufacture of chocolate. Milk chocolate has become a very popular product due to its unique blend of cocoa and milk flavours. Nowadays the consumption of milk chocolate greatly exceeds that of dark varieties. This is in line with other widespread applications of milk products being used as valuable food ingradients.

Traditionally milk and concentrated milk products have been used for food production in typical milk products such as yoghurt or cheese, or as food additives in various products, e.g. bakery, confectionery, sausage, dressings and mayonnaise, etc. Due to the recent developments in modern food processing milk-based ingredients can be produced over a wide range of compositions. This includes special powders made from whole milk, skimmed milk and buttermilk, and also various components and mixtures of these, including whey and milk proteins. The availability of this extended range of products has greatly enhanced the use of milk-based ingredients, especially in convenience and dietetic foods and in chocolate production.

Due to these tailormade products, research on milk-based ingredients has become much more orientated in the direction of their processing properties and nutritional value. The knowledge and experience gained so far has greatly improved the quality of the end products. In the same way the development of milk chocolate has been improved by the application of new technology for milk chocolate production. A wide range of milk powders and milk fat is now available, each having specific properties and composition. This chapter has therefore been divided into several sections. These are milk composition, tailormade milk products for chocolate production, milk fat products, and finally special products based on milk and chocolate ingredients, e.g. crumb.

4.2 The composition of milk

Being the total source of nutrients for the newborn mammal, milk must be a very special food with a complete and very complex composition. In Table 4.1 the average composition of cow's milk is shown. Beside the main components, fat, protein, lactose, and minerals, milk contains many different substances of

C

Table 4.1 Average composition of cow's milk

	(g/l approximately = % × 10)	% by mass of dry matter
Water	870 g/l	—
Fat	37 g/l	28.5
Protein	34 g/l	26.2
NPN	2.5 g/l	1.9
Lactose	47 g/l	36.2
Minerals	7 g/l	5.4
Na$^+$	0.5 g/l	—
K$^+$	1.5 g/l	—
Ca^{2+}	1.3 g/l	—
PO$_4^{3-}$	2.1 g/l	—
Citrate	2.0 g/l	—
Vitamins		
Vitamin D	0.4 μg/l	—
Riboflavin	1.5 mg/l	—
Ascorbic acid	20 mg/l	—
Nicotinic acid	1.2 mg/l	—

high nutritional value: vitamins, trace elements, non-protein nitrogens (NPN), etc. This complex composition is necessary to supply the newborn mammal with the complete set of its required nutrients. For this reason milk, as an ingredient, increases the nutritional value of food products.

Each milk-based product has its own processing properties as well as imparting distinct physical characteristics to the final product. This means the typical properties of proteins, fat, carbohydrates and minerals have to be recognized and standardized for food application. Variations in the concentration of the constituents have a profound effect on the processing and physical properties, and these may be utilized in order to obtain optimized basic ingredients in, for example, milk chocolate manufacture, as will be shown later.

Milk proteins have a wide potential as food ingredients. The composition of this milk protein fraction is given in Table 4.2. Cow's milk contains about 3–4% protein. This protein is a complex mixture predominantly composed of caseins (75–85%) and whey proteins (∼ 18%). Besides this, it contains a fat globule protein complex (∼ 1%) and the so-called non-protein nitrogen compounds (∼ 5%).

Caseins, whey proteins and the fat globule protein complex differ widely in their structures and potential uses. Besides this, caseins contain about 28

Table 4.2 Composition of milk proteins

	%	Number of components
Caseins	75–85	28
Whey proteins	∼18	15
Fat globule membrane proteins	∼1	25–30
NPN-compounds	∼5	?

Table 4.3 Important compositional and structural parameters of caseins and whey proteins

Caseins
  High proline content
  Phosphoseryl residues
  No disulphide bonds
  Distinct hydrophilic and hydrophobic domains
  Relatively flexible open chain structure
  Formation of high molecular weight aggregates (micelles)
  Detergent-like structure

Whey proteins
  High helical structure content stabilized by
      (i) ionic and hydrogen bonds
     (ii) hydrophobic interactions
    (iii) disulphide bonds
  Hydrophilic molecular surface
  Typical globular protein structure

different components, whey proteins about 15, and fat globule membrane 25–30, (all including genetic variants.)

A detailed analysis of these protein mixtures shows that single components, e.g. kappa-casein from the caseins and β-lactoglobulin from the whey proteins, are responsible for special properties of milk protein products. For this reason it is very important that, during production of dairy-based ingredients, the processing parameters are standardized in order to obtain high-quality food additives with reproducible physical properties.

When deciding which ingredient to use it must be remembered that the caseins are heat-stable open-chain proteins, whereas the whey proteins are globular proteins which can be denatured by heat treatment. Table 4.3 shows some of the underlying compositional and structural parameters of caseins and whey proteins which accounts for the different physico-chemical behaviour of these two fractions.

Table 4.4 Composition of lipids in cow's milk

	% of total fat
Triglycerides	96 –99
Diglycerides	0.3 –1.6
Monoglycerides	0.02–0.1
Phospholipids	0.2 –1.0
Sterols	0.2 –0.4
Cerebrosides	0.01–0.07
Free fatty acids	0.1 –0.4

Fat-soluble vitamins: A, D, E, K

Traces:
  Hydrocarbons
  Sterolesters
  Waxes
  Squalene

These differences can be used in the production of tailormade milk protein products. The heat stability of the caseins and the changes in the structure of whey proteins during processing are of great importance and related to the characteristics of the finished product.

Like the proteins, the lipid constituents of milk have a very complex composition. Table 4.4 shows the various components of the lipid fraction of milk.

The major components in milk fat are triglycerides. Besides these, small amounts of mono- and diglycerides, phospholipids, cerebrosides and gangliosides as well as fat-soluble vitamins are present. The surface-active constituents like mono- and diglycerides and phospholipids, e.g. lecithin, can have a great impact on the physical properties of the fat component.

Since the impact of lecithin on the rheological properties (viscosity, yield value) of chocolate masse passes through an optimum (see Chapter 9), the amount of lecithin already incorporated by the dairy ingredient has to be taken into account when working out total lecithin content. Although milk fat has a setting temperature in the range of 15–20 °C (59–68 °F) the butyric acid, content, esterified in triglycerides, results in a high proportion of liquid fat in the temperature range 15–30 °C (59–86 °F), either in the pure fat or in mixtures with cocoa butter. The hydrogenation and fractionation of butter fat enables the processing properties of milk fat additives to be optimized. In this case the compatibility of the milk fat fractions with cocoa butter with respect to the fat setting properties is normally the limiting factor. The milk fat content and type has to be adjusted according to the processing used and organoleptic properties of the product, e.g. mouthfeel.

The lactose content of milk raw materials is also of importance in milk chocolate manufacture. Lactose has a low solubility and forms different crystalline structures. Depending upon the drying conditions, the structure of lactose ranges from amorphous to α- or β-crystalline structure. When tailormade milk-based ingredients are produced, the crystalline structure of the lactose is of great importance, as it has an influence on the particle size distribution in the final chocolate product. Besides this, lactose has a distinct flavour-binding capacity which may affect the quality of the end product.

The minerals in milk are also important in chocolate production. In particular, calcium and phosphate content, which can be present in several different forms in milk, must be taken into account. Calcium phosphate is found in a soluble form and as so-called colloidal calcium phosphate. Phosphate is bound to caseins via phosphoseryl residues.

Because differing amounts of phosphates can be present, and the formation of calcium phosphate–lactose complexes during special drying procedures can influence the mineral balance in dairy-based ingredients, processing and product changes can be seen to occur.

4.3 Processing properties of milk ingredients

When considering the processing and physical properties of milk-based ingredients, various factors are important. The milk proteins, some components of the milk fat, and the milk fat triglyceride structure have a great influence on these properties, especially in combination with other ingredient materials such as cocoa powder.

In Table 4.5, important physical properties of milk proteins are shown. The table shows the complexity of possible reactions and interactions between milk proteins and other ingredients like sugar and cocoa powder. The relative importance of single factors may vary depending upon the ingredients used and the processing procedures employed, but most of them are important to a certain degree when manufacturing an optimal or at least an acceptable product.

Table 4.5 Physical properties of milk proteins

Solubility (PSI)	Rheological properties
Dispersibility (PDI)	Gelation properties
Water absorption	Elasticity
Water binding capacity	Adhesion
Fat binding capacity	Cohesion
Emulsification properties	Surface activity
Foaming properties	Texture-forming ability
Colour, taste, texture	

These properties are very different in caseins, in whey proteins and in fat-globule membrane proteins. These observed differences are due to their differing structural features. Caseins, having a flexible open-chain structure, show a high fat- and water-binding capacity and are very suitable emulsifiers and stabilizers for multiple-phase systems. The globular whey proteins can have similar properties, but their actual behaviour depends very much on the applied heat treatment during manufacture. The degree of denaturation of the original structure greatly influences its physical properties. It is therefore apparent that it is more difficult to manufacture standardized whey products than caseins with regard to these properties. In the case of whey protein, each processing step has to be standardized and the different types of whey have to be carefully analysed for composition and pretreatment changes.

4.4 Milk-based ingredients

Traditionally whole milk, skimmed milk, and buttermilk powder produced by different techniques have been used in chocolate manufacture. According to EEC food regulations and guidelines, for example, milk chocolate has to contain between 14–20% total milk solids derived from whole, partially skimmed or skimmed milk and milk fat.

Table 4.6 Methods for the production of tailored milk-based ingredients

Acid and/or heat precipitation
Reverse osmosis
Ultrafiltration, diafiltration
Ion exchange chromatography
Gel filtration
Electrodialysis
Precipitation by solvents or complexing agents

Today the development of new technologies provides the opportunity to supply tailormade products to the user of milk-based products. In Table 4.6 some of these technologies are listed.

Acid precipitation under moderate heating conditions is the principal stage in isolating high-quality casein. In order to obtain soluble or dispersible caseinates, a further step is necessary using for example ammonia, sodium or calcium hydroxide. The physical characteristics of the resultant products are highly influenced by the alkali(s) employed.

Heat precipitation of whey proteins has long been used for the manufacture of insoluble lactalbumin. The advent of new technologies like ultrafiltration or ion exchange, allowing the isolation of native whey proteins, has resulted in a decline in traditional lactalbumin production. The combined effect of acid and heat on skimmed milk yields a so-called 'coprecipitate'. It contains caseins and whey proteins in an almost insoluble form.

Recent methods to manufacture total milk protein isolates by precipitation are the so-called total milk proteinate (TMP) and the soluble lactoproteinate (SLP) processes. The composition is close to that of coprecipitates but the solubility and other physical properties of the products manufactured by these methods are said to be superior.

Reverse osmosis allows the concentration of total solids without any heat treatment. This means that there will be no denaturation of heat sensitive whey proteins during the concentration stage. *Ultrafiltration* has become the most popular technology for the manufacture of protein or protein- and fat-rich products from whey and skim milk or from whole milk. Low molecular weight compounds like minerals, lactose or NPN-components can be removed to a differing degrees. Using sweet whey as the basic ingredient, powders having the following composition can be obtained: protein 30–80%, lactose 4–47%, and minerals 3–8%. *Ion exchange* treatment of whey gives protein isolates with a protein content of at least 90%. Using ion exchange for demineralization it is possible to remove more than 99% of the ash. *Electrodialysis* is used to demineralize milk or whey products. One important aspect of this technology is that, due to the physico-chemical principles involved, certain ions can be selectively removed: sodium, potassium and chloride ions are removed more extensively than calcium, phosphate or organic acid ions.

In Table 4.7 the average compositions of a variety of milk-based products are listed.

Table 4.7 Average composition (%) of milk-based products (powder)

	Water	Protein	Fat	Lactose	Ash
Products containing total milk proteins					
Whole milk	2.0	25.4	27.5	38.2	5.9
Skimmed milk	3.0	35.8	0.7	51.6	7.9
Buttermilk	3.0	34.3	5.3	50.0	7.6
Coprecipitate	4.0	87–91	0.6–1.0	0.5	3.5–14.0
Caseinates	4.0	87–91	1.0–2.0	0.5	3.5– 5.0
Whey protein products					
Sweet whey	4.0	12.0	1.3	73	7.9
Acid whey	4.0	12.0	0.8	69	11.5
Whey protein concentrates (UF)					
A	4.6	36.2	2.1	46.5	7.8
B	4.2	63.0	5.6	21.1	3.9
C	4.0	81.0	7.2	3.5	3.1
Whey protein isolate	5.0	92.0	0.3	0.3	2.8
Demineralized whey					
(90% demin.)	3.8	14.0	0.7	83	0.9
(70% demin.)	4.0	12.0	1.5	79	2.7

Today whole milk powder dominates the usage of milk-based products in continental European chocolate production. The situation could change if suitable replacements, like some of those listed above, can be shown to give a processing, product or commercial advantage. By selecting the correct ingredients and processing methods it is possbile to produce a product which has the best physical and processing properties for the use that it is required. Milk chocolate production serves as a good example, since tailored milk-based and cocoa ingredients are combined to produce a quality product.

For example the composition and physical properties of whey protein products can be modified (see Table 4.7) according to requirement. The physical characteristics are very much related to the heat applied during processing. In milk chocolate production, the change in fat- and water-binding properties of whey proteins during heating are of particular importance.

4.5 Milk powders in milk chocolate production

There are some special requirements with regard to the composition and structure of the milk powders used in milk chocolate production. These are related to the quality of the raw milk and changes which take place during the manufacturing process. The raw milk quality predominantly influences the properties of the milk powder and depends very much on the storage conditions on dairy farms, e.g. storage time and temperature. It is important that only powders made from high-quality milk should be used for chocolate production.

The manufacturing process of the powders has the greatest impact on their suitability for milk chocolate. It consists of the following steps: preheating (homogenization), concentration by evaporation, and drying. The latter process can be performed by roller or spray drying. Each step has an effect upon the resultant powder. The preheating regime is usually pasteurization with a temperature/ time combination of 71–74 °C (160–165 °F) for 40–45 seconds or 85 °C (185 °F) for 8–15 seconds. This step is necessary to ensure the bacteriological stability of the powder. If a high degree of whey protein denaturation is required, the milk can be heated to 105–130 °C (220–265 °F) for 30 seconds. This changes not only the physical characteristics of the whey proteins but also those of the caseins, since certain interactions between the protein components, e.g. via disulphide interchange, take place.

A homogenization step before drying seems to reduce the amount of surface fat in the powder. It results in a reduction of viscosity and yield value, conditions which are advantageous for milk chocolate production. The flavour binding ability also seems to be enhanced.

During the concentration stage, a heat modification of the protein system can be obtained. The evaporation is performed in multiple-effect evaporators under reduced pressure. A concentration to 50% total solids with temperatures of 75 °C (167 °F) to 55 °C (131 °F) appears to be favourable for economic reasons.

More severe heating conditions during evaporation and drying can lead to caramelization and Maillard reactions. Caramelization has an even more pronounced effect on surface fat concentration, viscosity and yield value than homogenization. Also, the lipase activity is more markedly reduced. This activity does not seem to pose any problems in chocolate production since the water content of chocolate is around 0.5–1.0%. Additional heating can, however, lead to organoleptically active compounds which are responsible for particular flavour characteristics of the end product. It depends on the customer whether this caramel taste is accepted or in fact required.

After evaporation, drying is performed by spray or roller driers. Entirely different types of milk powders are produced depending upon which drier was used. The roller drying process, which is the more severe with respect to applied heat, is to be preferred in chocolate manufacture for several reasons. The main advantage of roller-dried compared to spray-dried powders is their high content of free surface fat (>95% $v.$ < 10%). This results in favourable rheological properties during the manufacturing steps of mixing and conching. The energy consumption is reduced considerably and the amount of cocoa butter can be reduced under constant rheological conditions.

Since the amount of free fat present is so important to the rheological properties both during manufacture and in the end product, a combination of skimmed milk powder and butter oil is now being used by some manufacturers.

Using spray-dried powders, the influence of different types of spray-drying equipment (feed system, atomizer and dryer operation) on the microstructure of the resultant powder has to be taken into account. The differences involved include bulk density and particle size distribution: both can be changed drastically by the recycling of fines.

Roller and spray dried powders also differ in their organoleptic properties, due to the Maillard reaction. Roller-dried powders tend to have a somewhat spicy and salty taste, whereas spray-dried ones have a distinct milky flavour. During conching of chocolate masse containing milk-based ingredients, the formation of Maillard products above 65 °C (149 °F) can be used by a manufacturer to create his special flavour profile. Since individual attitudes differ towards flavour perception, it is eventually the customer who decides which type of powder is preferred.

4.6 Milk and chocolate crumb

Special products used in the manufacture of milk chocolate are so-called milk and chocolate crumb. Crumb is manufactured from milk and sugar with or without the addition of cocoa mass respectively. The addition of cocoa mass stabilizes the crumb against oxidative rancidity. An essential part of the crumb-making process is the Maillard reaction, which occurs when (milk) proteins are heated in the presence of sugars and moisture. The organoleptically active compounds generated add a typical flavour to the end product. This reduces conching time to a considerable extent because of the earlier flavour generation.

The main determining factors for flavour development are:

 (i) Content of amino groups and reducing sugars (lactose)
 (ii) Optimal moisture content
(iii) Temperature/time relationship
(iv) Optimal pH level, in the range 5.5–7.5.

Two crumb manufacturing process employed today are the vacuum method and the atmospheric crumb method. The main features of these processes, together with their effect on the resultant product, will now be outlined.

4.6.1 Batch vacuum method

Milk is cooled to 4 °C (39 °F) and put into storage tanks. In a preliminary heat treatment, milk is heated to about 74 °C (165 °F) and concentrated in multiple effect evaporators to 30–40% total solids. Sugar is added to the condensed milk and the mixture is cooked at 75 °C (167 °F) under vacuum with rapid boiling until a concentration of 90% solids is obtained. At this point the sugar begins to crystallize and this continues in the subsequent kneading step.

When producing chocolate crumb, the cocoa mass is added at this stage. The stiff paste is spread in thin layers in trays and dried under vacuum to a water content of less than 1.5%. Variations made during this final stage give the manufacturer the most versatility in choosing a specific flavour for the finished product. This is due to the Maillard reaction's dependence on time and temperature. Product variability is often, however, associated with fluctuations during the final drying stages.

The dried crumb 'cake' is removed from the trays and broken to a kibble form in which it can be stored. If the crumb is to be used immediately it is ground to a fine powder and mixed with the other chocolate ingredients.

Because of the high costs associated with this batch process, continuous vacuum methods have been developed.

4.6.2 *Atmospheric crumb process*

In the mid-1970s an atmospheric process, designed to save time and energy, and therefore reduce costs, was developed.

At the initial stage sweetened condensed milk or reconstituted milk, sugar and water are thoroughly mixed and heated to 74 °C (165 °F), and cocoa mass is added and mixed until completely blended with the other ingredients. Some water is added to enhance the solubility of the sugar and to obtain a mixture containing about 70% total solids. A constant temperature is critical for a uniform end-product quality.

In the next stage the premix is concentrated to 94% total solids at 125 °C (257 °F) in a scraped surface evaporator. The Maillard reaction takes place during the two to four minutes that the paste is in the evaporator. Its temperature is continuously monitored at this stage. Crystallization of sugar occurs in a specially designed crystallizer. The moisture content of the crumb ranges from about 3.5% to 4.5% at about 55 °C (131 °F) when it leaves the crystallizer.

In the final stage, the crumb is dried to a moisture content 1.5–3% on an open conveyor. Longer storage times require a moisture content below 1%.

This atmospheric process is said to be superior to the vacuum method as regards cost, time and reproducibility of the quality of the finished product.

4.7 Summary

In Table 4.8 the important parameters for the production of high-quality ingredients for milk chocolate manufacture are summarized.

The complex set of parameters illustrates the possibility of improving the quality of milk-based ingredients for milk chocolate production. The opportunity to provide tailormade milk products with respect to composition and physical characteristics enables the manufacturer to produce a variety of different products.

Currently the traditional process using whole milk powders as the principal ingredient for chocolate production is often used. But economic, processing

Table 4.8 Important parameters for milk-based ingredients in milk chocolate manufacture

Raw milk quality	Storage conditions (time, temperatures)
	Bacteriological quality
Drying process	Preheat treatment
	Homogenization
	Evaporation conditions
	Drying conditions (spray-, roller-drying)
Composition	Whole milk
	Skimmed milk
	Buttermilk
	Milk, chocolate crumb
	Milk concentrates (UF)
	Coprecipitates of whey and casein proteins
	Caseinates
	Whey protein concentrates
	Demineralized wheys
	Recombined products, e.g. caseinate/milk powder
	blends, skimmed milk/butter oil blends

and product requirements may result in the use of tailormade materials manufactured using new recombination technologies. A great deal of current research is in fact being carried out in this direction.

References

1. Christiansen, B.J. *Manuf. Conf.*, (Nov. 1983) 49.
2. Jebson, R.S. The use of fractions of milk fat in chocolate. *XIX Int. Dairy Congr.* 1E (1974) 761.
3. Mann, E.J. *Dairy Industries Intern.* **45** (1980) 10.
4. Mehrens, H.-A. and Reimerdes, E.H. Aufarbeitung und funktionelle Eigenschaften von Milchproteinen (Isolation and functional properties of milk proteins). In *Milk Proteins for Food*, GE Behr Verlag, Hamburg, FRG (1987) 48.
5. Moor, de H. and Huygebaert, A. *IDF Bulletin*, **147** (1982) 58.
6. Reimerdes, E.H. and Mehrens, H.-A. Struktur und Funktion der Milchproteine (Structure and function of milk proteins). In *Milk Proteins for Food*, GE Behr Verlag, Hamburg, FRG (1987) 21.
7. Rosenstein, J. In *Silesia Confiserie Manual No. 3*, Vol. 2 (1983), Silesia Essenzenfabrik Gerhard Hanke KG, Abt. Fachbücherei Neuss, FRG, p. 698 ff.

5 Cocoa mass, cocoa powder, cocoa butter

J. KLEINERT

5.1 Introduction

Cocoa beans constitute the raw material for the production of cocoa mass, cocoa powder and cocoa butter. To obtain a high-grade product the cocoa beans must on the one hand come from fully ripe cocoa pods and on the other they must be properly fermented and dried. First-rate products cannot be made with poor-quality cocoa beans, even using the latest manufacturing methods.

5.2 Cocoa mass

The principal stages in the production of high-grade cocoa mass are illustrated in Figure 5.1.

5.2.1 Cleaning the cocoa beans

When the fermented and dried cocoa beans arrive from their countries of origin, they always contain a wide variety of foreign materials such as dust, sand, wood, stones, glass, leather and fibrous matter. It is essential to completely remove these foreign materials to maintain the quality of the product, as well as to protect the processing machines.

In order to develop the full flavour, further thermal treatment is necessary, and this is obtained by roasting either the whole beans, the cocoa nib or indeed the cocoa mass. Roasting whole cocoa beans produces combustion gases from many of these foreign materials which are especially detrimental to the flavour of the cocoa mass. For this reason an increasing number of manufacturers are now employing two-stage processes for thermal preparation of cocoa beans (e.g. the Indru process, micronizing process, saturated steam process). The final roasting process is carried out afterwards on the crushed cocoa nib or cocoa mass.

5.2.2 Preliminary thermal treatment of cocoa beans

The roasting of whole cocoa beans has several disadvantages.

Variation in size of whole cocoa beans. All batches of cocoa beans contain both large and small beans, with the proportion of beans of any particular size

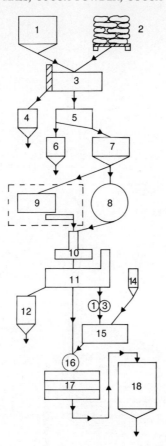

Figure 5.1 Schematic representation of the processing of cocoa beans into cocoa mass. (1) Cocoa bean silo; (2) cocoa beans in sacks; (3) preliminary cleaning unit; (4) bin for foreign materials; (5) stone and glass separator; (6) bin for stones and glass; (7) buffer receptacle for the cleaned beans; (8) preliminary thermal treatment of the cocoa beans, micronizing or Indru technology; (9) roasting of the cocoa beans by conventional methods (10) cocoa bean breaker; (11) removal of shell and germs from cocoa kernels (nibs); (12) bin for cocoa shell; (13) ribbed drums to disintegrate the cocoa nibs into finely fragmented nibs (1–4 mm); (14) receptacle for reaction solution; (15) ground cocoa treatment with water or reaction solution; (16) pregrinding the cocoa nibs; (17) fine grinding of the cocoa mass; (18) cocoa mass tank, heated, with mixing mechanism and circulating pump.

varying according to the source and the growth season. The degree of roasting which each receives will depend upon its size, as shown in Figure 5.3. If the roasting time is optimized for medium-sized cocoa beans, the small beans are inevitably overheated as a result. This has an unfavourable effect on the taste of the cocoa mass. In addition the large cocoa beans are not roasted sufficiently, which is also detrimental to the final chocolate. These differences can be demonstrated by measuring the aroma index (1), as shown by the UV absorption curves in Figure 5.4.

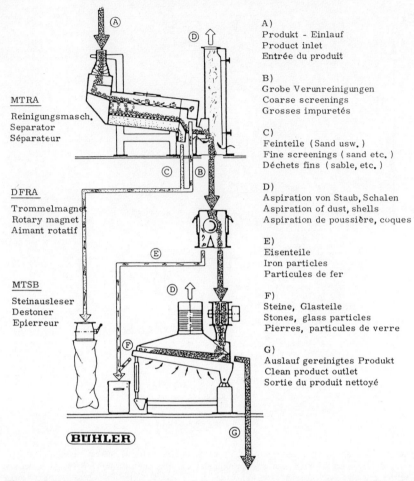

A)
Produkt - Einlauf
Product inlet
Entrée du produit

B)
Grobe Verunreinigungen
Coarse screenings
Grosses impuretés

C)
Feinteile (Sand usw.)
Fine screenings (sand etc.)
Déchets fins (sable, etc.)

D)
Aspiration von Staub, Schalen
Aspiration of dust, shells
Aspiration de poussière, coques

E)
Eisenteile
Iron particles
Particules de fer

F)
Steine, Glasteile
Stones, glass particles
Pierres, particules de verre

G)
Auslauf gereinigtes Produkt
Clean product outlet
Sortie du produit nettoyé

MTRA
Reinigungsmasch.
Separator
Séparateur

DFRA
Trommelmagnet
Rotary magnet
Aimant rotatif

MTSB
Steinausleser
Destoner
Epierreur

BÜHLER

Figure 5.2 Schematic representation of a unit for the preliminary cleaning of the cocoa beans with an additional device for the separation of denser foreign materials: stones, glass, metal (Buhler Brothers Ltd).

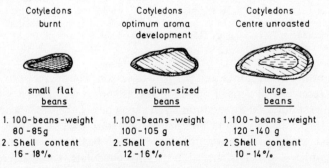

Cotyledons burnt	Cotyledons optimum aroma development	Cotyledons Centre unroasted
small flat beans	medium-sized beans	large beans
1. 100-beans-weight 80 -85g	1. 100-beans-weight 100-105 g	1. 100-beans-weight 120-140 g
2. Shell content 16 - 18%	2.Shell content 12 -16 %	2.Shell content 10 - 14%

Figure 5.3 Schematic representation of the effect of roasting small, medium-sized and large cocoa beans for the same period (J. Kleinert).

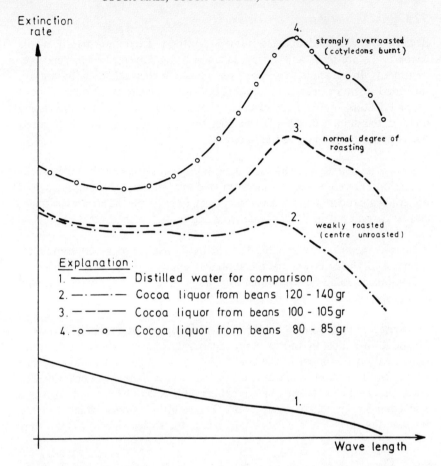

Figure 5.4 Absorption graphs of the water-soluble components steam distilled from three kinds of cocoa beans. For purposes of comparison, the filtrate level of distilled water is also included on the graph (J. Kleinert).

Proportion of broken and crushed cocoa beans. Broken and crushed cocoa beans occur in all bulk samples. Their proportion fluctuates considerably depending upon the origin of the beans and on the standard of fermentation. This damage results in a loss of cocoa butter when the beans are roasted because liquefied cocoa butter escapes from damaged cell clusters of the broken beans and is absorbed by the shells, which have a very low fat content. This can be illustrated by comparing the fat content of the shell of unroasted cocoa beans with that of shells which have been through a crushing machine. Shells from raw cocoa beans usually contain, in the dry state, less than 2% fat compared with more than 5–6% from the shells of some roasted beans.

5.2.3 *Principles of thermal pre-treatment*

Because of these disadvantages, instead of being roasted immediately, whole cocoa beans are more and more frequently being subjected to thermal pre-treatments. In principle these processes involve a thermal shock by hot air, saturated steam or infra-red radiation, during which the moisture content of the cotyledons should not fall below 3.5%. Irrespective of the processing method, the damaged cocoa beans must be separated from the whole cocoa beans prior to thermal pre-treatment.

Hot air treatment. By passing sufficient hot air over the cocoa beans, the shells can be loosened completely from the cotyledons. However, this treatment does not eradicate rodent hair or insect fragments. In principle the hot air treatment can be carried out in conventional roasting units.

Saturated steam treatment (2). The shell can also be loosened completely from the cotyledons by passing saturated steam over them. This also slightly reduces bacteriological contamination. A special unit for applying saturated steam has been developed by Buhler Brothers Ltd (2).

Infra-red treatment (3), (4). Infra-red technology is used in the micronizing (3) and infra-red drum (Indru) (4) processes. In the micronizing process, infra-red radiation treatment of the cocoa beans takes place on a fluidized bed conveyor. The energy is concentrated on the surface of the cocoa beans in such a way that particles of dust, remains of fruit pulp, rodent hair, insect fragments and other foreign matter burn visibly. The resulting vapour which accumulates on the surface of the cotyledons bursts the shells without the moisture

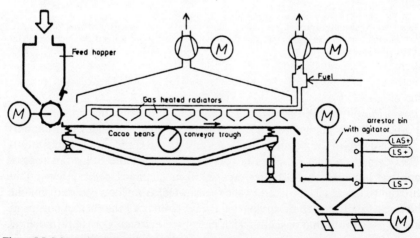

Figure 5.5 Schematic representation of the operating principles of the micronizing unit according to Newton (refs. 3, 4). (Micronizing Co. (UK) Ltd and G.W. Barth GmbH).

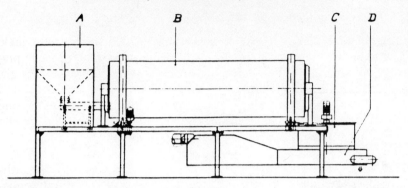

Figure 5.6 Schematic diagram of an infra-red drum unit according to Kleinert (ref. 5). (A) Pretreatment container; (B) infra-red treatment; (C) spark extinguisher and bean-buffer; (D) cooling apparatus. (G.W. Barth GmbH).

content of the cotyledons being substantially reduced. The operating principles of micronizing technology are illustrated in Figure 5.5. The high surface temperature also brings about a drop in the amount of microbiological contamination, especially yeast and other fungi.

In the infra-red drum Indru process, in contrast to the micronizing process, the cocoa beans are exposed to infra-red radiation in a rotating drum fitted with an internal spiral conveyor, as is illustrated in Figure 5.6. The cocoa beans are added in a continuous stream so that they form a single layer in the drum. The cocoa beans roll off the drum as it rotates and are heated more evenly than in the fluidized conveyor bed. The additional rubbing action of the cocoa beans causes a grinding down of the remains of any fruit pulp, to which hair rodent and insect fragments adhere. This is beneficial to the process.

A great deal of importance is attached to infra-red technology, in particular in connection with the processing of small flat and indented cocoa beans. As a result of the improved loosening of the cocoa bean shells from the cotyledons, there is a distinctly better yield of cocoa nib than was obtained by traditional methods. Furthermore, the loss of fat through the transfer of cocoa butter from the nib to the shells is now very small.

5.2.4 Breaking of the cocoa beans and separation of the shell

The cocoa beans are normally broken while they are still hot following thermal pre-treatment or roasting of the intact beans. The shell and in certain circumstances the bean germ is then separated as far as possible from the broken cocoa bean kernels (nibs). It is crucial to ensure that the cocoa nib is as free as possible from fine sand, dried pulp remains and other hard material. Cocoa bean shells, except those from washed beans, always have adhering pulp which contains fine sand. This is very detrimental to the operation of

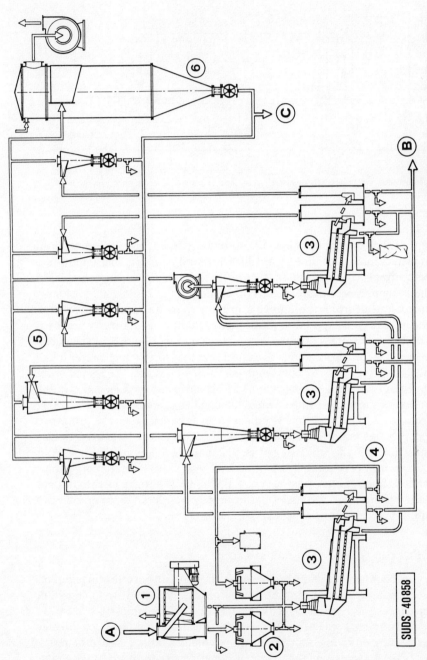

SUDS - 40 858

Figure 5.7 Schematic representation of a modern cocoa bean breaker and sifting unit, the Buhler SUDS 40 858 Winnowing Plant (3 t/h); (*A*) Cocoa beans; (*B*) nibs; (*C*) shells/dust; (1) preliminary sieve MKZM; (2) impact crushers SBC; (3) separator 'classifier' MTRA; (4) return of uncrushed beans; (5) cyclone separators MGXB; (6) air-jet filter MVRS (Buhler Brothers Ltd).

subsequent preliminary and fine grinding units. The importance of separating the shell as fully as possible cannot be overemphasized.

Only whole cocoa beans should be fed into the breaker (2), which is why the broken beans are separated through a preliminary sieve (1) after thermal treatment. This type of process is illustrated in Figure 5.7. The separated broken cocoa beans are fed directly into the first shaking sieve 3, on the one hand to relieve the first breaker 2, and on the other to avoid an excessive build-up of fine matter. So that the smallest possible amount of finely fragmented nibs is produced, the cocoa beans are flung individually at a very high speed against radially arranged interchangeable baffle plates. The shell is removed from the broken cocoa beans in the shaking sieves. The air drawn off from the cocoa bean breaker unit is purified in order to prevent taint in other products and pollution of the atmosphere.

The very hard germ which makes up about 0.9% of the dry cocoa beans is not specifically separated from the broken cocoa kernels in this unit although it is generally detrimental to the processing. A clean separation of the cocoa germ from the broken cocoa nib, however, requires a special device (grain cleaning machine), and is today undertaken only by a very small number of cocoa processing plants.

The cocoa nib produced can be ground down directly to cocoa mass. To an increasing degree a preliminary grinding process is now included at this stage. This will depend on the end use. For example, for the production of certain chocolate types specific reaction solutions are used. Alternatively, when manufacturing cocoa powder, alkaline solutions are employed. A thermal treatment for the development of colour and aroma is carried out in conjunction with this, which also serves for the eradication of all sorts of micro-organisms. It is possible to easily adapt other types of machine so as to gain the advantages of the Buhler breaker process (2).

5.2.5 Flavour development and alkalizing of cocoa nib

With the aid of a microscope it is possible to see that the cocoa butter and cocoa starch in the nib are contained extensively in the cell clusters (Figure 5.8a). Once this is ground down to cocoa mass, fat surrounds the solid particle forming a water-repellent layer (Figure 5.8b). In order to increase the speed of any reactions, it is therefore more sensible to carry out any flavour development or the alkalizing on finely divided nib particles and not on the cocoa mass. As the cocoa nib (especially when it is thermally pre-treated) tends to be very coarse (Figure 5.9), reaction conditions can obviously be optimized by grinding the nib to a particle size of 1–4 mm.

In contrast to cocoa mass during the treatment of finely fragmented cocoa the aqueous solution can start an unhindered reaction with the rest of the cocoa bean components. At the end of the process any undesirable substances still remaining after steam treatment can be removed without

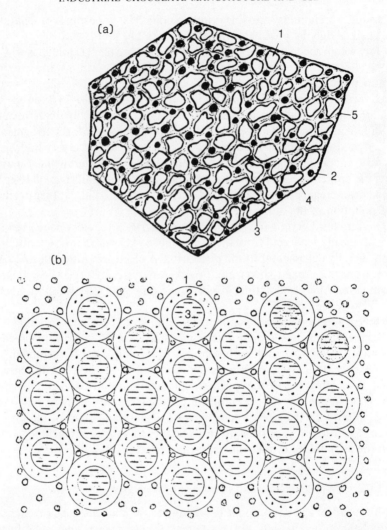

Figure 5.8 Schematic representation of the difference between treating finely fragmented cocoa nib and cocoa mass with reaction solutions. (Top) cocoa nib (1) Cocoa butter in cell cluster; (2) cocoa starch; (3) protein capsules; (4) cellulose wall. (Bottom) cocoa mass. (1) Water-based reaction solution; (2) fat cell (cocoa butter), water-repellent; (3) fat-free cocoa particle.

problem. This is because they do not have to pass through a fatty barrier, as in the case of the cocoa nib (Figure 5.8*a*). With cocoa mass foul-smelling reaction components are taken up partially by the cocoa butter, the cocoa starch and the cocoa protein. This makes their extraction with water vapour much more difficult. Following treatment the ground cocoa is dried, and afterwards thermally treated, i.e. roasted, according to the amount of flavour development desired. This method leads on one hand to the eradication of micro-

Figure 5.9 Cocoa nib from the breaking unit and after further grinding. (Right) finely fragmented cocoa nib, particle sizes 1 to 4 mm; (left) cocoa nib, particle sizes 1 to 10 mm. (Photo J. North).

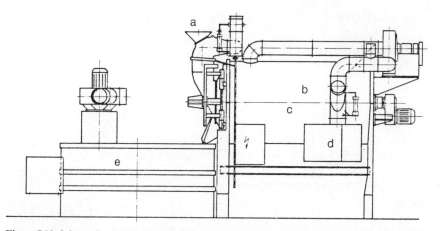

Figure 5.10 Schematic representation of a batch alkalizing or flavour development unit (Barth System, ref. 4). (*a*) Funnel tube; (*b*) reaction drum; (*c*) perforated pipe to carry the reaction solution; (*d*) gas or oil heating; (*e*) cooling pan (G.W. Barth GmbH).

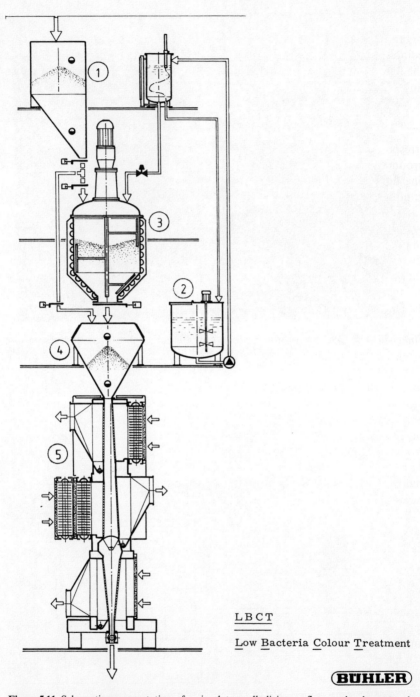

LBCT

Low Bacteria Colour Treatment

BÜHLER

Figure 5.11 Schematic representation of a circulatory alkalizing or flavour development plant (Buhler System, ref. 2). (1) Batch hopper; (2) solution preparation; (3) SLT reactor; (4) intermediate hopper; (5) STT dryer/roaster (Buhler Brothers Ltd).

organisms such as yeast and other fungi. On the other hand, flavouring and colouring components are developed. As can be seen in Figures 5.10 and 5.11, the finely ground cocoa can be processed by batch as well as by continuous-flow processes.

5.2.6 Grinding of cocoa nib to cocoa mass

Irrespective of whether cocoa powder, cocoa butter or chocolate is to be produced from nib or ground cocoa, the material must be ground down to a fine homogeneous mass. A wide variety of different machines is used for grinding, which may take place in one or many stages.

Pre-grinding. Nowadays blade mills are amongst a range of machines employed for the pre-grinding of the nib. During this process cocoa butter, already released from the cell clusters, melts as a result of the consequent rise in temperature, producing a liquid mass. The still relatively coarse-grained product is ground down very finely in one or two further stages according to the end use.

Fine grinding. When producing cocoa powder or cocoa butter it is important to release that fat completely from the cell clusters. It is also imperative that the proportion of very fine particles of cocoa solid material is kept as small as possible because fine particles will bind up the fat and result in a cocoa mass with very poor flow properties. Disc, roller, hammer and ball mills are employed for fine grinding. The choice of the fine-grinding unit depends to an increasing degree on the end use of the cocoa mass. In cocoa powder production, in particular, very fine particles of metal must be avoided, as they become a darker colour when combined with water, e.g. in whips and puddings. For this reason millstones are once again being employed to an increasing degree for the fine grinding of cocoa mass.

A system where a vacuum treatment is incorporated between the preliminary and fine grinding units is illustrated in Figure 5.12. Figure 5.13 shows the design of a modern disc mill for the fine grinding of cocoa liquor.

5.2.7 Alkalizing and cocoa liquor treatment

From the point of view of evenness of heating, roasting of cocoa mass would appear to be ideal. In raw cocoa beans, the shell is, however, the main carrier of contaminants and pesticides as well as of micro-organisms, and, in addition, it is not possible satisfactorily to separate the shell from the cotyledons in untreated beans. Grinding of raw cocoa beans can therefore result in excessive levels of pesticide residues and other undesirable matter, in particular fat-soluble contaminants. Proper thermal pre-treatment is thus essential for reasons of quality and hygiene.

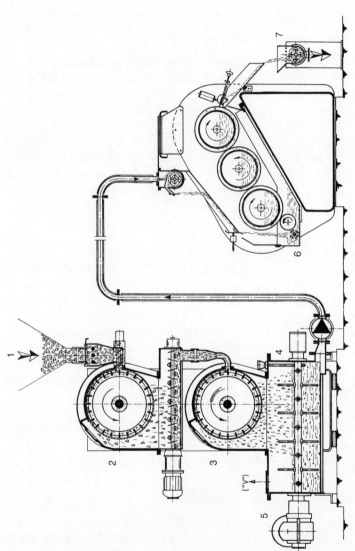

Figure 5.12 Schematic representation of a pre-grinding and fine-grinding unit together with vacuum device for the production of cocoa mass (Buhler System, ref. 2). (1) Feed-pipe for the cocoa nib; (2) pre-grinding, stage I; (3) pre-grinding, stage II; (4) liquefaction and deodorizing; (5) vacuum pump; (6) fine grinding; (7) liquefier (Buhler Brothers Ltd).

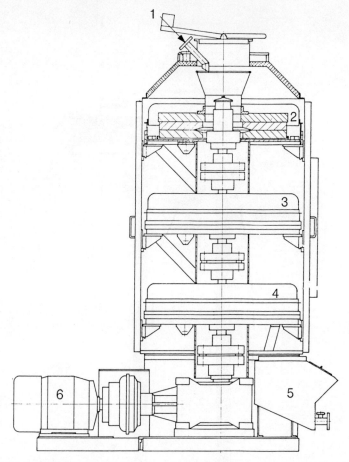

Figure 5.13 Schematic representation of a modern fine-grinding disc mill with millstones (Lehmann System, ref. 6). (1) Feed-pipe for cocoa nib; (2) grinding zone 1; (3) grinding zone 2; (4) grinding zone 3; (5) finely ground cocoa mass; (6) drive motor (Lehmann Maschinenfabrik GmbH).

If, however, cocoa mass is purchased instead of beans, this can only be treated by a thin-layer process. This presupposes that the mass has not already undergone irreversible thermal damage in previous processing. Cocoa beans, as has already been shown, should be thermally pre-treated in order to obtain a clean separation from the husks (shell). They may also be subjected to a further heat/chemical treatment process. These involve some risk of damaging the flavour and smell of the product. Once this has occurred, the quality can only be partially improved by thin-layer treatment. For this reason it is

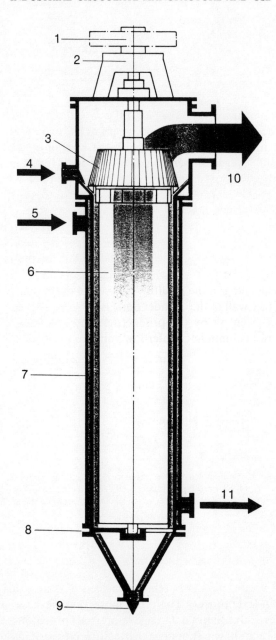

Figure 5.14 Schematic representation of the operating principles of the Luwa Thin-Layer Evaporator used for flavour development, moisture removal and roasting of cocoa mass. (1) Rotor drive; (2) upper rotor bearing; (3) exhaust vapour filter; (4) cocoa mass inlet; (5) heat inlet; (6) rotors; (7) heating jacket; (8) lower rotor bearing; (9) cocoa mass outlet; (10) heat outlet; (11) exhaust vapours outlet (Buss–Luwa AG and Petzholdt GmbH).

advisable to control any heating processes which take place before thin-layer-treatment by using aroma index evaluation (1).

Various firms have developed specific units for the flavour development of cocoa mass, with and without reaction solutions. There are also specific units for roasting cocoa mass.

Luwa Thin-Layer Evaporator (7). Figure 5.14 illustrates the principle of operation of the evaporator. The cocoa mass to be treated is fed in over the heating jacket into the cylinder, and with the help of a separating ring is distributed evenly over the inner wall of the column. Here it is caught by the blades of a high-speed rotor and turbulently sprayed in a thin film mass over the whole of the heated inner surface. The thin film of cocoa mass flows down a screw-shaped path on the inner wall of the cylinder, during which time the low-molecular-weight volatile components evaporate. In front of each rotary blade a bow wave forms, which changes gradually into a highly turbulent zone as it passes between the rotor blades and the inner wall of the cylinder. With substances of high viscosity this turbulence results in good heat transmission. This intensive blending in the resulting bow wave prevents the formation of crusts on the inner wall of the cylinder and also avoids overheating the mass.

The vapours given off by the products during treatment pass upwards through the Luwa column in counterflow and into the exhaust-vapour filter. Particles contained in the airstream are ejected by the high-speed rotor and flow back into the evaporation zone. Afterwards the exhaust vapours from the solid and liquid components of the mass are fed into a condensation unit. The treated cocoa mass reaches the bottom end of the evaporation zone within a few seconds. It is extracted with a pump, cooled in a static mixer to below 80 °C (176 °F) and stored in a tank with a circulation pump and a heated jacket ready for further processing.

Petzomat Thin-Layer Process (8). The Petzomat unit consists of one or more spray columns, as illustrated in Figure 5.15. The cocoa mass is homogeneously moistened either with water or with a particular treatment solution in a static mixer. The unit is fed by a controllable pump. The mass flowing from the top to bottom is processed very intensively in the turbulent current of rising hot air in the annulus which forms between the stator and the rotor. The mass, present in the form of very fine droplets, gives off its moisture, together with undesirable components such as acetic acid, carbonyl compounds, amines, etc., into the circulating air. Afterwards the cocoa mass is cooled and stored in a tank until required for further processing.

Convap ^{R}process (9). In this alternative device the cocoa mass to be treated is fed into the lower end of the mounting standing column, as illustrated in cross-section in Figure 5.16. Because of the centrifugal effect of the rotor, the mass is transferred onto the heated surface of the inner cylinder and then removed

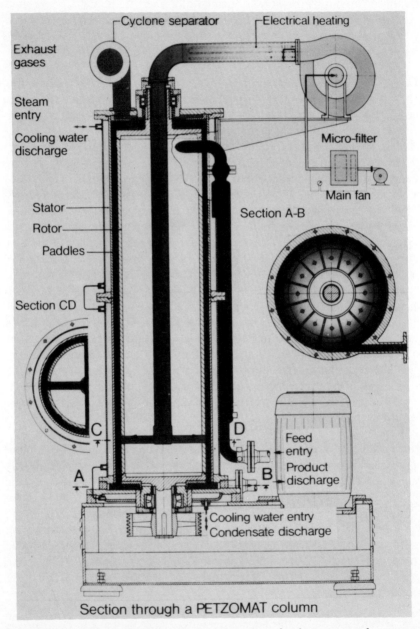

Figure 5.15 Schematic representation of a Petzomat column for the treatment of cocoa mass (Petzholdt GmbH).

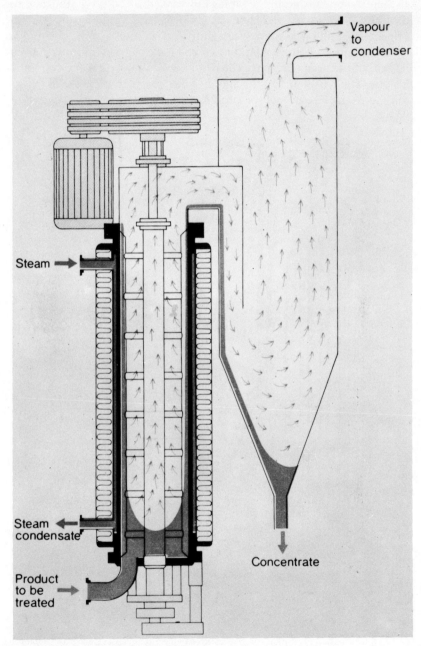

Vapour
to
condenser

Steam →

Steam ←
condensate

Product
to be
treated →

Concentrate

Figure. 5.16 Schematic representation of the operating principle of the Convap ᴿThin Film Evaporator for moisture removal as well as the preliminary flavour treatment and roasting of cocoa mass (The Contherm Division of Alfa-Laval Inc.).

from there by specially ground scraping knives. In this way, a very thin layer of mass with a short thermal diffusion path is formed giving a very uniform heat treatment. In addition, because of the short residence time of the material on the heated surface, deposits cannot build up on the metal. This results in very good heat transfer into the mass. The hydraulic drive provides a continuously adjustable control of the rotor speed, which enables it to be optimized according to the viscosity of the product being treated.

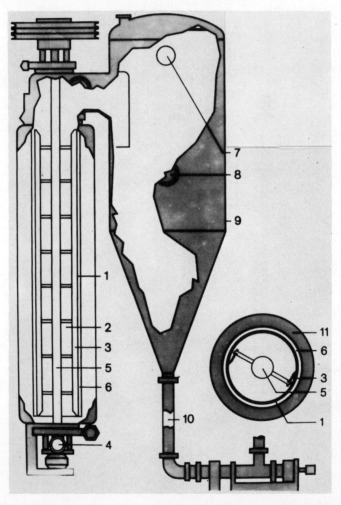

Figure 5.17 Schematic representation in cross-section of the Bauermeister Cocoavap Unit for the drying, flavour development and roasting of cocoa mass. (1) Cylinder; (2) water vapour; (3) stripping device; (4) product inlet; (5) rotor; (6) heating space; (7) exhaust vapours outlet; (8) inspection window; (9) separator; (10) product outlet; (11) insulation (Gebrüder Bauermeister & Co).

As the cocoa mass is transported upwards, it is heated to the prescribed treatment temperature. The heat consumption fluctuates with the evaporation temperature employed, which itself can be independently regulated by adjusting the pressure.

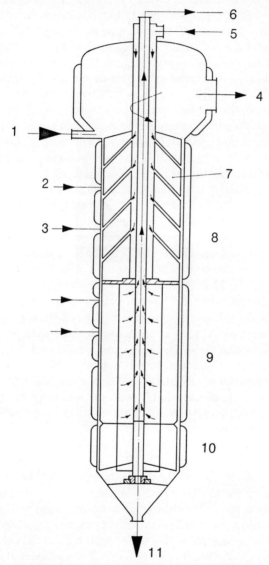

Figure 5.18 Schematic representation of the operating principles of the Lehmann KFA Unit for the removal of moisture and other volatiles and for the roasting of cocoa mass. (1) Solids inlet; (2),(3) metered addition; (4)(removal by) suction; (5) low-pressure chamber; (6) vacuum; (7) rotor with nozzle for gas addition; (8) drying zone 1; (9) roasting zone 2; (10) cooling zone; (11) mass outlet (Lehmann Maschinenfabrik GmbH).

At the heat of the column there is a baffle plate to remove the vapour extracted from the cocoa mass. This condensate can be collected in addition to the treated cocoa mass.

Cocoavap process (10). The Cocoavap process is a further development of the Alfa-Laval technology, and the system is shown in cross-section in Figure 5.17. The Cocoavap unit requires only hot air for roasting cocoa mass. If moisture removal and flavour development of the cocoa mass are also required, the Cocoavap unit can be fitted with a vacuum stage.

The cocoa mass to be treated is fed by a pump into the reaction cylinder. Immediately after entry, the cocoa mass is distributed in a thin film over the heated surface where rotating stripping devices convey it continuously upwards. The extraction of components of medium to high volatility takes place under vacuum, so as to avoid roasting. The temperature and pressure can be adapted within a specific range to the requirements of the cocoa mass. In the Cocoavap, as in the Alfa-Laval Convap, there is a short thermal diffusion path leading to good heat transmission into the cocoa mass.

Lehmann KFA process (6). This thin-layer flavour development unit (Figure 5.18) has a column containing three treatment zones. Hot gas is supplied to the cocoa mass which is fed in at the head of the column and then into the drying zone. The treatment of the cocoa mass takes place on the inside of the heated stator. The roasting zone is separated from the head of the column by a disc, and does not have hot air flowing through it, but is kept under low vacuum to remove the volatiles formed during roasting. In the drying, as well as in the roasting zones, there are dispensing points for adding reaction solutions such as water, sugar syrup, protein solutions, etc., for the formation of specific aroma and flavour components.

In the Lehmann KFA unit the raw cocoa mass is pre-dried before the roasting takes place under adjustable moisture and temperature conditions. The result should be that all particles in the cocoa mass remain in the unit for the same length of time.

Carle–Montanari Process (11). The PDAT reactor functions as a batch machine. The cocoa mass is metered into the reactor from a supply tank at a temperature of 60–80 °C (140–176 °F). When the filling process is complete, the reactor is evacuated and the dispersion system is filled with about twice as much inert gas (nitrogen or carbon dioxide) as the volume of cocoa mass. The extraction of air and its replacement by inert gas reduces the risk of oxidation. Subsequently water or a specific reaction solution is added as the temperature is raised to the desired level. After a fixed period the inert gas is released, and the vacuum slowly increased to remove the volatiles present. Acids and other undesirable substances are thereby removed from the cocoa mass. The PDTA reactor is suitable for deodorizing as well as for roasting and alkalizing cocoa

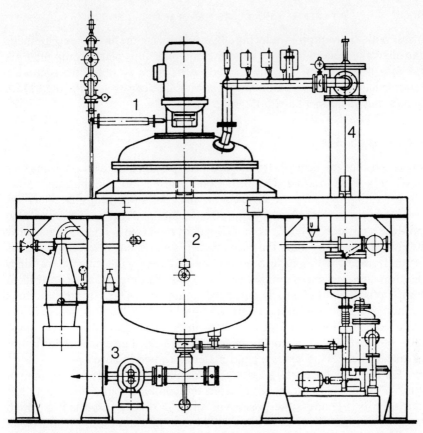

Figure 5.19 Sketch of the Carle-Montanari PDAT Reactor (ref. 11). (1) Cocoa mass inlet; (2) PDAT reactor; (3) discharge pump; (4) water vapour condenser (courtesy D. Ley).

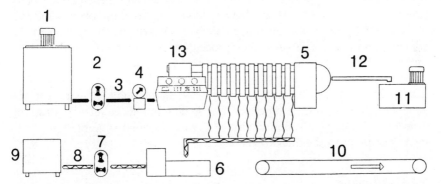

Figure 5.20 Schematic representation of the operating principle of a modern horizontal cocoa butter press (ref. 10). (1) Cocoa mass container; (2) cocoa mass pumps; (3) pipe for cocoa mass; (4) metered dosing device; (5) hydraulic cocoa press; (6) cocoa butter scales; (7) cocoa butter pump; (8) cocoa butter pipe; (9) cocoa butter container; (10) cocoa cake conveyor; (11) hydraulic pumping unit with control; (12) hydraulic pipes; (13) electric control panel (Gebrüder Bauermeister & Co).

mass. With the additon of reducing sugar solutions some flavour development can also take place. The water vapour containing any volatile components is fed into a condensor and the treated cocoa mass is stored in a suitable intermediary tank to await further processing. The construction of the PDTA unit is illustrated in Figure 5.19.

5.3 Cocoa powder

Cocoa powder can be made from cocoa mass prepared from cocoa beans that have only been roasted, i.e. from non-alkalized (non-solubilized) cocoa. However, to provide an attractive colour and good suspension characteristics in milk, cocoa powder is usually made from solubilized cocoa mass. In order to produce a powdery product from the high-fat cocoa mass, it must be partially defatted, as happens when the cocoa butter is extracted. The vertical presses formerly used for this have been replaced by more efficient modern high-performance horizontal presses (Figure 5.20). The presses are constructed from individual horizontally arranged containers. The bottoms of these containers are made from special steel mesh which is replaceable. A specifically designed pump fills the press chambers with the finely-ground hot cocoa mass (90–100 °C) (194–212 °F) and by pressing in the steel plungers the cocoa butter is squeezed out. This leaves a solid material known as cocoa press cake.

After pressing, the cocoa press cake, normally with a residual fat content of 10–20%, is ejected automatically from the containers, and falls on to the conveyor belt beneath the press. The press cake is transported in rough fragments either into a silo with cool air passing through it, or into pallet containers which are stored in refrigerated rooms. The cooled press cake fragments are further broken down in a toothed roll breaker with two spiked rollers moving in opposite directions. The pre-refined crushed cocoa press cake can also be treated with lecithin and flavouring solutions prior to milling.

Nowadays pinned disc mills are chiefly employed for the milling of the cooled press cake. The operating procedure of a modern cocoa powder milling, sifting and cooling unit is illustrated in Figure 5.21.

With its extensive range of colours, cocoa powder is a much-sought-after ingredient which can be used in many ways in the food industry, e.g. the production of coating compound, filling compound, cake mix, drinking powder etc. Cocoa powder is also still sold in small quantities in retail packages direct to the final consumer.

5.4 Cocoa butter

Cocoa butter can be produced from crushed cocoa nibs, i.e. from dehusked cocoa bean cotyledons which have been ground down to cocoa mass. Alternatively, whole cocoa beans with the shell still on can be used to produce cocoa butter by the expeller process, as illustrated in Figure 5.22b.

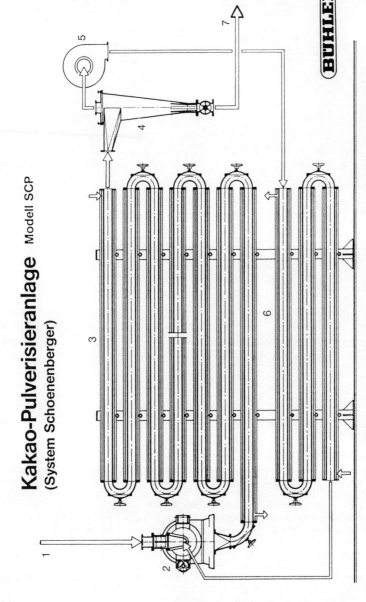

Kakao-Pulverisieranlage Modell SCP
(System Schoenenberger)

BÜHLER

Figure. 5.21 Diagram of a cocoa milling, sifting and cooling unit (Buhler System, ref. 2). (1) Broken cocoa press cake inlet; (2) broken cocoa press cake milling unit (cocoa mill); (3) cocoa powder refrigeration unit; (4) cocoa powder cyclone separator; (5) aspirator; (6) air and cocoa dust cooler; (7) to cocoa powder bagging point (Buhler Brothers Ltd).

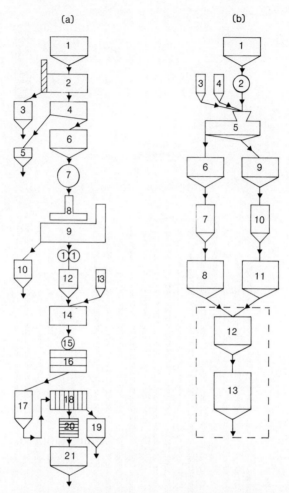

Figure 5.22 Schematic diagram of the manufacture of cocoa butter by the hygienic press process (*a*) and the less hygienic expeller process (*b*). *Press process*: (1) Raw bean silo; (2) preliminary cleaning; (3) foreign materials; (4) stone and glass separator; (5) container for stone and glass; (6) buffer container; (7) thermal pre-treatment; (8) impact mill; (9) shell separator; (10) cocoa bean shell; (11) breaker to give finely fragmented cocoa nib; (12) buffer container for the fragmented nib; (13) reaction solution; (14) flavour treatment of the fragmented cocoa nib; (15) pre-grinding of the nib; (16) fine grinding into cocoa mass; (17) cocoa mass tank; (18) cocoa butter press; (19) cocoa press cake; (20) cocoa butter filtration; (21) press-cocoa butter. *Expeller process*: (1) Silo for raw beans and high-fat waste material; (2) preheating of the material for pressing; (3) cocoa fines; (4) cocoa bean shell, fines rubbish and dust; (5) expeller press; (6) expeller press cake; (7) fat extraction from the expeller press cake with solvent; (8) extracted cocoa fat; (9) expeller press fat; (10) filtration of expeller fat; (11) purified expeller fat; (12) sanitization of the expeller fat by a total 'refining' treatment; (13) clean and hygienic expeller fat.

5.4.1 Press cocoa butter

This is obtained from cocoa mass, the production of which was described in great detail in section 5.2. Figure 5.22 illustrates the different processing stages. The cocoa butter is partially pressed out of the cocoa mass into large containers, as in the case of cocoa powder (described in the previous section). The cocoa butter from the press is put on the market in liquid or in solid form in a purified form as natural cocoa butter, or it may be deodorized and marketed as such (see 5.4.4).

During the pressing it is important that hard layers do not build up, which would form a barrier to the cocoa butter. Also it is desirable to produce a fat which is as clear as possible. To achieve this the pressure must be increased slowly. Depending upon the residual fat content in the cocoa press cake, a pressure of 400–500 bars is necessary to extract the cocoa butter from the mass. When producing cocoa powder with 21–22% fat, approximately 41–42% cocoa butter is extracted. In the case of the so-called highly de-fatted types of cocoa powder, the proportion extracted increases to 48% or even more.

5.4.2 Expeller fat (cocoa fat)

As can be seen in Figure 5.22b, the raw beans are fed directly into the expeller press together with all their contaminants (vermin, pesticides and harmful micro-organisms), sub-grade beans, and other waste materials (cocoa dust and particles, mouldy produce etc). In the expeller press the residual fat content of these materials is reduced to less than 10%. After loosening up the conglomerates, the expeller-press residual matter (expeller-press cake) with a residual fat content of 8–10% is de-fatted with an organic solvent to leave a fat content of less than 1%. The practically fat-free residual matter contains all the original contaminants in enriched form and for this reason should never be supplied to the food and consumption industry as a half-finished product at a bargain price. The expeller fat, with the fat-soluble components from the shell and dirt, especially the excreta and the pesticides, could nevertheless be put on the market as filtered or deodorized cocoa butter.

Although this would never be used by reputable manufacturers it is possible that it could be obtained by less scrupulous operators. Manufacturers should take care to ensure that any fat used has not been produced from contaminated material.

5.4.3 Quality criteria

The following definitions from *Codex Alimentarius* are relevant.

Press cocoa butter, natural or deodorized. This is cocoa butter from cocoa beans which are of perfect quality, have been properly pre-cleaned, and from which the shell has been compeletely separated. The resulting crushed cocoa nib is ground down to mass either directly or following alkalizing and the cocoa butter is pressed out hydraulically. After filtration the cocoa butter can be put on the market either directly or after deodorization.

Fully refined cocoa butter. This is cocoa butter obtained by mechanical extraction from cocoa beans, crushed nib, mass, cocoa dust as well as cocoa particles, which undergoes full treatment according to the standard processing technique.

Extraction fat. This is fat which is obtained by means of a solvent extraction from cocoa press residual matter as well as from rancid cocoa and from waste material, which afterwards undergoes a full treatment according to the standard processing technique. Cocoa fat differs analytically from pure press cocoa butter. The product is put on the market in its pure form as well as in a mixture with press cocoa butter, e.g. fully refined cocoa butter.

5.4.4 *Cocoa butter deodorization (13)*

Since, to an increasing degree, buyers are stipulating a cocoa butter which is as neutral as possible in smell and taste, it is increasingly being deodorized. In order to remove the substances which give the cocoa its characteristic aroma and taste, the cocoa butter is subjected to a specific water vapour treatment under vacuum at 160–170 °C (320–338 °F). The deodorization process can be carried out by batch methods as well as by a continuous process. Figure 5.23 illustrates a modern continuous cocoa butter deodorization unit.

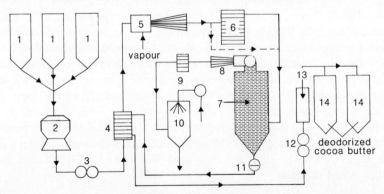

Figure 5.23 Representation of a modern continuous cocoa butter deodorization unit. (1) Raw cocoa butter tanks; (2) centrifuge for clarifying the cocoa butter; (3) cocoa butter pump; (4) cocoa butter preheater; (5) water vapour inlet; (6) heat retention and flavour exchange; (7) vacuum tank for the separation of the flavour volatiles; (8) steam jet vacuum pump; (9) water vapour cooler; (10) water vapour condenser; (11) air lock and discharge of cocoa butter; (12) conveyor pump for deodorized cocoa butter; (13) cocoa butter after cooler; (14) storage tanks.

5.5 Quality control

Quality control tests for cocoa beans, nib, shell, mass, powder and butter are summarized below.

5.5.1 Cocoa beans

Visual inspection. The visual inspection includes recording the appearance and size of the whole beans as well as of the cross-section of at least 100 cocoa beans for evidence of damage by mould and moths as well as the proportions of slaty and totally purple beans.

Flavour control. This test is very important for foreign flavours and permits a rapid response.

Analytical tests. (i) Determination of the weight of 100 beans; (ii) determination of the moisture content, which should not exceed 7%.

5.5.2 Cocoa nib

Estimation of the proportion of shell in a representative sample taken from the cross-section of the main channel of the breaking and winnowing machine: (i) visual selection of the small proportion of shell particles and small germinating roots and weighing the quantities; (ii) sedimentation process.

5.5.3 Cocoa bean shell

(i) Determination of the amount of cocoa nib in a representative cross-section sample from the main channel of the breaking/winnowing machine; (ii) visual selection of nib particles and weighing of the quantities.

5.5.4 Cocoa mass

(i) *Microbiological tests*

Total count of aerobic bacteria	< 20 000/g
Enterococci	< 100/g
Staphylococci, coagulase positive	< 10/g
Enterobacteria, fermenting lactose	< 10/g
E. coli	< 1/g
Mould fungi	< 100/g
Yeast fungi	< 100/g
Salmonella bacilli, not detectable in 50 g	

(ii) *Analytical tests*

Moisture
Total fat
Total protein
Minerals
Heavy metals, Cd, Pb
Pesticides, chlorinated hydrocarbons

(iii) *Physical tests*

Amount of coarser material over 50 μm (0.002 inch)
Flavour (aroma) index
pH value at 20 °C (68 °F) in 10% suspension
Filth test

(iv) *Sensory tests*

Appearance
Flavour
Taste (40 g sucrose, 60 g cocoa solids mixed homogeneously)

5.5.5 *Cocoa powder*

(i) *Microbiological tests*

Total aerobic bacterial count	< 20 000/g
Enterococci	< 100/g
Staphylococci, coagulase positive	< 10/g
Enterobacteria, fermenting lactose	< 10/g
E. coli	< 1/g
Mould fungi	< 100/g
Yeast fungi	< 100/g

Salmonella bacilli, not detectable in 50 g

(ii) *Analytical tests*

Moisture
Fat content
Protein content
Total ash
Water-soluble ash
Ash insoluble in water
Alkalinity, total ash
Alkalinity, water-soluble ash
Alkalinity, water-insoluble ash
Heavy metals, Cd, Pb
Pesticides, chlorinated hydrocarbons

(iii) *Physical tests*

Amount of coarse particles over 30 μm (0.001 inch)
Flavour (aroma) index
pH value at 20 °C (68 °F) in 10% suspension
Filth test
Colour, e.g. Minolta appliance
Colouring power in milk

(iv) *Sensory tests*

Appearance
Flavour
Taste (40 g sucrose, 60 g cocoa powder mixed homogeneously)

5.5.6 *Cocoa butter*

(i) *Sensory tests*

Appearance in solid and melted state
Flavour
Taste

(ii) *Analytical tests*

Iodine value
Unsaponifiable matter
Gas chromatographic analysis
Triglycerides
Pesticides, chlorinated hydrocarbons

(iii) *Physical tests*

Melting point
Characteristic figures from the cooling curve
Time interval T30 to the reversal point (TWP)
Reversal point temperature (TWP)
Freezing point temperature (TEP)
Temperature quotient, $\Delta T : \Delta t$ in °C/min.
Refraction index nD^{40}
Polygravimetric penetration at 20 °C (40° cone with 50g, 100g, 200g and 400g)
Flavour (aroma) indices.

References

1. Kleinert, J. 'Praktische Erfahrung bei der sensorischem Beurteilung von Süsswaren–Der Aroma-Index'. *Sensorische Erfassung und Beurteilung von Lebensmitteln*, LWT-Edition 3, Foster Verlag AG, Zürich (1977).

2. Buhler Ltd. Uzwil, Switzerland (cocoa bean precleaning unit, saturated steam technology, LBCT technology, cocoa bean breaking and sifting unit, SCS–8 cocoa grinding unit, cocoa milling and sifting unit).
3. Newton, D. Micronizing Co. UK Ltd., Framlingham, nr. Woodbridge, Suffolk 2 IP 13 9 PT.
4. G.W. Barth Ltd, Martin-Luther-Strasse 44, D-7140, Ludwigsburg, West Germany (Infra-red drum technology, tornado technology).
5. Kleinert, J. *Thermal Cocoa Refinement* (chocolate technology project, 1983), Central College of Technology for German Confectionery Industry, Solingen–Grafrath, West Germany.
6. F.B. Lehmann Ltd. Daimlerstrasse 12–13, D-7070 Aalen, West Germany (cocoa grinding unit, KFA–cocoa mass refinement unit).
7. Buss-Luwa Ltd, Anemonenstrasse 40, CH-8047, Zürich, Switzerland (Refinement of cocoa mass, Luwa system).
8. Schmitt, A. Spray/thin layer process for the treatment of cocoa mass. *Confectionery* **1/2** (1986) 41–44, Petzholdt Ltd, Engineering Works, D-600, Frankfurt 1, West Germany.
9. Alfa-Laval Ltd, Tumba, Sweden (Convap Process).
10. Bauermeister & Co. Ltd, Friedensallee 44, Hamburg, West Germany (cocoa bean breaking and sifting unit, Cocoavap-cocoa mass treatment).
11. Carle & Montanari Ltd, Via Neera 39, 1-20141, Milan, Italy (cocoa precleaning units, cocoa breaking and sifting units, cocoa mass refinement units, cocoa nib grinding units, cocoa mass treatment units).
12. Kleinert J. *Codex Alimentarius*, Kakao- und Schokoladeprodukte. Fachschrift: *Zucker- und Süsswarenwirtschaft* **35**(6) (1982) 183–184.
13. Kleinert J. Kakaobutter-Desodorierung. *Ullmann's Encyclopaedia of Technical Chemistry*, 4th edn., Chemistry Publishing House Ltd, D-6940 Weinheim, West Germany, 673–688.

6 Particle size reduction

E.A. NIEDIEK

6.1 Introduction

Particle size reduction (or grinding) cannot be looked at in isolation from the other stages of the chocolate masse manufacturing process. The particle size reduction process chosen depends upon the processes which both precede and succeed it, and affects both the economic viability of the chocolate manufacturing process and the final quality of the chocolate masse. The differences in the quality of chocolate found in the trade are due not only to differences in the recipe; they are also dependent on the production process used. The consumer's quality criteria for chocolate vary according to geographical location, and within a single market there are considerable differences. Producers are able to cover many of these various different requirements by supplying a range of products.

Chocolate is a product which has been developed empirically. It was only once chocolate was being mass-produced that the traditional combined processes of mixing, grinding and conching were split into separate parts and studied individually. The production techniques common today were developed by constant testing and modification and by research which was guided by developments and trials in the chocolate manufacturing area. The practical implications of chocolate research can be seen in the analysis of each individual section of the chocolate manufacturing process. Measuring techniques are constantly improving and give an ever-increasing insight into the reactions taking place during chocolate manufacture. However, it must be remembered that analytical and monitoring techniques even today are still inadequate, and development is still largely empirical. However, improved analytical techniques frequently show us important parameters which had previously been neglected. The latest analytical results in the chocolate area can thus make previous decades of research work look unimportant, however significant they may have appeared when they themselves were new.

A typical example of this is the description of flow characteristics of liquid chocolate. For decades the OICC method for describing flow behaviour using the Casson model was an important and useful aid to the chocolate industry. Now it has been shown that using the 'Hyrheo method' (1, 2) to measure absolute values for flow characteristics gives a better correlation with the processing conditions than with the Casson Yield value figures. Although it will take some time before the new method is generally accepted throughout the industry, the findings obtained with the old OICC method are no longer up to date. This also applies to the influence of particle size reduction on the flow characteristics of the final product. This relationship will have to be re-evaluated using the most up-to-date measuring and analytical techniques.

Consequently, in the following sections on particle size reduction there is a clear distinction between statements which are based upon current techniques and those founded on the older ones, which may be less relevant.

6.2 Current types of particle size reduction process

Three types of process are presently in use:

(i) Combined particle size reduction of components
(ii) Partly combined particle size reduction of components
(iii) Separate particle size reduction of components.

All three processes can be carried out either in one stage or in several stages. The type of process to be used is dependent on the final quality required and on the raw materials to be used. In addition, a distinction is made between so-called 'wet grinding' and so-called 'dry grinding'. In wet grinding, the particles to be ground are suspended in a liquid medium (in this case melted fat). In contrast, in dry grinding the particles being ground are suspended in a gas, generally air.

When grinding the raw materials to finished chocolate fineness, it is sensible to first consider the grinding (or milling) of the cocoa nib as a separate entity. Roughly or finely ground cocoa mass for chocolate production can be considered as a part-processed product. There are in fact companies which limit themselves to this kind of product. If a chocolate masse producer owns his own cocoa mass preparation plant, this is generally regarded as a separate process which may be carried out at a remote location.

6.3 Grinding cocoa nib into cocoa mass

6.3.1 Effect of the roasting process

Only the nib of the cocoa bean is ground, after the bean has been shelled. Removal of the shell breaks up to cocoa nib into coarse pieces and a relatively small proportion of fine material. This mixture is called crushed cocoa nib.

Dependent on the roasting process chosen, shelling (winnowing) can take place either before or after roasting (see Chapter 5). In grinding cocoa nib into cocoa mass there are two degrees of difficulty in grinding. Cocoa beans which are unroasted but only pre-dried are more difficult to grind down than roasted cocoa nib.

The reasons for this are easy to understand. One thing which affects the degree of difficulty of grinding is the unavoidable remains of cocoa bean shells, germ and the relatively high water content in the cocoa nib. In unroasted cocoa beans, the shells do not come away from the bean very easily and the shell content is larger. Unroasted beans are generally only dried down to a moisture content of 2.5–3% whilst the moisture content after roasting is 1.5%. The proportion of germ is equally great in roasted and unroasted beans.

The shell and germ content account for several per cent of the weight of cocoa nib. Shells and germ have rather woodlike characteristics and are particularly difficult to grind. If the milling machine is adjusted to grind the shells and germ, this also grinds the rest of the cocoa nib structure correctly. In practical terms the grinding problem of cocoa nib can be illustrated by imagining just a few kg of germ and shells mixed with 1,000 litres of cocoa butter, with the requirement to grind the germ and shell part to a particle size of between 70 μm and 15 μm (according to subsequent use).

An additional cocoa nib treatment process (e.g. alkalizing), in the production of cocoa mass for making cocoa powder, changes the grinding qualities of the cocoa nib, usually making cocoa nib easier to grind.

A further influence on the level of shell content is the type of bean. The size and shape of the cocoa bean put different limits on the unavoidable shell content in the crushed cocoa nib.

6.3.2 Cocoa bean grinding (3)

In grinding terms we can distinguish four groups of substances: shell, germ, cell walls, cell contents.

(i) *Shell.* In uncrushed form, shells are about 200–250 μm thick and are plate-shaped. Although they are relatively brittle, pressure does not cause them to break. Pressure alone makes them into thin platelets with very high tensile strength. The pressure thus created easily exceeds the flow limit of the best grinding equipment and thus leads to irreversible wear on the surfaces of the machinery, e.g. the surfaces of grinding rollers. In order to grind down shells, therefore, in addition to pressure a cutting action is also needed, which disintegrates the compressed platelets. In practice, however, this disintegration is not very effective, and normally only tears the platelets. This, however, makes them easier to grind during subsequent processing.

Despite their relatively small number and their thinness (about 50 μm, with diameters of up to 500 μm) these platelets have a definite effect on the flavour even at low concentrations in the fat.

For further particle size reduction the latest techniques involve ball mills, with balls several millimetres in diameter. Between the balls the damaged platelets are easily ground below 20 μm.

(ii) *Germ.* This also has woodlike characteristics and is brittle. The grinding qualities correspond to those of the shell particles. Grinding is also carried out as for shell particles.

(iii) *Cell walls.* The thickness of the cell wall is about $2\,\mu$m, and the cell diameter is 20–$30\,\mu$m. Cell walls are brittle and in the grinding process they fracture easily into plate-shaped particles. The smallest particles found after grinding have a particle size of about $0.2\,\mu$m.

(iv) *Substances contained in the cell.* The contents of the cell represent the greater part of the particles in the finely ground cocoa mass. As the cell diameter of 20–$30\,\mu$m alone means that they must naturally be smaller than 20–$30\,\mu$m, they do not need to be ground. Presumably they unavoidably go through an unnecessary grinding process.

The question is often raised as to whether cocoa mass is too finely ground by present-day particle-size reduction techniques. The largest proportion of the finest particles is obtained from the cell contents (starch, protein) which are produced by nature. Present milling processes can only affect the fineness of the small proportion of coarse substances such as shell, germ and cell walls. In practice it is not easy to over-refine cocoa nib.

6.3.3 Pre-grinding techniques

The reader is referred to references 4, 5, 6, 7, 8, 9, 10, 11, 12, 13, 14, 15, 16, 17, 18, 19, 20, 21, 22, 23, 24, 25 and 26. The following equipment can be used for pre-grinding cocoa nib:

  (i)   Fluted rollers (these used to be common decades ago)
 (ii)   Hammer mills
(iii)   Blade mills
 (iv)   Disc mills
  (v)   Extruders
 (vi)   Tumbling ball mills.

The aim of pre-grinding is to pre-grind shells and germs to a particle size of below 200–$300\,\mu$m, thereby avoiding overtaxing the subsequent grinding equipment.

Fluted roller mills carry out this initial step at low temperatures; the pre-milled product is not yet in a paste form as the fat is not fully melted. Fluted rollers (typical capacity 300 kg per hour) are no longer efficient enough. It is questionable whether it will be possible in the future to build fluted rollers with a higher throughput.

Traditionally, some manufacturers when making milk chocolate used a light coloured cocoa mass which was only dried and not roasted. Roasting and hot milling, however, put the cocoa protein through a heat treatment and de-naturing process. In the grinding process for chocolate manufacturing, the flavour-determining reactions between the proteins and the sugar work differently when the cocoa protein has been de-natured. This kind of low-heat-treated pre-ground mass is normally only produced by fluted rollers.

Hammer mills work in a different way. Apart from grinding, by intensive stirring and mixing they turn the rough cocoa nibs into a coarse viscous cocoa mass. Sieves in this kind of mill restrict the particle size of the mass produced. This kind of mill needs higher temperatures, e.g. 60°–80 °C (140°–180 °F) than the fluted roller mill. The temperature of the mass leaving the mill is not a precise indication of the temperatures within the mill itself which, locally, must be higher than may be desired. Some cooling takes place by evaporation of moisture which also takes with it some undesirable odours. This effect is accentuated by grinding under a vacuum.

The grinding equipment in these hammer mills may take the form of hammers or pins, for example.

Blade mills (27, 28, 29) differ from normal hammer mills by constant working in a sump. The rotating blade forms a 'cylinder' of material against the sieve; the finest grade sieve lets through only those particles which in one direction are smaller than e.g. 150–250 μm. The coarse particles which are difficult to grind accumulate in the 'cylinder'. The higher the concentration of coarse particles the better the grinding, because it increases the probability of the particles being struck by the blade and against each other. The blade mill can also be operated in a vacuum in order to remove moisture and some unpleasant flavours by evaporation.

Disc mills may have steel or carborundum elements. The nib is poured in centrally between two grinding discs and is carried along in the narrow grinding slots to the edge of the disc. Various disc movements are possible:

(i) One disc remains still, the second one rotates. The older, more simple carborundum disc mills work on this principle.
(ii) Both discs rotate in the same way, but at different speeds. Here the speed of grinding in the grinding slot and the throughput can be adjusted independently from one another. Disc grinders can operate in one stage or in several stages, and they may also have an automatic slot adjustment.

Extruders can crush or grind dried cocoa nib to the particle size of pre-ground mass. The grinding energy causes the fat to melt and turns the mass to liquid. Extruders are still at an experimental stage, and only when cheaper extruders become available will this operation become viable. An extruder may also have a vacuum area for removal of moisture.

Tumbling ball mills have a grinding volume which turns and is filled with ball bearings or grinding rods. These mills operate similarly to agitator ball mills, the movement of the grinding bodies is caused by rotating of the whole grinding volume instead of by stirring (see Chapter 17).

6.3.4 Fine grinding techniques

(i) *General.* Fine refining of cocoa mass is interesting for both production of cocoa powder and for chocolate masse (6, 14, 26, 30, 31, 32, 33, 34, 35, 36, 37, 38, 39, 40, 41, 42, 43, 44, 45, 46, 47, 48, 49, 50, 51).

The fineness requirements of cocoa mass for the production of cocoa powder are greater than those for the production of chocolate. The fineness of cocoa mass determines the possible fineness of the cocoa powder to be produced from it.

As the proportion of coarse particles in cocoa mass consists mainly of germ and shell, these are the most difficult components to refine in terms of the effort and they affect the viability of the refining process for chocolate manufacturing. This applies in particular for dark-chocolate processing. Ground cocoa mass is in general more liquid than it is in its coarse ground state. Thus finely-ground mass when mixed with sugar in the kneader permits the fat content to be kept down relative to coarsely-ground material. Experience shows that reducing the fat at the kneading stage also means less fat needs to be added in the final recipe to give a chocolate of the same viscosity. Chocolate refining equipment is expensive and it must be protected from wear and tear as far as possible. If the shells and germ in the cocoa mass are already ground so finely that the rollers in the chocolate refiners are not overtaxed, then the wear and tear of the refiners is considerably reduced. When making high-fat couvertures it is not always possible to put all the cocoa mass in the recipe at the pre-refining stage. To do this would give a mass which is too liquid to be ground on a roll refiner. In this case it is necessary to add the fine-milled cocoa mass to the conche together with the roll-refined product. In the production of chocolate masse when using separately refined components, or part of components, this also may require finely-ground cocoa mass to be added directly in the conche.

The following section describes techniques for refining the cocoa mass.

(ii) *Use of triple mills.* Experience has shown that, of all carborundum disc grinders (52), only the triple mill is able to supply the required fineness of the end product over a long operating period without expensive supervision.

The name triple mill comes from the three-stage milling by successive pairs of discs. Only one of the discs is driven, and the pressure of the discs on one another can be adjusted. Because these disc mills are relatively expensive, they have limited use, but recently disc mill have become more popular. When operating they can use between 50 and 100 kWh per tonne, according to the shell content.

The capacity of these triple mills is about 1 000 kg per hour but varies according to the required fineness. The life of the grinding disc can be up to several years if care is taken to prevent possible metallic contamination of the cocoa mass before it goes into the disc grinders.

(iii) *Agitator ball mills.* Agitator ball mills (53, 54, 55), also known as attrition mills, are today used for refining the greater part of the world's cocoa harvest. Agitator ball mills have a grinding container filled with steel ball-bearings, in which the ball-bearings are moved around by a stirrer while the cocoa mass to be ground flows through. There are considerable differences between the agitator ball mills produced by different suppliers in relation to size of the grinding volume, the form of agitator used, and diameter of ball bearings. In order to obtain the required degree of fineness for cocoa powder and chocolate production, experience shows that the pre-ground mass has to be subjected to the techniques described previously and has to be refined in several stages in the agitator ball mills. Several stages means that the ball-bearing diameter becomes smaller from one mill to the next, and there could be up to three mills in succession. The ball-bearing diameter in the first mill would be around 15 mm and in the last mill about 2–5 mm; with decreasing size of ball-bearing the agitator rotation may also increase. Important parameters for grinding in the agitator ball mill are the ball bearing content in relation to the grinding volume (a few per cent deviation from the optimum value can severely affect the grinding); capacity or rate of flow; and rotation of the agitator.

Important construction characteristics are (i) the separation of the product which has been ground and the ball-bearings themselves (sometimes done by a sieve), (ii) clutch-controlled drive aid before the stirrer, (iii) refill possibilities for ball bearings, ease of replacing worn parts, etc.

The ball-bearings in the grinding volume impact against each other and there is also a rolling effect just as between rollers. Both of these mechanisms produce grinding.

The disadvantage of agitator ball mills compared with roll refiners is their broader particle-size spectrum. Particles to be ground can pass through unmilled, but they may also be broken several times unnecessarily. Thus the required grinding energy of 70 to 100 kWh per tonne is higher than in the roll refiner process. The output is also normally similar to the triple mills at about 1 000 kg per hour. The investment costs are lower than with triple mills but the cost of maintenance and wear and tear are greater.

According to the final particle size required, an equipment wear rate of 20–100 g per tonne of cocoa mass produced must be taken into account. Up to now, this finely distributed metal content in the cocoa mass and in the finished products has not been considered detrimental. However, a systematic investigation of this problem has not yet been carried out.

(iv) *Fitting a double roll refiner before the agitator ball mill.* The high dilution of the difficult-to-grind shell and germ particles in the cocoa mass and the lack of definite fineness criteria results in relatively high costs for agitator ball mill systems. The costs can be reduced considerably if double rollers about 50 μm apart are used to pre-refine out and grind up the shells and germ particles to this size. They come out of the double roller system as plate-shaped agglomerates, which are then easily ground in the agitator ball mill (56).

This kind of double roller system is not yet directly available from machinery manufacturers, and until now three-roll refiners (57, 58), converted by the manufacturers, have been used. Fixed spacers were then fitted into these refiners between the rollers, so that the gap between the last two rollers was about 50 μm. This kind of converted cocoa mass refiner has been used for about ten years, with relatively low wear and tear. Even with a two-shift system, it has only been necessary to overhaul the refiner once each year. The total grinding energy for this sytem is 30–50 kWh per tonne. The costs for wear and tear, investment and maintenance are lower than with triple disc mills or with the sole use of agitator ball mills.

This grinding procedure will probably be accepted in the long term. What is at present unsatisfactory about the use of roll refiners for this purpose is their relatively high price and the open mode of operation (relatively unhygienic). This grinding technique also demands more supervision from staff than the triple mills.

6.3.5 Standardization of cocoa mass

Like every natural substance, cocoa fluctuates in its content and components. In cocoa the fluctuations in the fat content are significant in the range 52–56%; in addition, the moisture content of the cocoa mass at the end of the grinding process generally fluctuates between 1–2% by weight. The possibility of standardization of the fat content does exist by decanting a part of the cocoa butter and thus reducing the fat content to 45–50% by weight. By mixing various mass obtained in this way, a batch of standardized mass can be obtained. Low-fat cocoa mass is also of interest in the manufacture of high-fat couvertures rich in cocoa solids.

With the aid of infra-red measurement techniques we can measure the moisture content of cocoa mass and corrections can be made by varying the process (generally by adding water).

6.3.6 Storage of part-processed cocoa products

At room temperature, 20 °C (68 °F), with an average relative humidity of 70%, cocoa beans absorb moisture of up to 5–6% weight until they reach equilibrium, so that cocoa nibs after drying, roasting and grinding are hygroscopic. This should be considered with respect to storage hopper design, particularly at the pre-refining stage. Quite often before grinding, cocoa nib can absorb so much water (overnight or over the weekend) that it can no longer be ground as usual and the grinders become hot. The moisture limit when obtaining a fully ground product is about 3% by weight. Systematic investigations have been published about changes in quality of the components of cocoa through storage and processing. These include fermentation of the harvested cocoa, storage, transport, processing, the effect of metal worn off the machinery, oxygen and temperature whilst grinding, as well as long storage in the liquid form in stainless steel tanks. All of these have some effect on the fat quality and the proteins in the cocoa (59). These require further investigations, as does the question of pesticide residues (e.g. from warehouse

Table 6.1 Particle size distribution of a range of different chocolates

(a) Particle count per 100 g block of full milk chocolate masse
    Size of sieve (calculated from sieve residue)

	500 μm	400–500 μm	300–400 μm	200–300 μm	100–200 μm
Type 1	8	7	21	1000	5000
Type 2	7	6	98	360	6200
Type 3	2	8	31	240	4300
Type 4	13	9	52	210	6900
Type 5	4	10	124	850	20000
Type 6	39	45	65	460	6800

(b) Particle count per 100 g block of dark chocolate masse
    Size of sieve (calculated from sieve residue)

	500 μm	400–500 μm	300–400 μm	200–300 μm	100–200 μm
Type 1	56	58	70	760	6100
Type 2	22	5	9	460	4530
Type 3	8	7	32	155	4750
Type 4	7	14	15	90	1350

(c) Milk chocolate masse (lower milk content)

Size of sieve	Average residues (% by weight)		
(μm)	USA (reference)	UK	Mainland Europe
> 800	0.003	0.002	
> 500	0.025	0.015	0.010
> 400	0.050	0.030	0.020
> 300	0.200	0.090	0.070
> 200	0.500	0.450	0.400
> 100	10.000	9.000	10.000

(d) Milk chocolate masse (lower milk content) (lower size range)

Size of sieve	Average residues (% by weight)		
(μm)	USA (reference)	UK	Mainland Europe
> 100	0.3	0.12	0.12
> 70	0.8	0.32	0.3
> 50	2.1	1.16	0.46
> 40	4.9	3.0	0.9
> 30	11	10.3	2.8
> 20	20	19.5	7
> 10	37	35	35
> 8	45	43	40
> 6	58	50	55

fumigation) in cocoa. In grinding and roasting, such constituents may either the partly eliminated or they may react with other constituents.

6.4 Refining of chocolate masses

6.4.1 *Differences in refining requirements*

Both the use to which the final product is to be put and the market in which it is to be sold will affect the final particle size and hence the suitability of one or other of the grinding processes (12–14, 24, 25, 60–85). There is no published systematic research on the effect of solid particle fineness on the quality of flavour and the flow behaviour of chocolate masse. Up-to-date analytical methods can, however, be used to measure the fineness of products and their suitability for various applications and markets (86).

Here one should take into account that, after refining, conching also has a greater or lesser effect on the fineness, according to the process chosen. Apart from the particle size, the particle shape is also important. There is, however, no published systematic research information on this parameter. Table 6.1 gives a summary of some of the available particle size data. Chocolate also includes products which may contain other sweeteners such as sorbitol, fructose or palatinite, in place of sucrose. In refining these substances one should take into account their special characteristics. The use of a mixture of components before refining can have a significant effect on the refining process. For this reason exploratory tests should be carried out before such mixtures are used.

6.4.2 *Grinding properties*

Carbohydrates (sucrose, lactose, dextrose and maltose) behave in a brittle way under mechanical stress. They are normally used in their crystalline form, but under the stress of roll refining their structure inevitably changes to some degree into a soft amorphous form. Under the influence of moisture and temperature this amorphous portion generally crystallizes again during the chocolate manufacturing process (87). This has an important effect on processing.

(i) *Grinding of sugar into icing sugar or micro-milled sugar.* If a sugar crystal is broken by impact or pressure (roll) refining, then the broken face of the crack in the crystal has sonic velocity related to the solid sugar. Here the energy conversion at the break is so great that a temperature of about 2 300 °C (4 200 °F) is created. The heat output given off in a mill can be measured from the wavelength emitted. This wavelength is unique for a particular temperature, which can then be determined from the calibration curve. The sugar does not burn at this temperature because the break moves so quickly that the heating affects only a few layers of molecules into the broken surface (one heating point stays on the broken surface for less than one-millionth of a second). Heat conduction into the broken piece almost instantaneously cools the molten surface, which sets in an unordered molecular structure (amorphous condition) (83, 88–94). In grinding granulated sugar into icing sugar, the amorphous surface layer accounts for less than 2–3% weight of the total sugar. This amorphous layer is, however, extremely hygroscopic. The amorphous part, at 20 °C and 70% relative humidity, absorbs about 15% weight of moisture (compared with about 0.3% in the total icing sugar). The amorphous surface layer then crystallizes suddenly and releases the water it has absorbed. In practice, the water absorption starts straight after milling, from the surrounding air, and then continues in the storage stage before the mixer. If a long enough time elapses for water to be absorbed to the crystallization stage, about 3 kg of water per tonne of sugar is suddenly released. This cannot usually be removed quickly enough from the storage container. The air in the icing sugar can only absorb a negligible amount of water, so the remainder is absorbed by the surface of the sugar. This causes the crystals to stick together into lumps (see Chapter 3). To store freshly-ground icing sugar, it must first be crystallized on the surface by deliberate moistening and drying.

It is interesting that the temperature shock on the surface during sugar grinding can also promote chemical reactions, which experience has shown to give icing sugar its typical and normally detrimental secondary flavour. Coffee drinkers do not generally use icing sugar to sweeten their coffee. If, for comparison, sugar before grinding and sugar after it has been ground into icing sugar are dissolved into water at a concentration of 10%, the solution of icing sugar tastes unpleasant. This is transferred to a greater or lesser extent into the chocolate flavour. This is a critical aspect of chocolate making.

(ii) *Grinding of sugar mixed with other components to obtain finished fineness chocolate masse.* Amorphous sugar particles are produced both in impact milling and in roll refining. If a sugar particle passes quickly under high pressure through the gap between the rolls it comes out as a thin transparent plate. As in metal being rolled into sheets, the sugar particle undergoes a plastic shock-like distortion. The melting point of about 170 °C (340 °F) for sucrose must be exceeded temporarily, and analysis of the plate structure shows that, directly after the pressure is applied, it is largely amorphous, although it is capable of very quickly absorbing water from the environment (air or other components such as cocoa mass or dried milk) and recrystallizing. However, amorphous substances, intensively absorb not only water, but also flavours from the environment. The difference between combined and separate refining of components lies in the fact that the temporary amorphous structure of the sugar in the chocolate can absorb flavours on to itself. Thus variations in the refining process can lead to differences in the final flavour of the chocolate. The warmer the amorphous sugar, the less water is needed for recrystallization. For this reason it is very likely that the amorphous particles will recrystallize to a large extent during conching.

In the amorphous state, sugar is highly soluble. This can be significant when determining how much Maillard reaction takes place between the proteins in the cocoa, and those in the milk and sugar. Recrystallization with its subsequent release of moisture should change the proportion of water-soluble fines during conching (95). At about 85 °C (185 °F) amorphous particles such as sucrose are soft and can be kneaded.

(iii) *Dextrose and fructose.* A problem also arises with the milling of dextrose in that its crystalline water content is released at 65 °C (150 °F). Even when the measured refining temperature is lower than this, the moment pressure is applied, temperatures are produced which release the water of crystallization (96). Fructose is even more susceptible to pressure. For this reason, it is difficult to produce fructose-based chocolate.

(iv) *Palatinite and sorbitol.* In the raw-ingredient state these are also in crystalline form. They react similarly to sugar, but are much more sensitive to high temperatures.

(v) *Grinding of full cream milk powder.* Milk powder is composed of protein, fat and lactose. The fat content is about 25% weight, the remaining 75% being water (about 3%) lactose (about 39%) minerals (about 6%) protein (about 27%). We can assume that in milk powder all the constituents are evenly distributed. The factor which affects the consistency is the protein, which is not a brittle substance (97).

Under mechanical pressure, the dried milk particles distort in a plastic manner. Only under extremely high impact speeds (close to sonic velocity), preferably with simultaneous grinding of the sugar, can one succeed in using impact milling to reduce milk powder to the same fineness as chocolate. If a mixture of milk powder and cocoa fat, for example cocoa mass, is pumped into an agitator ball mill, then grinding is satisfactory. A low-fat mixture passing through a normal five-roll refiner does not reach a satisfactory particle size distribution. Only when this mixture has about 10% by weight of sucrose added to it, does satisfactory refining occur in the roll refiner.

(vi) *Grinding of hazelnuts, soya beans, etc.* Small additions of hazelnut kernels or part-processed soya products, etc., are sometimes used in some markets to improve the flavour quality and reduce the recipe costs. Their grinding properties are similar to those of cocoa beans kernels, i.e. cocoa nib.

(vii) *Grinding of chocolate crumb.* Crumb (ref. 98 and Chapter 4) is a dehydrated mixture of milk, cocoa mass and sugar. The traditional vacuum drying process produces a relatively high proportion of amorphous constituents. This can be seen from the abnormal hygroscopicity of the material. This amorphous condition does not easily change to the crystalline state. In conching, the amorphous state tends to lead to formation of lumps. In taste terms, however, this combined ingredient can have advantages over a process in which the individual components are mixed together. In part this is because during the drying process the Maillard reaction, which helps to determine the flavour, can be carried out in a more controlled way. Crystallization of amorphous particles before refining can be aided by moistening the masse.

6.5 The roll refiner

The most important and most frequently used machine for grinding chocolate masse is still the five-roll refiner (Figure 6.1). Refiners (99, 100–103) generally have five successive rollers and an operational width of up to 2 m (6.5 ft). The roller diameter is about 400 mm (16 in). The two lowest rollers have to form the first film and determine the output. The lowest roller runs at the lowest speed of rotation, whilst those above it go increasingly faster. The film of product being refined always sticks to the faster-moving roller and moves upwards, becoming ever thinner, until it reaches the scraper on the last roller. The change in thickness of the layer of film corresponds to the increase in rotation speed of the roller.

There are four manual adjustments possible on this kind of refiner to adjust the fineness. The most important of these is the size of the working gap between the two bottom rollers. The second roller is fixed firmly in position, but the first roller is suspended and movable. By means of hydraulic pistons this first roller can be moved at both ends and pressed against the second roller. For a given

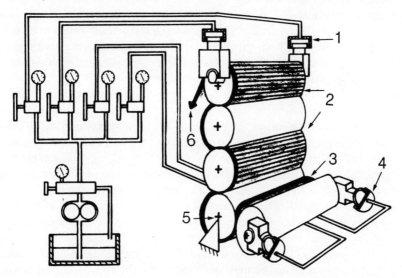

Figure 6.1 Diagram of a five-roll refiner. (1) Roll stack pressure; (2) chocolate film; (3) chocolate feed; (4) feed roll pressure; (5) fixed roll; (6) chocolate from scraper.

roll diameter, the width of the working gap between the drums affects the refining output. The grinding of the product in this first refining space is negligible. The first refining gap is about $100\,\mu m$ (4×10^{-3} inch) wide when preparing enrobing and moulding chocolate and about $60\,\mu m$ (2.5×10^{-3} inch) for the production of block chocolate. The coarsest particles in the feed material should not exceed $150\,\mu m$ (6×10^{-3} inch) if possible. In practical terms, the feed to the final refiner should have less than 10% of the particles larger than $125\,\mu m$ (5×10^{-3} inch).

The film formed on the second roller at about $100\,\mu m$ (4×10^{-3} inch) now passes to the second refining slit and is then reduced by a factor between 1.5 and 1.7 times, which means to about $60\,\mu m$ (2.5×10^{-3} inch). In the second refining slit, there is a further corresponding reduction in fineness ($38\,\mu m$, 1.5×10^{-3} inch) and in the fourth space between the rollers the $20\,\mu m$ (0.8×10^{-3} inch) layer thickness is reached before the scraper. This thickness corresponds to that in high-fat couverture chocolate production (although this varies from market to market; see Chapter 13). In European block chocolate production the final slit between the rollers is about $10\,\mu m$ (0.4×10^{-3} inch) starting from a space of about $50\,\mu m$ (2×10^{-3} inch). The proportion of coarse particles in the end product in this case must be much smaller. The increase in roll to roll speed corresponds to a ratio of between 1.5 and 1.7.

A further possibility of adjusting the fineness is by varying the pressure of the rolls on one another. It used to be assumed that the mixture was so liquid that hydraulic pressure held the two drums apart (similar to aquaplaning), so it was believed that by varying the pressure, the space between the rolls could be reduced to obtain finer material. In fact, the product between the rollers does not have the kind of flow characteristics which can build up hydraulic pressure. On the contrary, the product has the quality of sea-sand after a wave has run off it. The space between the particles is only partially filled with liquid. Capillary pressure is created, which solidifies the particle agglomeration. Surface forces transfer the film on to the faster-moving roll. Just as fine sea-sand containing sufficient moisture is hard enough to support vehicles, the rollers peel off the film of chocolate masse and grind the particles in it. With steady rotation of the rollers the chocolate moves continuously through the refiner (provided there are no obstructions) and it can only be adjusted by altering the slit width between the rollers or the speed of rotation. Only by varying the speed of rotation can the fineness be changed when the gap between the rollers is fixed. Changing the pressure of the rollers relative to one another helps to stabilize the system and prevent the formation of an irregular film of chocolate masse, but does not change the overall particle size reduction.

The roll temperatures are a further influence on the process; faults in the temperature setting lead to incorrect gaps between the rollers. The texture of the starting product in the first space between the rollers also affects the refining quality on the fifth drum. Kneading of the feed material in the mixer makes the product more liquid. In general, the minimum viscosity has not yet been reached before refining. But if once the product reaches the fifth roll there is only a relatively small volume of masse present, then it shear-thins and becomes more liquid. This is of particular importance if the product first goes through a pre-refiner and then passes directly to a fine-grinding refiner.

When making chocolate using the roll refiner, one of two procedures is normally employed. In the first the sugar and sometimes some of the other ingredients are pre-milled. The complete recipe is then placed in a mixer which works the material until it has the correct flow properties to pass through a five-roll refiner. Here it is ground to its final fineness (although further changes in particle size may take place in the conche). In the other method, only the cocoa itself is normally pre-crushed. Granulated sugar is added to the mixer, together with cocoa mass and any other ingredients. The ex-mixer material is then passed through two roll refiners. The first refiner provides an amount of milling approximately corresponding to that which is performed by the other mills in the first process, while the second refiner once again produces a masse at its final fineness. Here it is very important that the ex-first-refine material is liquid enough to pass up the second refiner. Initially, five-roll refiners were used to perform the first stage of milling, but more recently a two-roll machine has been manufactured for the purpose. This has been found to give adequate milling, and it is claimed to be easier to operate and give a higher throughput on the second refiners.

The surface of the rolls on the refiner can be damaged by high temperatures and it is important that it should switch off automatically if there is no product present between the rolls. Nowadays controls are also available for adjusting the amount of material being fed into a refiner.

Despite the fact that these refiners are well tried and tested, there is an attraction in developing alternative particle size reduction techniques (see Chapter 17). Refining machinery is relatively expensive, and, from a hygiene point of view, operates unsatisfactorily, because transport of material and refining are open to the atmosphere. Automatic control of the fineness would also be desirable, as manual control and supervision would then no longer be necessary.

6.6 Refining for production of chocolate masses for different uses

The aim of chocolate production is to give the product the optimum flow properties for further processing. The yield value required differs according to the potential use (e.g. hollow moulding or enrobing). The particle size affects both the yield value and the plastic viscosity.

In order to reduce recipe costs, the most expensive constituent (the cocoa butter) should be used in the smallest possible proportion which will give correct flow characteristics. In practice this is achieved when, after refining, the mixture of components undergoes intensive mixing and shearing whilst still in its liquid state. As one cannot extract fat from very liquid masse, the refined chocolate components at the start of conching must be so low in fat that it is possible to achieve the optimum work input into the chocolate (Chapter 8). This is not possible when using an agitator ball mill—only with masse which contains no cocoa butter does the ball mill have benefits. It is also very difficult to get optimum mixing.

Investigations have shown that, to obtain the best flow characteristics, the sugar must be included at the initial mixing of the ingredients.

(i) *Refining of dark cooking chocolate.* Cooking chocolate is the kind of product which is only coarsely ground. Here it is sufficient to stir finely milled sugar into a fine cocoa mass. Alternatively, of course, it is also possible to refine it in agitator ball mills or on roll refiners.

(ii) *Refining of dark (high-fat) couvertures.* The raw materials are granulated sugar, lecithin and very finely milled cocoa mass. The water content in the cocoa mass should be 1.5%. Two different processes are possible.

In the most common method (a), the granulated sugar is ground down to icing sugar in an impact mill and is mixed with finely ground cocoa mass in such a way that the mixture can easily be refined and conched. Alternatively, a pre-mix of cocoa mass and granulated sugar is pre-refined (fixed space between the rollers) before being fully ground on a roll refiner. With the correct consistency of the feed material, refining does not give any problems. Additional ingredients and perhaps also some of the finely ground cocoa mass are poured into the conche. The cocoa mass is often pre-treated with respect to the flavour (Chapter 5). Where permissible, cocoa powder and other types of fat may also be incorporated.

In more up to date (but seldom used) method (b), finely milled icing (pulverized) sugar is mixed with the other ingredients either in a pre-mixer or it may be added directly into the conche and mixed with the other components. The conching operation is in this case affected by the sequence in which the other components are added.

(iii) *Refining of milk (high-fat) couvertures.* Here the process is similar to refining of dark couvertures. Both methods can be employed, but using a pre-refiner is preferable to milling the icing sugar. Pre-milling of milk powder is similarly preferred as this makes the final refining easier. When using an agitator ball mill and method (b) it has been found to be preferable to firstly add the milk powder, cocoa butter and cocoa mass. The very finely milled sugar is added later (104–106).

(iv) *Refining for producing block (dessert) chocolate.* With block chocolate, the flavour formed during the processing is of prime importance. But there is also a connection between flow qualities in the melted state and the 'melt' of the chocolate in the mouth. The process used exclusively in continental Europe is that of pre-refining followed by a second roll refine to final fineness.

(v) *Refining of high-fat mass without cocoa butter (chocolate-flavoured coatings)*. In this case it is not so vital to reduce the fat, and agitator ball mills are frequently used (107, 108).

References

1. Windhab, E., *Zucker and Süsswarenwirtschaft* (April) **4** (1986) 141–145.
2. Windhab, E., *Voraussichtliche Veröffentlichung*, Behr's Taschenbucher (1986).
3. Gooch, J.U., Lysons, A., Roscoe, G.E., Griffiths, W.A., Moran, D.P.J., Lobf, P.L., Harris, R.V., Hardy, R., Laws, D.R.J., Scholefield, J., Rhodes, M.J.C., Coton, S.G., Chinery, P.I., Lister, D., Turner, P.T., Rees, J.A.G., Gates, M.Y. and Paine, F.A. *Rep. Prog. Appl. Chemi.* (1972) 57.
4. British Food Manufacturing Research Association, Brit. Pat. 1 216 296, 1970.
5. British Food Manufacturing Research Association, *Confectionery Production* **36**(3) (1970) 143–145, 167.
6. Anon, *Food Trade Rev.* **39**(11) (1969) 43–46.
7. Anon, *Con. Prod.* **35**(9) (1969) 601–603.
8. Anon, *Int. Choc. Rev.* **25**(12) (1970) 434–437.
9. Anon, *Manuf. Conf.* **50**(11) (1970) 27–30.
10. Anon, *Food Manuf.* **45**(3) (1970) 40–44.
11. Anon, *Int. Choc. Rev.* **26**(10) (1971) 292–293.
12. Anon, *Candy Ind.* **136**(1) (1971) 36–38.
13. Anon, *Food Proc. Ind.* **41**(487) (1972) 41–43, 45.
14. Anon, *Kakao and Zucker* **24**(4) (1972) 136–138.
15. Anon, *Int. Choc. Rev.* **21**(1) (1972) 15.
16. Boller, G. *Manuf. Conf.* **60**(5) (1980) 53–60, 62.
17. Keeney, P.G., Abstracts of papers, American Chemical Society (1971) 162: AGFD 17.
18. Lauth, K. West German Patent Application, 1 507 740 (1969).
19. Marshalkin, G.A. Klimovtseva, Z.G., Mel'nikov, E.M., Balashoro, A.E., USSR Patent, S V 1 007 636 (1983).
20. Rohan, T.A., Veen Pvan, West German Patent Application, 1 919 870 (1969).
21. Seyfert, R., *Süsswaren* **26**(11) (1982) 362, 364.
22. Swiechowski, C., *Przeglad Piekarski i Cukierniczy* **22**(1) (1974) 16–18.
23. Szegrandi, A., *Candy Ind. and Conf. J.* **135**(5) (1970) 42, 48, 50, 53.
24. Taubert, A., West German Patent Application, 2 002 958 (1971).
25. Tourell, A.G., US Patent, 3 663 231 (1972).
26. Vogeno, W., *Süsswaren* **16**(2) (1972) 50–52.
27. Anon, *Kakao and Zucker* **29**(2) (1977) 58–60.
28. Daffey, L.F., *Conf. Prod.* **42**(11) (1976) 502–503.
29. Rix, W. and Teschner, K. European Patent. Application EP 0 157 896 A1 (1985).
30. Bühler, A.G., West German Patent Application, 1 507 473 (1969).
31. Carle & Montanari SpA. Brit. Pat. 1 243 909 (1971).
32. Wiener & Co BY, Brit. Pat. 1 556 425 (1979).
33. Anon, *Gordian* **71**(9) (1971) 278.
34. Goerling, P. and Zuercher, K. German Federal Republic Patent Application, 2 036 202 (1972).
35. Goryacheva, G.N., Kleshko, G.M., Osetrova, E.Y., Simutenko, V.V., Khisyamet-dinova, R Ya. *Khlebopekarnaya i Konditershkaya Promyshlennost'* **10** (1979) 30–31.
36. Hanks, W.T. *Manuf. Conf.* **50**(9) (1970) 30–33.
37. Jetter, J. *Edesipar* **21**(3) (1970) 72–76.
38. Jetter, J *Edesipar* **21**(2) (1970) 43–47.
39. Niediek, E.A., *Industries Alimentaires et Agricoles* **93**(9/10) (1976) 1067–1078.
40. Niediek, E.A., *Int. Choc. Rev.* **28**(10) (1973) 244–256.

41. Niediek, E.A., *Int. Choc. Rev.* **28**(4) (1973) 82–95.
42. Niediek, E.A., *Int. Choc. Rev.* **28**(2) (1973) 30–36; (3) 72–77.
43. Niediek, E.A., *Int. Choc. Rev. for Sugar and Confectionery* **27**(10) (1974) 433–436; (11) 480–482.
44. Rapp, R., *Süsswaren* **25**(3) (1981) 73–76.
45. Reudenbach, R., *Kakao and Zucker* **25**(10) (1973) 459–464.
46. Rink, R. and Selinger, I. *Przemysl Spozywczy* **27**(10) (1973) 471–472.
47. Vakrilov, V., *Nauchni Trudove*, Vissh. Institut po Khranitelna i Vkusova Promyshlennost **23**(2) (1976) 149–156.
48. Wahjudi, J. and Plassmann, P., *Gordian* **73**(4) (1973) 128, 130. 133.
49. Wahjudi, K. and Plassmann, P., *Süsswaren* **17**(9) (1973) 476, 481–484.
50. Wyczanski, S., *Vyroba Kakaa* (1968) Warsaw WPL.
51. Zorzut, C.M. and Chichester, C.O., *Rivista Italiana delle Sostanze Grasse* **48**(5)(1971)236–241.
52. Goll, R. and Weber, C., West German Patent Application, 1 607 612 (1970).
53. Anon, *CCB Rev. for Chocolate, Confectionery and Bakery* **6**(1) (1981) 12.
54. Goeser, P.G., West German Patent 1 272 698 (1968).
55. Szegvardi, A., *Manuf. Conf.* **50**(5) (1970) 34–37.
56. Niediek, E.A., *Int. Rev. of Sugar and Confectionery* **31**(3) (1978) 13–18.
57. Bolliger, W., Swiss Patent 494 070 (1970).
58. Niediek, E.A., *Süsswaren* **22**(2) (1978) 28–40.
59. Mohr, W., Zurcher, K. and Knezevic, G., *Gordian* **71**(6) (1971) 184–188; (7&8) 224–226.
60. Taneri, C.E., Wing, D.H., Campbell, M.O., Welch, R.C., Sartoretto, P., Siebers, B.H., De Roeck, F.D., Palmer, R.M., Clarke, W.T., Mitchell, D.G., Tanner, R., Graham, A.S. and Thomas, A., Chocolate Symposium of PMCA, in *Manuf. Conf.* **56**(6) (1972) 45–96.
61. See ref. (3).
62. Anon, *Voeding en Techniek* **2**(38) (1968) 973.
63. Anon, *RTIA* **18**(184) (1970) 75, 77, 79.
64. Bauermeister, J., *Industrie Alimentari* **17**(5) (1978) 424–428.
65. Burdick, A., *Candy and Snack Ind.* **136**(10) (1971) 18–20, 22.
66. General Mills Inc., US Patent 3 544 328 (1970).
67. Kleinert, J., *Int. Choc. Rev.* **27**(5) (1972) 111–127; (7/8) 183–199; (10) 255–272.
68. Knetchtel, H., *Candy Ind. and Conf. J.* **135**(8) (1970) 5, 9, 49.
69. Krauss, P., *Manuf. Conf.* **52**(4) (1972) 38–41.
70. Niediek, E.A., *Ernährungswirtschaft* **18**(4) (1971) 104, 206–216.
71. Niediek, E.A., *1st Int. Congr. on Cocoa and Chocolate Research*, 1974, 273–286.
72. Niediek, E.A., *Proc. 6th Eur. Symp.* (European Federation of Chemical Engineering, Engineering and Food Quality Symposium) 1975, 366–379.
73. Niediek, E.A., *Süsswaren* **21**(21) (1977) 652–655.
74. Rumpf, H. and Niediek, E.A., West German Patent Application, 1 532 376 (1970).
75. Samans, H., *Kakao und Zucker* **30**(5) (1978) 134–137.
76. Sander, E.H., *Food Eng. Int.* **2**(8) (1977) 45–47.
77. Simon, E.J., *Int. Choc. Rev.* **24**(8) (1969) 340–349.
78. Simon, E.J., *Int. Choc. Rev.* **24**(10) (1969) 413–421.
79. Simon, E.J., *Candy Ind. and Conf. J.* **133**(10) (1969) 5, 30–36, 40, 56, 59.
80. Simon, E.J., *Revue des Fabricants de Confiserie, Chocolaterie, Confiturerie, Biscuiterie* **44**(11) (1969) 51–58, 71, 76; (12) 49–58.
81. Simon, E.J., *Candy Ind. and Conf. J.* **133**(8) (1969) 22, 24, 75.
82. Simon, E.J., *Int. Choc. Rev.* **24**(4) (1969) 140–155.
83. Ubezio, P., *Industrie Alimentari* **15**(9) (1976) 137–140.
84. Zucker, F.J. and Bruchmann, H.D. West German Patent Application 1 693 367 (1971).
85. Zuercher, K., *Int. Rev. of Sugar & Confectionery* **27**(7) (1974) 288–289.
86. Niediek, E.A., *Int. Choc. Rev.* **27**(12) (1972) 330–339.
87. Niedick, E.A. and Barbernics, L., *Gordian* **80**(11) (1980) 267–269.
88. Heidenreich, E. and Huth, W., *Lebensmitt.-Ind.* **23**(11) (1976) 495–499.
89. Mosimann, G., West German Patent Application, 1 925 170 (1969).
90. Mosimann, G., Swiss Patent, 490 810 (1970).
91. Mosimann, G., Swiss Patent, 518, 066 (1972).
92. Niediek, E.A., *Ernährungswirtschaft* (1971) 298, 300, 302, 304, 307.

93. Niediek, E.A., *Süsswaren* **15**(10) (1971) 400–406.
94. Niediek, E.A., *Süsswaren* **15**(8) (1971) 273–274, 278.
95. Niediek, E.A., *Gordian* **70**(6) (1970) 244–51; (7) 300–301.
96. Voelker, H.H., *Süsswaren* **15**(22) (1971) 918–920.
97. Milchwirtschaftliche Forschungs- & Untersuchungsgesellschaft GmbH, Brit. Pat. 1 273 943 (1972).
98. Rusoff 11, US Patent, 3 622 342 (1971).
99. Anon, *Revue des Fabricants de Confiserie, Chocolaterie, Confiturerie Biscuiterie* **54**(1) (1979) 38–42.
100. Anon, *Süsswaren* **23**(3) (1979) 16, 18.
101. Gaupp, E., *Süsswaren* **23**(10) (1979) 20, 25–26.
102. Niediek, E.A., *Gordian* **72**(4) (1972) 121–131.
103. Niediek, E.A., *Süsswaren* **22**(13) (1978) 21–25.
104. Niediek, E.A., *Süsswaren* **15**(3) (1971) 91–96.
105. Niediek, E.A., *Süsswaren* **15**(4) (1971) 124–132.
106. Rolle, P., *Gordian* **74**(4) (1974) 130–134.
107. Wiener en Co., Netherland Patent Application, 6 901 227 (1969).
108. Van der Molen, J.H., Netherlands Patent Application, 6 811 128 (1970).

7 Chemistry of flavour development in chocolate

J.C. HOSKIN and P.S. DIMICK

7.1 Introduction

The most notable attribute of chocolate which accounts for its universal appeal is its unique flavour. Like most natural foods, the mix and balance of the numerous compounds which contribute to the final flavour depend on a host of factors such as genetics, environmental conditions, harvesting and processing. With chocolate, more than any other product, the chemical complexity of flavour development is evident when one realizes the numerous parameters which may influence its development. Its complexity is equally obvious when considering that, even today, this desirable flavour has not been duplicated by the flavour chemist.

Among numerous factors, the organoleptic properties of chocolate may be dependent on the variety of the cacao used as starting material in manufacture. For example, the Criollo- and Trinitario-based varieties are known to have fine chocolate flavour, often described as a mild nutty and a full chocolate flavour, respectively. Thus the beans from these varieties are considered 'fine grade' cocoa. The 'bulk' cocoas which make up the vast majority of the world's production are from Forastero varieties. Although they are of good quality, they have less fine chocolate flavour when compared to the Criollo type (see Chapters 2, 13). It should be noted, however, that even within a varietal type, great variations in flavour potential exist because of the lack of pure strains. Therefore formulations ultilizing a single bean variety or those using blends may have dramatic variations in final flavour. Cultivation and husbandry practices may also have an impact on the flavour quality. However, the limited time and effort spent by investigators on this subject indicates that this apparently is a minor influence.

To date, much of the understanding of chocolate flavour is based on gas chromatographic and mass spectral data of the volatile aroma fraction of cured and/or roasted cocoa beans and the resulting formulated chocolate. More than 400 compounds have been found in cocoa following fermentation, drying, roasting and conching (1). However, the direct correlation of the gas

Technical Contribution No. 2696 of the South Carolina Agricultural Experiment Station, Clemson University. Technical Contribution No. 7658 of the Pennsylvania State University Agricultural Experiment Station.

chromatographic profile data of the aroma fraction with the flavour has been elusive.

7.2 Fermentation

Once fruit is harvested, biochemical mechanisms which contribute precursors for flavour begin. Even though the fermentation process facilitates the removal of the mucilaginous pulp and prevents germination of the beans, it primarily results in the development of the necessary flavour precursors. However, the

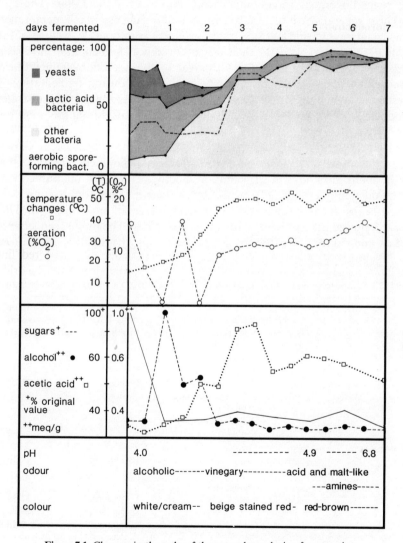

Figure 7.1 Changes in the pulp of the cacao bean during fermentation.

beans must have reached maturity; otherwise, no amount of processing can produce the desired flavour. Early work with cacao cell cultures showed that only when the cells have 'matured' could a chocolate or cocoa flavour result from further processing.

Fermentation is therefore required because unfermented beans may develop little chocolate flavour when roasted. Likewise, the outcome of excessive fermentation may also result in unwanted flavour. Thus, the first major post-harvesting phase to have an impact on flavour development is that of fermentation. During this phase of curing, the mucilaginous pulp surrounding the beans undergoes an ethanolic, acetic and lactic fermentation. The acid and heat generated kill the beans, with a resulting change in cell membranes. This facilitates enzyme and substrate movement with notable swelling of the bean. Excellent reviews of the chemistry of cocoa bean fermentation have been presented (2). A summary of the changes in the microflora, temperature, atmosphere and selected compounds in the pulp during fermentation is illustrated in Figure 7.1 (3). Reactions between the enzymes and substrates producing flavour precursors which occur in the beans during this stage are not fully understood.

The problems of defining the flavour mechanisms which occur during fermentation are difficult and as yet have not been definitively elaborated due to the many variations in fermentation methods. There are undoubtedly as many different methods of fermentation as there are countries that produce cocoa beans. It can be said that the beans undergo an anaerobic hydrolytic phase followed intermittently with an aerobic phase. The timing, sequence of events and degree of hydrolysis and oxidation may be highly variable from

Table 7.1 Characterization of the principal enzymes active during the curing of the cacao bean (3)

Enzyme	Location	Substrate	Products	pH	Temperature (°C)
Invertase	Bean testa	Sucrose	Glucose, fructose	4 5.25	52, 37
Glycosidases (β-galactosidase)	Bean cotyledon	Glycosides (3-β-D-galactosidyl cyanidin, 3-α-L-arabinosidyl cyanidin)	Cyanidin, sugars	3.8- 4.5	45
Proteinases	Bean cotyledon fragments	Proteins	Peptides, amino acids	55	4.7
Polyphenol oxidases	Bean cotyledon	Polyphenols (epicatechin)	O-quinones, o-diquinones	6	31.5, 34.5

fermentation to fermentation. The presence and concentration of the many flavour precursors which occur during this stage are dependent on enzymatic mechanisms. A summary of enzyme-catalysed reactions and their products is presented in Table 7.1. Not only are flavour precursors produced, but colour changes also occur. The hydrolysis of the polyphenol compounds by the glycosidases (4) results in bleaching and at the same time influences the flavour (5).

In early fermentation the major saccharide is sucrose; however, it is soon hydrolysed into monosaccharides (namely, glucose and fructose) as fermentation progresses (6). One interesting point during hydrolysis reactions is the increase in fructose as compared to glucose after approximately 2 days of fermentation. The increase in starch concentration during fermentation (1%) might imply that glucose originating from hydrolysed sucrose is utilized to produce starch (7). Fructose, less utilized because it requires conversion, would then be found in higher concentrations compared to glucose. However, it has been suggested that sugars are available from the fermenting pulp or other sugar sources. The reason for this change has not been determined, although it could be very important relative to the establishment of reducing sugars for the interaction with amino compounds in carbonyl–amino reactions. Thus, the sugars present in the cotyledon, associated shell and dried pulp are very important flavour precursors.

The major nitrogen-containing precursors formed during the anaerobic phase of fermentation are the numerous amino acids and peptides (8). These amino compounds are available for non-oxidative carbonyl–amino condensation reactions which occur during this and later heating phases, such as drying, roasting and grinding. Although some of the protein is degraded to flavour precursors, the purity of the remaining protein decreases due to polyphenolic–protein interactions.

During the aerobic phase, many oxygen-mediated reactions may occur, one

Table 7.2 Concentration of monocarbonyl classes in raw mouldy and non-mouldy cocoa beans (11). Reprinted from *J. Food Sci.* **35** (1970) 37–40. Copyright © by Institute of Food Technologists

| | Micromoles per 30 g of beans | | | |
Samples	Methyl ketones	Saturated aldehydes	2-enal	2, 4-dienal
Mouldy Bahia A	9.44	1.77	0.138	0.388
Non-mouldy Bahia A	2.15	0.70	0.115	0.001
Mouldy Bahia B	10.73	0.72	0.360	0.290
Non-mouldy Bahia B	5.55	0.21	0.061	0.028
Mouldy Accra A	9.82	0.02	0.450	0.125
Non-mouldy Accra A	1.55	0.17	0.110	0.020
Mouldy Accra B	9.65	0.83	0.450	0.265
Non-mouldy Accra B	6.43	1.30	0.180	0.400

E

Table 7.3 Concentration of phenols in cocoa mass from various regions of cocoa production as related to taste panel evaluation of the smoke flavour defect (12). Reprinted from *J. Food Sci.* **43** (1978) 743–745. Copyright © by Institute of Food Technologists

Producing region	μg phenol/10 g cocoa mass[a]	Smoke flavour intensity
Bahia	14.2	High
Bahia	10.0	Medium
Bahia	10.0	Medium
Bahia	7.4	Medium
Bahia	4.5	Slight–medium
Bahia	2.6	None
Bahia	2.2	Slight
Nigeria	0.6	Slight
Ivory Coast	2.2	Slight
Ivory Coast	14.3	High
Costa Rica	1.6	Trace
Dominican Republic	3.0	None
New Guinea	2.6	Trace
Malaysia	6.2	Medium
Malaysia	11.8	High
Mexico	0.0	None
Venezuela	0.0	None

[a] Values reflect concentrations of all reactive phenols in the samples based on standard curve for phenols.

of importance being that of the oxidation of protein–polyphenol complexes formed during the anaerobic phase (9). These mechanisms reduce the astringency and bitterness, as the oxidized polyphenols can complex with proteins and peptides, modifying further protein degradation reactions (10).

7.3 Drying

Following the fermentation process the beans are dried. This drying process is also instrumental in flavour precursor development. Indicators of good drying practices relating to flavour quality of the beans are good brown colour and low astringency and bitterness. Freedom from off-flavours such as excessive acidity and hammy/smoky flavour are also indicative of proper drying.

It is during the drying phase of cocoa curing that the characteristic brown colour of chocolate develops. Major oxidizing reactions occur with polyphenols catalysed by the enzyme polyphenol oxidase. The death of the bean, with loss of membrane integrity, allows previously restricted enzymatic reactions and results in brown colour formation.

Modern technology has changed the dependence on the sun for drying and includes artificial or mechanical drying processes which are not without drawbacks. With mechanical driers, where high temperatures may be used, case-hardening may result. Excessive heat and rapid drying may not allow for adequate loss of the volatile acids, e.g. acetic acid, and therefore are detrimental to quality. There also may be situations where the water activity within the bean in such that the drying process is merely a continuation of the fermentation process and many reactions can proceed. A high level of water activity, as a result of incomplete drying or rain soaking, may also result in mould contamination, particularly noticeable for beans in transit.

High concentrations of strongly flavoured carbonyls may result from heavy mould growth and thus alter bean flavour (11). Total carbonyls may double in concentration, with monocarbonyls increasing by 500% and methyl ketones exhibiting a notable increase (Table 7.2). Although virtually all the carbonyls are present in non-mouldy beans, their presence in high concentrations in mouldy beans can affect the final flavour (12). Artificial heating beneath the bed of beans is sometimes used; however, if smoke being generated from the burning fuel comes in contact with the beans, an off-flavour known as a hammy or smoky flavour can result (Table 7.3). Another proposed source of the hammy off-flavour is over-fermentation (13).

7.4 Roasting

Although the previous processes and those following roasting are important for flavour development, it is the controlled high-temperature heating step of roasting that magnifies the complex interaction between the flavour precursors and results in chocolate flavour. Prior to roasting, the beans may taste astringent, bitter, acidy, musty, unclean, nutty or even chocolate-like, depending on the batch examined (14). After roasting, the beans possess the typical intense aroma of cocoa, although remaining unpalatable. The process of roasting understandably deserves considerable attention.

7.4.1 An introduction to browning reactions

The roasting process, normally heating the beans to 110–220 °C (230–428 °F) depending on bean type, is needed for the reduction of moisture, release of the beans from the shell and development of chocolate flavour. One of the most important and complex reactions involving flavour during the roasting process is called the non-enzymatic browning, Maillard browning or the carbonyl–amine reaction. The reactions may be conveniently separated into three stages: initiation, intermediate and final. Although this separation is convenient, it should not be interpreted that such a separation exists. With the numerous and interacting steps of the Maillard reactions it is difficult to separate all of the individual reactions even within a stage.

The simplest, and therefore the best understood, step is the initial step. Reducing sugars and amino acids form addition compounds, which in turn form glucosylamines or frustosylamines, depending on the inital reducing sugar. One of the key reactions is the rearrangement of these glycosylamines into isomerization products. If glucose is the starting sugar in the reaction, it would be converted to aminated fructose. Although the transformed compounds are not detectable by colour or flavour change and may be reversible at this stage, the initial reactions are important because most of the later reactions cannot occur without the rearranged end-products.

Intermediate reactions involve the previous compounds as well as sugar and amino acid degradation (Strecker). They are essential for the proper flavour of a cocoa product and involve interactions of numerous compounds. When heated, mixtures of carbonyls and amino acids can produce a variety of aromas including chocolate, butterscotch, caramel, potato, cheese and 'bread-like'. It is the later stages of this intermediate phase which allow the generation and interaction of many compounds which form numerous flavourful notes. The aldehydes, ketones and other carbonyls react to form the pyrazines, furans and other compounds found in chocolate. Cocoa products have been known as the best source of pyrazines, and as many as 80 have been isolated and identified. Although the compounds identified in cocoa products number in the hundreds, it is likely that as analytical equipment continues to be improved, further examination may indicate that there are thousands or more. Not only is flavour or aroma generated during this stage, but a yellow colour (in model systems) starts to appear. This stage is permanent and cannot revert to the initial stage but may progress into the final stage.

The appropriateness of the term 'browning reactions' comes from the final stage. Only then can the insoluble dark brown pigments, called melanoidins, be produced. It is easily understood why less is known about this stage since many, or perhaps all, of the preceding compounds can condense and polymerize into very large compounds. Unlike most large molecules, the melanoidin pigments are made up of many compounds which may interact in any order and arrangement.

7.4.2 A closer look at browning reactions

The importance of amino acids and sugars to chocolate or cocoa flavour has been suggested by investigations of the heating of those compounds. Odours from mixtures of leucine and glucose, threonine and glucose, and glutamine and glucose, when heated to 100 °C (212 °F), have been described as 'sweet chocolate', 'chocolate' and 'chocolate' respectively. Valine and glucose, upon heating to 180 °C (356 °F), have been described as 'penetrating chocolate'. The complexity of such a basic flavour system is obvious when considering that a change in temperature elicits a change in description. For example, when heated 20 °C (36 °F) higher, the leucine–glucose mixture was described as 'moderate breadcrust' (15). The number of potential compounds is equally impressive: simply heating glucose (250–500 °C or 482–932 °F) has produced 96 compounds (16). The interaction of amino acids and sugars is evident during roasting and can be seen in Accra beans, for example, in which free amino acids decreased by 50% (17) and 92% of the reducing sugars were destroyed (18). The generation of aroma indicates that reactions have proceeded past the initial stage.

The isomerization products formed during the initial phase are primarily addition compounds formed from amino acids and sugars. During the intermediate stage they are then dehydrated, fragmented and transaminated, forming complex compounds, depending on temperature and pH. On the acid side, generally hydroxymethylfurfural and other furfural products are formed. If the pH is neutral, the result of the reaction is reductones. All of the intermediates are very complex, and little is known about their structure and the exact nature of their formation in food systems. In the end, however, a host of compounds, depending on the substrates and the pH, will polymerize and in turn contribute to the final chocolate flavour. Some of the most important compounds are pyrazines, pyrroles, pyridines, imidazoles, thiazoles, and oxazoles.

An important interaction in the intermediate stage of the Maillard scheme is a specific type of amino acid degradation called Strecker degradation. Amino acids are tasteless and odourless: even the sulphur-containing ones have no flavour unless they are contaminated. An example of Strecker degradation is the interaction of the amino acid glycine with glyoxal. Glyoxal is a 1, 2-dioxo compound and acts as an acceptor during transamination. Dioxo compounds may result from the formation of the dehydroreductones, sugar fragmentation and dehydration reactions. With Strecker degradation, glyoxal forms a nitrogen-containing moiety needed to form ringed compounds after reacting with glycine. The amino acid is decarboxylated and deaminated, producing an aldehyde which, in this example, is formaldehyde. This nitrogen-substituted 1, 2-dioxo compound (now 2-amino-ethanal) could combine with a duplicate molecule. When the two compounds react and are further oxidized, pyrazine is formed. This is one of the major pathways for pyrazine formation during the browning reaction.

The actual pyrazine structure itself is dictated by the side groups on the dioxo compound. For example, if one starts with pyruvaldehyde and valine, the end products will be 2-methyl-propanal and 2, 5-dimethyl pyrazine, which has been described as having a nutty flavour (19). The high concentrations of precursors make chocolate products an abundant source of pyrazines, at least 80, which have an important role in chocolate flavour (20). The concentration of pyrazines in roasted beans varies among varieties. Ghanian beans may have about 698 μg per 100 g of beans whereas Tabasco beans (Mexico) may contain as little as 142 μg per 100 g (21). Because of this variation and importance in determining the flavour of the final chocolate, investigation of the pyrazine concentrations is a very important area of research.

As important as the carbonyl–amine reactions are, one must realize that, with the exception of model systems, it is unlikely that a food system will consist of only amino acids and sugars. Other compounds such as peptides, proteins, vitamins, fats and their oxidation products and other derivatives can enter the reactions and influence the final product. With the many compounds found in chocolate, it is virtually impossible to identify all reactions and pathways needed to produce chocolate flavour.

The production of aldehydes from amino acids plays an important role in the flavour balance of the final chocolate. As already mentioned, they are converted via Strecker degradation of amino acids, which also produces pyrazines. The amino acid structure dictates the resulting aldehyde and also the amine and acid which could be produced from the amino acid degradation (Table 7.4).

Figure 7.4 The degradation products of amino acids which have been found in cocoa products (20)

| Amino acid | Degradation products | | |
	Amine	Aldehyde	Acid
Alanine	Ethylamine	Acetaldehyde	Acetic acid
Glycine	Methylamine		Formic acid
Valine	Isobutylamine	2-methylpropanal	2-methylpropanoic acid
Leucine	Isoamylamine	3-methylbutanal	3-methylbutanoic acid
Isoleucine		2-methylbutanal	2-methylbutanoic acid
Threonine			2-hydroxypropanoic acid
Phenylalanine	2-phenethylamine	2-phenylacetaldehyde	2-phenylacetic acid
Tyrosine			2-(4-hydroxyphenol) acetic acid
Methionine		Methional	

By utilizing what is known about Strecker degradation during the roasting process, a method to evaluate cocoa bean quality by comparison of headspace volatiles has been suggested (22). Cocoa beans to be evaluated are placed in an Erlenmeyer flask sealed with a cork and septum. The sample flask is heated for a pre-set time and temperature in an oven and the headspace volatiles are then analysed by gas-liquid chromatography. Two major peaks increase as the time and/or temperature of roast is increased. The compounds are isovaleraldehyde (from valine) and isobutyraldehyde (from leucine), and if one plots their concentration over time of roasting, one can correlate the concentration with the subjective evaluation of the flavour quality. For example, Caracas and Trinidad beans have been considered to be of good quality with high flavour intensity, and concurrently, chemical analyses indicate high concentration of isovaleraldehyde and isobutyraldehyde. The Arriba, Bahia and Accra beans fall in the middle, and Costa Rican, Sanchez and Tabasco beans have low concentrations of these aldehydes. This is a very simple method to evaluate bean quality. Thus, the aldehydes are also very important in the total flavour development of chocolate.

Another reaction, which is important as a generator of carbonyls when the moisture content is high in the bean, is the breakdown of lipids mainly through oxidation via moulds. The increase in 2-enals and 2, 4-dienals has been found in mouldy beans when compared to non-mouldy beans. The unsaturated aldehydes formed have very low flavour thresholds and are potent flavour components.

Up to this point we have been discussing some of the flavour-generating reactions through the roasting process. There are probably about 400–500 compounds that have been identified from various volatile and non-volatile fractions of chocolate. These compounds include hydrocarbons, alcohols, aldehydes, ketones, esters, amines, oxazoles, sulphur compounds, etc. It is very difficult to elaborate the flavour of a product like chocolate when so many different parameters (for example, fermentation, drying, roasting, refining and conching) influence the final product. During further processing such as refining and conching, additional flavour changes may occur. Refining is essentially a grinding process during which the broken cocoa beans (called nibs) are further reduced in particle size. That process is inherently one of high temperature because of the high-pressure rollers used. The reactions associated with the roasting process may therefore continue during refining but would not be expected to affect flavour to the degree of roasting.

7.5 Conching

Conching (Chapter 8) can be described as the working of chocolate flake and crumb into a fluid paste. Before the invention of roller refiners, chocolate paste was coarse and gritty. The use of a machine called a 'conche', so named because of its resemblance to the conche shell, changed the texture. After pounding for days in a conche, the chocolate and sugar particles were thought to be reduced in size, giving the conched chocolate a smooth mouth feel. Thus, the conche was initially used to refine the texture. Since preconche chocolate can now be adequately refined without this time-consuming process, the question must be asked, 'Why do we continue to conche?'

Although conching allows the chocolate mass to be further mixed, it is the modification in flavour that becomes important. Typically, conched chocolate is described as mellow compared to unconched. The bitterness is reduced, perhaps allowing other flavour notes to be more pronounced. The nature of the flavour change during conching has not been completely explained at the chemical level even though much has been learned about the total volatiles, free fatty acids, pyrazines and sulphur compound concentrations.

The air space immediately surrounding a conche in operation has an odour of acetic acid. Therefore, the first chemical change suspected to occur during the conching of chocolate was the loss of short-chain volatile fatty acids. Acetic acid is present because it is an end product of the fermentation stage although, to a lesser extent, several other acids are also found. When the role of volatile fatty acids in the conching phase was investigated, it was found that the concentrations decreased (23). However, the major change in

Table 7.5 Volatile fatty acid concentrations at final conche time for dark semi-sweet chocolate (23).

Conche type	Total VFA (μg/g choc)	Volatile fatty acids (VFA) (% of total)			
		Acetic	Propionic	Isobutyric	Isovaleric
Pug mill	909	96.5	1.1	0.8	1.6
Longitudinal	721	97.0	1.2	0.5	1.3
Vertical	634[a]	97.4	0.9	—	1.6
Horizontal rotary	614[a]	100.0	—	—	—
Pre-conche	823	95.7	1.8	0.6	1.8

[a] Significantly different at $p < 0.05$

concentration previously implicated by the literature was not noted (Table 7.5). The lowest boiling point of the volatile fatty acids is 118 °C (244 °F) for acetic acid. Since conching does not reach this temperature, it would seem difficult to significantly reduce the VFA concentration. The very low boiling points of the associated aldehydes, almost 100 °C lower than the acids, might make them a more significant class of compounds because their loss is more likely to occur and have an effect on the final flavour of the chocolate.

The major difference between the boiling point of the VFA and the corresponding aldehydes suggests that the aldehydes would decrease. Surprisingly, an analysis of total carbonyls and monocarbonyls during conching showed no significant differences (24). In order to ascertain the importance of conching as a continuation of the roasting process, analyses were performed on chocolates made from beans subjected to a low roast and conched for a longer period (up to 44 h). No statistical differences were found in the concentration of total carbonyls and total monocarbonyls. In addition, higher roasting temperatures result in higher concentrations of carbonyl compounds. That is, the chocolate from the high-roast beans averaged 23.90 $\pm$ 2.21 μM/10 g of fat as compared to 19.02 $\pm$ 1.05 μM/10 g fat in total carbonyls for the low-roast chocolate. The average monocarbonyl concentrations were 5.06 $\pm$ 0.81 and 3.36 $\pm$ 0.19 μM/10 g of fat for the chocolate from the high and low roast, respectively. Investigations of the roasting of cocoa beans revealed that the Strecker degradation was not complete at the end of the roasting process and suggested that it continued during conching (25). However, it was also determined that the concentration of amino acids and reducing sugars exhibited no change as a result of conching (longitudinal, 48 h at 71 °C or 160 °F). Thus, the modification of the flavour during conching could not be attributed to the Strecker degradation of the free amino acids.

Nonetheless, the analysis of headspace volatiles does suggest that a chemical change did occur during conching (24). Volatiles decreased by almost 80% during the first hours of conching before levelling off (Figure 7.2). That point seemed to coincide with taste panel responses indicating that no further

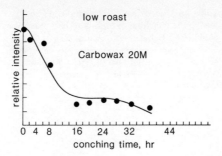

Figure 7.2 Relative intensity of head space volatiles of a dark semi-sweet chocolate from low roast beans during conche (24).

flavour changes occurred. The same phenomenon of reducing the amount of volatiles has been previously noted (26). Conching would be expected to affect the volatiles since it is a relatively long mixing process, usually heated or the cause of temperature change itself, which also increases the loss of volatiles. However, it should be noted that major changes in volatiles are not always responsible for a detectable flavour change.

Other chemical investigations have been undertaken comparing the differences between conched and unconched chocolate. A relatively rapid loss of phenols during the conching of a smoky Bahian liquor has been noted (12). What occurs in terms of phenol concentrations during the conching of non-smoky beans has been studied (27). Chocolate made from Accra and Bahia beans, which were previously roasted at 149 °C (300 °F) and 163 °C (325 °F), was conched at 82 °C (180 °F) for 0–44 h ('low-roast'). In addition, chocolate made from Accra and Bahia beans, previously roasted at 204 °C (400 °F) and 218 °C (424 °F), was conched (Suchard) at 82 °C (180 °F) for 0–24 h ('high-roast'). Chocolate samples were removed from the conche at specified times, and refrigerated until analysed. The concentration of phenol compounds decreased from 21.3 μg/100 g at 0 h of conche to 10.9 μg/100 g after 44 h in the low-roast chocolate (Table 7.6). The concentration of phenols decreased from 10.3 μg/100 g at 0 h to 6.0 μg/100 g after 24 h in the high-roast chocolate. Unfortunately, the relationship between the phenols and the carbonyl–amine reactions during the conching process is not known, although it seems apparent that these ring compounds would play a major role. It is well established that the polyphenols, through oxidation and enzymatic mechanisms (in the tanning process), form complexes with amino acids, peptides, and proteins. The decrease in phenols noted during conching may be due to irreversible protein–phenol interactions. This is turn would reduce the astringency characteristic of phenols and thus provide a more mellow chocolate after conching.

Understanding the conching process of chocolate requires consideration of its role in producing a smooth mouthfeel. This effect is so obvious that

Table 7.6 Concentration of phenols in chocolate during the conching of high- and low-roast bean blend (40% Bahia −60% Accra) (27)

Hours of conche . (Suchard)	μg phenol/100 g chocolate	
	Low-roast	High-roast
	$n=4$	$n=2$
0	21.3	10.3
2	21.5	
6	19.1	
8		11.3
12	17.8	9.4
24	16.5	6.0
32	16.7	
38	13.2	
44	10.9	

unconched chocolate appears visually coarser than the conched product. This phenomenon is easily explained by simply considering that cocoa butter is coating all the particles of cocoa mass and sugar rather than physically modifying or smoothing the sugar crystals (28). Since the temperature of the chocolate masse determines the degree of liquidity of the cocoa butter and, therefore, the length of time needed to conche, it may explain the relationship between time and temperature. The question of how a physical process can effect a flavour change needs to be considered.

It is likely that everyone is familiar with the adage 'a bitter pill is hard to swallow'. If that same pill is coated with butter or cocoa butter, it would be no surprise to find that the bitter pill goes down much more easily or would be perceptibly less bitter. Perhaps that is analogous to what happens during conching. The bitterness of the cocoa mass is muted by the coating of the particles with cocoa butter. A similar effect must occur with the sugar particles, since the clean sweetness of the sugar is not as obvious in conched chocolate. Although the conche is an inherently inefficient machine, it may be impossible to replace because physically working the chocolate masse is difficult to accomplish. It may not be possible to increase efficiency much beyond fine-tuning the process unless a completely different method is devised.

If the flavour change caused by conching is simply due to the coating of the particle with cocoa butter, then controlling the particle size might allow for maximum conching efficiency. Because larger particles require less cocoa butter per unit volume for coverage, the larger particles might be easier to conche. It is important to consider the range of particle sizes the mouth determines to be gritty.

It should seem obvious that the more liquid the cocoa butter, the faster conching and coating can proceed. The upper limit on temperatures starts changing the flavour by imparting more of a burnt flavour. Because sugar is also coated, it might be advantageous to conche the sugar separately at a higher temperature and then add it back to the cocoa mass.

7.6 Conclusion

The importance of chocolate flavour can only increase in the future. Chocolate flavour is appreciated by virtually everyone, and that is very meaningful when considering that many strong flavours are not universally acceptable. It has an ability to mask or mute flavours regardless of whether they are desirable and undesirable. That ability makes it a prime candidate for increasing the palatability of under-utilized and newly developed food sources. It is also likely that the continued investigation of the chemical and sensory nature of chocolate will result in vastly new and different products.

References

1. (a) Carlin, J.T., Lee, K.N., Hsieh, O.A.-L., Hwang, L.S., Ho, C.T. and Chang, S.S. *J. Amer. Oil Chem. Soc.* **63** (1986) 1031–1036. (b) Hoskin, J.C. and Dimick, P.S. *Proc. Bioch.* **19**(3) (1984) 92–104. (c) Hoskin, J.C. and Dimick, P.S. *Proc. Bioch.* **19**(4) (1984) 150–6. (d) Maniere, F.Y. and Dimick, P.S. *Lebensm.-Wiss. u.-Technol.* **12** (1979) 102–7. (e) Keeney, P.G. *J. Amer. Oil Chem. Soc.* **49**(10) (1972) 567–72. (f) Ziegleder, G. *Süsswaren* **10**(1984) 422–6.
2. (a) Lehrian, D.W. and Patterson, G.R. *Biotechnology*, **5** (1983) 529–75. (b) Lopez, A.S. *Proc. Cacao Biotechnology Symp.*, ed. Dimick, P.S. The Pennsylvania State University (1986) 19–53.
3. Ref. 2(b).
4. Forsyth, W.G.C. and Quesnel, V.C. *J. Sci. Food Agric.* **8** (1957) 505–509.
5. (a) Wadsworth, R.V. *Tropical Agric.* (*Trinidad*) **32** (1955) 1–9. (b) Rohan, T.A. *J. Sci. Food Agric.* **9** (1958) 542–551.
6. (a) Berbert, P.R.F. *Rev. Theobroma* **9** (1978) 55–61. (b) Knapp, A.W. *Cocoa Fermentations*. John Bale, Curnow (1937) (c) Reineccius, G.A., Anderson, D.A., Kavanagh, T.E. and Keeney, P.G. *J. Agr. Food Chem.* **20** (1972) 199–202.
7. Ref. 6 (c).
8. (a) de Witt, K.W. *Report Cacao Res.* (*Trinidad*) **30** (1957) 228–238. (b) Ref. 6(b). (c) Roelofsen, P.A. *Adv. Food Res.* **8** (1958) 225–296. (d) Timbie, D.J. and Keeney, P.G. *J. Food Sci.* **42**(6) (1977) 1590–1599. (e) Zak, D.L. and Keeney, P.G. *J. Agric. Food Chem.* **24** (1976) 483–486.
9. Forsyth, W.G.C. and Quesnel, V.C. *Adv. Enzymol.* **25** (1963) 457–492.
10. Ref. 8(c).
11. Hansen, A.P. and Keeney, P.G. *J. Food Sci.* **35** (1970) 37–40.
12. Lehrian, D.W. Keeney, P.G. and Lopez, A.S. *J. Food Sci.* **43** (1978) 743–745.
13. Kaden, O.F. *Gordian* (1950) 1–16.
14. Ref. 1(d).
15. Mabrouk, A. In *Food Taste Chemistry*, ed. Boudreau, J.C. ACS Symp. Ser. **115** (1979).
16. Fagerson, I.S. *J. Agric. Food Chem.* **17** (1969) 747–750.
17. Rohan, T.A. and Stewart, T. *J. Food Sci.* **31** (1966) 202–205.
18. Rohan, T.A. and Stewart, T. *J. Food Sci.* **31** (1966) 206–209.
19. Maga, J.A. *CRC Crit. Rev. Food Sci. Nutr.* **16** (1982) 1–48.
20. Hoskin, J.C. Ph. D. thesis, The Pennsylvania State University (1982).
21. Reineccius, G.A. Keeney, P.G. and Weissberger, W. *J. Agric. Food Chem.* **20** (1972) 202–206.
22. Ziegleder, G. *Rev. Choc. Conf. and Bakery* **7**(2) (1982) 17–22.
23. Hoskin, J.M. and Dimick, P.S. *Proc. 33rd Ann. Prod. Conf. of PMCA* (1979) 23–31.
24. Maniere, F.Y. and Dimick, P.S. *Lebensm.-Wiss. u.-Tech.* **12** (1979) 102–107.
25. Refs. 17 and 18.
26. Mohr, W. *Rev. Int. Choc.* **14** (1959) 371.
27. Hoskin, J.C. and Dimick P.S. *Proc. 36th Ann. Prod. Conf. of PMCA* (1982) 23–31.
28. Hoskin, J.M. and Dimick, P.S. *J. Food Sci.* **45** (1980) 555–557.

8 Conching

D. LEY

8.1 Introduction: principles of conching

The conching process is of paramount importance in the production of chocolate. During this process chemical and physical processes are taking place concurrently and should not be separated from each other. These include the development of the full desirable chocolate flavour and also the conversion of the powdery, crumbly refined product into a flowable suspension of sugar, cocoa and milk powder particles in a liquid phase of cocoa butter (and other fats as appropriate). For many years more attention was paid to the physical processes than to the chemical ones, because the former were easier to study visibly.

The development of the gas chromatograph, and improvements in the interpretation of the data it provided enabled us to understand some of the chemical reactions and changes taking place. Fundamental work in this field was carried out by Dr W. Mohr at the Institute for Food Technology and Packaging, Munich. This topic is reviewed in more detail in Chapter 7.

It is important that conching should not be studied in isolation. Processes which have been initiated during fermentation and roasting are being completed in the conche. Where damage to flavour has already taken place due to incorrect processing or poor beans, this cannot be overcome in the conche. Nowadays, it is considered that the conching cycle and roasting cycles are equally important in terms of chemical change. The higher temperature of roasting, however, enables these reactions to take place more quickly in this process than those that take place at the low temperature of conching. Some flavour components like pyrazines require the higher temperature provided by roasting, while other flavours develop during the longer reaction time in the conche.

This means that care should be taken during the roasting process to ensure that acetic acid, lower boiling aldehydes and the heat-sensitive pyrazines are not all driven off. In other words the moisture content of the finished roasted beans should not drop below 2–2.5%. This can only be obtained by carrying out all heat treatments under carefully controlled conditions. Heating above 100 °C will not leave enough moisture to evaporate during later stages to take with it the undesirable flavour components. Thin film processing is now being used increasingly to remove unwanted low-boiling-point acidic components (see Chapter 5). Usually this is carried out at temperature below 100 °C and by adding water. Here temperature control must again take place to ensure that no over-roasting effects occur. The added water is removed once more when this process is carried out correctly. Chemical reactions forming free amino acids cannot take place due to the low temperature and the short duration of the process, which is limited to the volatilization of the low-boiling-point components. The use of vacuum is to be avoided, as both beneficial and detrimental flavour components, which have a similar boiling point, are volatilized.

Conching is required to complete the flavour development started in the roasting and/or thin-film processing. In the conche the water content of the chocolate mass is lowered from about 1.6% to 0.6–0.8%. As the moisture is removed it takes with it many unwanted flavour components. In this way approximately 30% of acetic acid and up to 50% of low-boiling aldehydes are volatilized. This volatilization is considerably helped by means of the so-called 'dry conching' of chocolate masse with relatively low fat content. The partial removal of these volatilized acid components is necessary to give the finished chocolate a full 'rounded' flavour. The use of additional ventilation through the conche may aid moisture removal, but does not cause additional chemical changes. At one time it was incorrectly thought that this would oxidize the tannins and bring about flavour development.

The theory of amino acid development after roasting developed by Strecker seems far more probable. This suggests that a significant formation of free amino acid takes place during conching, which is connected with flavour development in the chocolate. In fact, during conching the amount of free amino acid developed corresponds to about one-third or half of that formed during roasting. Free amino acids together with reducing sugars are the flavour precursors from which the variety of flavours develop during heating by means of Maillard reactions. During roasting approximately 50% of the free amino acids which are created are also destroyed. The remainder are available in the conche as flavour precursors. In addition the further amino acids which are formed are slow to do so owing to the relatively low temperatures in the conche. It has been proved that the shearing stress in the conche not only serves the purpose of liquefying the masse, but also positively influences and accelerates the flavour development processes. This is probably due to the interactions which occur as the solid particles are forced against one another, in particular the sugar particles.

It cannot be overstressed that the chemical and physical processes occur together during the conching process, and they cannot be totally separated from each other.

8.1.1 Liquefaction and shearing forces

During the conching cycle mechanical and shearing forces are exerted in order to separate the agglomerates formed during grinding, to coat the single particles with fat and to disperse the cocoa butter phase throughout the chocolate. The physical tasks of the conche are to disperse, to dehumidify, to remove volatile components, and to homogenize, in order to improve the viscosity, improve the flowability, improve the texture and produce a chocolate with good melting characteristics. The efficiency of a conche, therefore, is very dependent on how well the kneading and shearing arms are arranged, and on the precision of the thermal controls.

A good consistency for the refined flake filling the conche is one which is powdery and crumbly, but does not flow under the influence of gravity. The effective viscosity drops as the temperature rises and also depends upon the range of stress applied, i.e. upon the speed and frequency of the mixing elements. A thorough mixing action is required which involves high shearing, repeated folding, utilization of back flows, mechanical subdivision of flows and permanent rearrangement of particles.

In general, shearing of a deformable material between parallel planes results in a linear shearing action. If the material is stationary in one plane, for instance at a wall of the machine, it is subject to a variable rate of shear, usually parabolically distributed. Given streamline behaviour, laminar shearing of this type causes the particles of a suspension to become coated with the liquid phase; the particles, whether solid or elastic, also tend to be rotated and to migrate into neighbouring streamlines. Agglomerates of particles are also broken down by this action. Usually the intensity of the dispersing action is expressed as the rate of shear, which is expressed as the relative velocity of any two planes divided by their distance apart.

Combined shear and compression forces are formed when the relative motion of the planes is not purely parallel (causing wedge-like streaming). Suitable combinations of the forces and motions produce highly effective back flows and circular mixing motions. This is illustrated in Figure 8.1.

These theoretical principles have been successfully applied in the design of mixing and shearing elements in conches. By means of continuous recirculation and changes of streamline directions shearing forces are produced, with the shearing process mainly taking place at the wall and scrapers. At this location, it is possible to measure the shearing action from the rate of relative velocity of the two planes divided by their distance apart. The power required

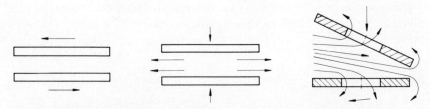

Figure 8.1 Illustration of the development of different types of shearing action.

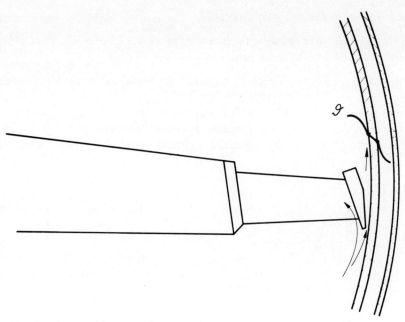

Figure 8.2 The product being heated by shearing as the scraper moves parallel to the wall.

for this shearing depends upon the rheological properties of the product. The particle size, the fat content and especially the water content greatly affect the frictional heat developed during the shearing process (Figure 8.2). The masse becomes smoother as the fat melts, which in turn causes a quicker coating of the particles. As this happens the energy requirement of the conche falls.

If the stationary wall (conche side) is being cooled rather than heated, sooner or later the mass becomes pasty, which greatly affects the power usage. In the early stages of the process, however, during 'drying conching', it is desirable to impart as much energy into the mass as possible, in order to produce as large a shearing stress as possible. On the other hand a slow temperature rise is required in order to evaporate the moisture satisfactorily from the mass (this also takes with it many of the organic volatiles). At first sight this seems contradictory, but if the temperature rise takes place too quickly the solid particles become coated with fat before the moisture escapes. This has two very detrimental effects. Firstly, the water-soluble volatile 'undesirable' flavour components are driven off at a greatly reduced rate. Secondly, the water thickens up the chocolate by its reactions with the hydrophilic material in it. No further size reduction of particles takes place in the conche, although some of the crystal surfaces may be subjected to 'smoothing and finishing'.

8.1.2 *Phases in the conching process*

In all, there are three phases in the conching process (Figure 8.3), each of which takes more or less the same time.

Dry phase: Shearing, moisture evaporation, removal of other volatiles

Pasty phase: Flavour development by means of shearing and heating, moisture removal, homogenizing

Liquid phase: Homogenizing by means of intense stirring, shearing.

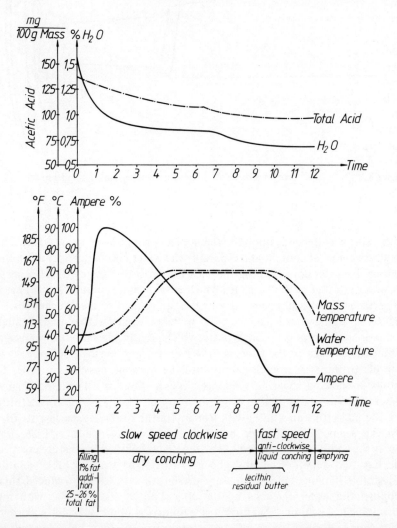

Figure 8.3 Machine operating conditions and changes in moisture and acidity during a conche cycle (time in hours).

Good processing involves a slow temperature rise up to a temperature level corresponding to the desired chocolate flavour. During the pasty phase it is necessary to maintain a steady conching temperature in order to obtain reproducible flavour results. Here it is advantageous to use a thermostatic water jacket on the conche which is able to react spontaneously to temperature changes.

As the conching progresses, the flow properties improve. This is principally due to the reduced water content, coupled with the coating of cocoa butter across the surfaces of the particles. This formation of a stable continuous fat film between and on the particles is seen as particularly significant. It is perhaps worth mentioning that moisture contents of less than 0.6% have been shown to produce no further improvement in the flow properties of the chocolate. However, a 0.15% reduction of moisture content obtained by intense dry conching is the equivalent to a 1.5% saving of fat (viscosity and flow properties being the same).

The final reduction in viscosity is usually made at or near the end of conching, by the addition of the emulsifying agent lecithin. This addition should take place at a moderate temperature and as late as possible. It is widely accepted that one part of commercial lecithin can substitute for 9 to 10 parts of cocoa butter so that, commercially, lecithin is a very important constituent of chocolate.

Not all countries allow other surface-active agents, such as ammonium phosphatides (YN) and polyglycerol polyricinoleate (PGPR), to be used (see also Chapter 12). Ammonium phosphatide (YN) in dark chocolate does not reduce the yield value as much as lecithin (126% of the value for lecithin) but it has the same effect on the plastic viscosity. In milk chocolate the plastic viscosity is increased (130% of the value for lecithin). However, yield value drops to half the value for lecithin. Polyglycerol polyricinoleate (PGPR) has a dramatic effect on yield values. Zero values (Newtonian flow) are possible compared with lecithin. The plastic viscosity is hardly affected in dark chocolate, whereas for milk chocolate it almost doubles. Combinations of the different surface-active agents may be added, and give different effects depending upon the proportions used.

If chocolate is solidified and re-melted several times there is an increase in both its plastic viscosity and yield values.

8.2 Types of conche

The Mexican grinding stones can be considered as the forerunners of present-day conches. As early as 1878, Rudi Lindt noted the flavour-enriching effect brought about by the continuous movement of the chocolate over long periods. Thus began the development of the longitudinal conche. Even when new types of conche were developed, many producers of high-quality chocolate preferred to retain the longitudinal type, because of its exceptionally good flavour development properties. It did, however, have several major disadvantages, notably its high energy consumption, small filling capacity, poor temperature control and especially its inability to 'dry conche', which resulted in long conching times. Consequently rotary conches have normally replaced the longitudinal conches.

In addition to these two types of conche, processes have been developed whereby the flavour development is chiefly carried out by 'conching' the cocoa mass alone (or with a small proportion of other ingredients). The traditional conching operation is then reduced chiefly to one of liquefying the product.

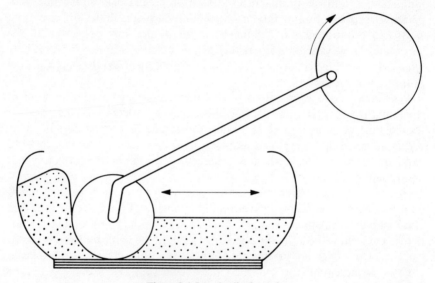

Figure 8.4 Longitudinal conche.

8.2.1 *Longitudinal conches*

One machine generally contained four conche troughs (Figure 8.4). The bottoms of the conche troughs and the grinding rollers were made of granite. A connecting rod moved the grinding roller to and fro at a frequency of 20–40 r.p.m. The top of each trough, which has a capacity of 100–1000 kg (220 lb–1 ton), was covered. The chocolate to be processed had to be filled into the troughs in a pasty or liquid state.

Because of the large angle of the shearing gap between granite trough and grinding roll, only these pasty and highly fatty masses can be fed into these conches. Powdery material compacts and may damage the machine.

As the particles are already coated with fat before conching starts, the removal of volatiles requires additional heating. In addition the induced shearing stress is low, so in order to produce the full flavour and liquidity development, conching times of up to 96 hours are required.

8.2.2 *Rotary (round) conches*

Historically there has been a large number of conche manufacturers. Because of the limited space in this chapter, however, this review is restricted to several of the important manufacturers of today:

Bauermeister, Hamburg, Federal Republic of Germany
Carle & Montanari, Milan, Italy
Richard Frisse, Herford, Federal Republic of Germany
Lloveras, Terrassa, Spain
Nagema, Heidenau, German Democratic Republic
Petzholdt, Frankfurt, Federal Republic of Germany
Thouet, Aachen, Federal Republic of Germany.

Bauermeister. Unlike other manufacturers, Bauermeister and Co. constructed a round conche about 1960. Its axis was inclined by 45° and its trough rotated (rotary conche). Inside a scraper was provided which could be operated independently of the rotary movement of the trough.

The centre axle was equipped with a worm, ensuring mass transport in the axial direction. Heating was frictional or additionally via an infra-red lamp. Air cooling was available. The batch capacity was 2–3 tons. The rotary conche could operate with dry, crumbly feed material and so dry conching could be carried out.

Figure 8.5 Bauermeister type TRC conche (Hermann Bauermeister GmbH)

Because of its small capacity, poor temperature control and very complicated construction, this conche type was only used for a very short period before being succeeded by the Bauermeister Dry Round Conche TRC (Figure 8.5). In this conche a higher capacity (up to 6 tons) was achieved. The inclined position and plough-like conching elements increase the rate of change of product surface and intensive aeration and mixing in the dry phase. The drive operates at two speeds: dry conching is carried out at lower speed and the liquid phase at the higher. The conching vessel is jacketed for rapid cooling or heating. The temperature increases during the dry phase due to frictional heat. This mixing action and conche design gives a very large product surface, from which volatiles can escape.

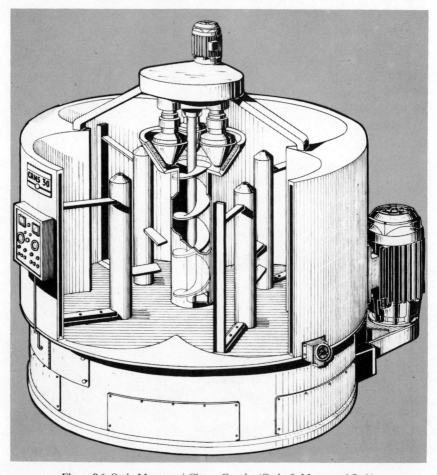

Figure 8.6 Carle-Montanari Clover Conche (Carle & Montanari SpA).

Carle and Montanari. The current Clover-Conche was also developed from an earlier model. The Carle–Montanari Conche consisted of a round, heatable outer trough and a conical inner trough (Figure 8.6). The chocolate masse is first agitated in the outer trough before being transported into the conical inner trough by a conveying worm. The tapered granite rollers provide an intensive shearing action by running against the conical inner wall and throwing the mass against the wall of the outer trough. Here scrapers provide further shearing stress. During this circulation the surrounding air flow provides a good ventilation for the removal of volatiles. This type of construction enables both dry and liquid conching to be carried out. The batch capacity of the present Clover Conche is 3–6 tonnes according to the model.

Richard Frisse. About 1960 it was realized that the maximum batch capacity, 2 tonnes, of the Double Round Conche (DRC) no longer met the requirements of industry. Therefore today's Double Overthrow Conche (DUC) having a batch capacity of 3–6 tonnes was conceived. The DRC had shown the great importance of keeping the wall–scraper surfaces as large as possible to provide a good shearing stress. The jacketed triple-trough assembly was therefore developed, with the outer two troughs having a smaller diameter than the central larger one (Figure 8.7).

The kneader-stirrer arms turn towards one another, but at different speeds. All three shafts are designed for very heavy duty work and are driven by helical

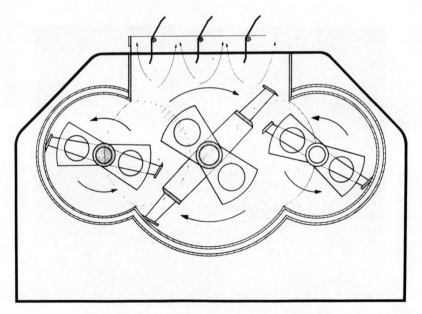

Figure 8.7 The Frisse Conche (Richard Frisse GmbH).

gears operating in an oil bath. The scrapers and the kneader-strirrer arms overlap on the shafts, ensuring powerful shearing in the masse and at the walls. The different direction of rotation of the stirrers gives the conche its well-known 'double-overthrow' effect. The low water content of the jacket gives precise temperature control, which is of great importance in flavour development.

Due to its construction the conche has a large, rapidly changing product surface area. There is no need for additional ventilation, which only has the effect of changing conditions in the room outside the conche. As with most manufacturers' equipment, the conching process in the DUC is controlled by means of microprocessors.

Lloveras. Today the Lloveras 'Mura' Conche, and the Low and Duff MacIntyre Conche (see Chapter 17) are usually employed where only a low capital budget is available for a limited throughout system. The machine itself consists of a double-jacket cylinder (Figure 8.8). Inside there is an axle fitted with a blade, which is fitted at an angle so as to scrape and raise the dough, thereby keeping it in constant motion. Baffles are fitted to prevent the dough all turning together with the machine without being homogenized, and enable the components to be thoroughly mixed. At the same time the front and back surfaces of the cylinder are scraped to prevent a build-up of the chocolate mass. The strong turbulence and the powerful airflow through the masse are designed to reduce the conching time.

Figure 8.8 L & D MacIntyre Refiner/Conche (Low & Duff (Developments) Ltd).

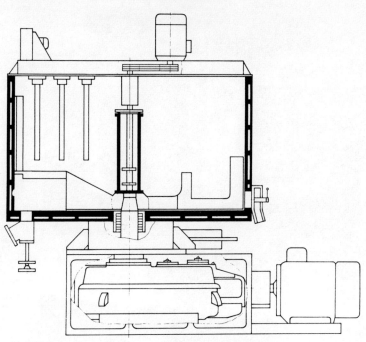

Figure 8.9 Nagema CRT Conche.

Nagema. The dry conche CRT 3000 (Figure 8.9) with a batch capacity of 3 tonnes is used for dry and liquid conching of chocolate masse. The round trough is machined so that there is a constant small gap between the scrapers and the inner contours. While dry conching, the stirrer runs at the lower speed which is designed to produce an intense circulation of the masse. The conche can also be ventilated. In the liquid phase the chocolate is homogenized by the stirrer running at a higher speed. The whole conching process is programmably controlled.

Petzholdt. The Petzholdt-Super-Conche PVS (Figure 8.10) is a conventional batch-operated conche, providing for both dry and liquid conching. The conche is designed to give maximum flexibility because the kneading-stirring action can be varied with the masse texture during the different phases of the operation. Of particular interest are the homogenizing and spraying arrangements, which are operated after the dry conching phase has been completed. It is during this final phase that the breakdown of microagglomerates and maximum surface exposure of the now liquid chocolate takes place. These conches have a capacity of 100–6000 kg (220 lb–6 tons) and can all be ventilated.

Thouet. The double round conche (DRC) (Figure 8.11) was developed for factories where large quantities of chocolate are produced in a very limited

Figure 8.10 Petzholdt PVS 4000 Super Conche (Petzholdt GmbH).

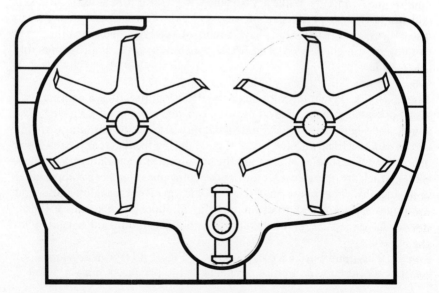

Figure 8.11 Thouet DRC Conche (Thouet KG Maschinenbau).

space. The batch capacity of the conche is up to 12 tons, and it has a very compact design. Although its triple assembly is similar to that of the Frisse conche, it operates in a very different way. For example, the outer troughs are the bigger ones and the small inner trough is used only to move the mass around the conche. Intense dry conching can be carried out together with a grinding and shearing action (similar to the longitudinal conche) in the liquid phase. The conche has a complete water jacket with a very large capacity, $4.0\,m^3$ ($140\,ft^3$).

The mixing and the two conching elements are driven separately, providing a uniform shearing stress and distribution of the masse in the conche. Additional heating and aeration are possible, and the whole conching process can be programmably controlled.

8.2.3 'Continuous' conches

Apart from the round batch-operated conches, several chocolate production methods have developed which claim to be continuous. In this case the refined masse is fed continuously, reciprocally or semi-continuously into a 'conching' process. The following manufacturers have been involved in this type of machine: Nagema, Petzholdt and Tourell.

The Nagema Konti-Conche 420 (Figure 8.12). The refined masse is continuously supplied to two shearing tanks. In these tanks the ingredients have their moisture and other volatiles removed under the influence of conditioned air, as well as being made into a paste-like material by the intense shearing action. At the end of this treatment the masse is converted into flowable state by the addition of lecithin. The final additions to the recipe are made in the subsequent dosing and weighing stage. Finally, the chocolate mass is homogenized, cooled and conveyed into storage tanks.

One feature of this type of machine is the very short retention time at each stage. Even if the cocoa mass is pre-treated, it is unlikely to achieve a conversion of free amino acids during the conching process comparable with that taking place in conventional conches which have precise temperature control and a longer reaction time (see 8.1). This applies to all systems operating by the continuous-flow method.

Petzholdt PIV intensive treatment unit. The PIV intensive treatment unit was designed by Petzholdt GmbH to try to eliminate the need for a prolonged conching. The chocolate components, such as cocoa mass, milk powder, etc., are thermally pretreated before refining to produce the majority of flavour development in this type of process. The PIV's role is mainly one of liquefying the chocolate.

A PIV unit consists of 2 or more reaction vessels and can be fed continuously. The unit is designed so that as one vessel is filling others are

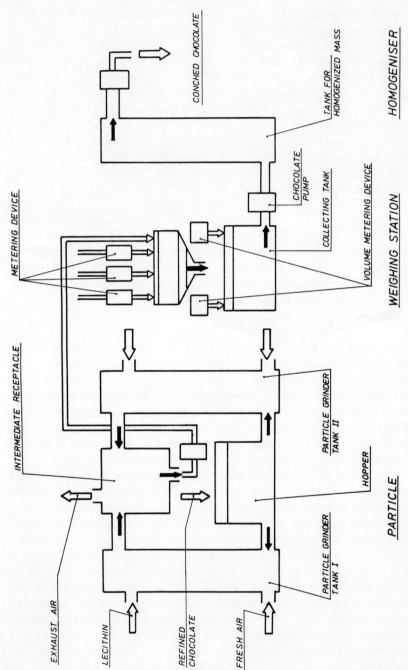

Figure 8.12 Nagema Konti-Conche 420.

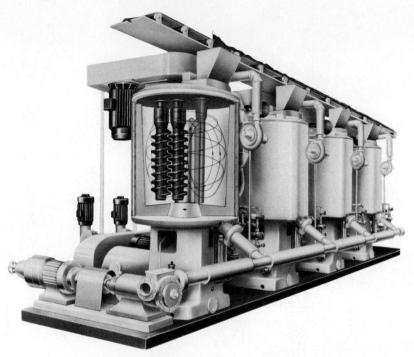

Figure 8.13 Petzholdt high-intensity short-dwell PIV 4000 conche (Petzholdt GmbH).

liquefying or emptying—thus a batch/continuous system is developed. The treatment time for 1000 kg, or 1 ton (one vessel), including feeding and discharge, is approximately 1 hour.

The double-walled vessel contains a stirrer system comprising two inter-leaving mixer screws, both turning in the same direction. They are suspended in the vessel from above and set up parallel to, but off-centre with respect to the central axis of the vessel (Figure 8.13). The lower 'anchor' stirrer system has scraper arms which reach up the sides of the vessel and which are angled so as to scrape material off the bowl wall. Shearing fields are generated when the inclined scraper arms on the 'anchor' stirrer presses the masse at right angles into the worm screws. The two interleaving worm screws then force the masse downwards towards the base of the bowl, where it is subjected to further shear from the material being driven at right angles to it, by the 'anchor' stirrer. The process time doubles if the cocoa mass has not been pre-treated. As with the Nagema plant, the reaction time is very short compared with that required for the formation of free amino acids.

Tourell. The Tourell Conching System (Figure 8.14) was developed by co-operation between Cadbury UK and the Tourell Conche Manufacturing Company. It consists of an arrangement of horizontal troughs with stirrers.

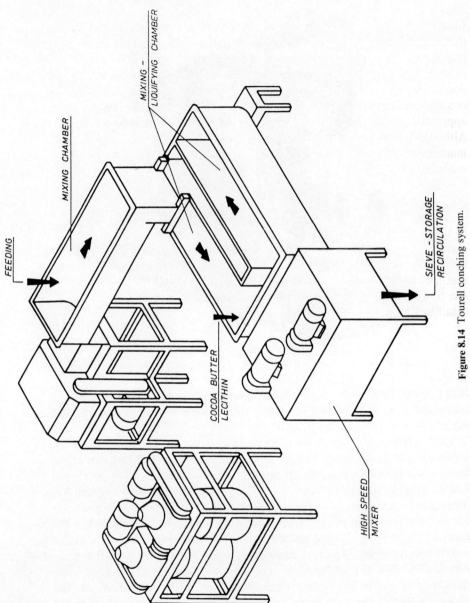

Figure 8.14 Tourell conching system.

These move the product through each trough in turn. The stirring paddles of the first trough are arranged on two parallel shafts and work in opposite directions. This results in a high mixing and shearing action which also conveys chocolate masse from the feed hopper to the discharge. In the second trough, the paddles provide only a low shearing and transport action. In the third trough, the masse is again exposed to high shearing effects caused by two counter-rotating shafts. Here lecithin addition takes place. The transit time is approximately two hours. The capacity of a Tourell line is about 3 tonnes/h. Although a substantial tonnage of milk chocolate can be produced in these machines, their use has been restricted to a relatively few large-scale chocolate manufacturers.

8.2.4 Recirculation chocolate manufacturing plants

In these systems the chocolate is held in a storage vessel from which it is pumped several times through a grinding machine (usually a ball mill). Flavour development is achieved by a combination of mixing and forced ventilation; manufacturers include Wiener and Bauermeister and Co. These are, however, not conches in a traditional sense, and are described in more detail in Chapter 17.

8.2.5 Cocoa mass treatment devices

In the introductory section, the significance of pre-treatment of cocoa mass with the resulting saving of conching time was noted. Thin-film treatment of cocoa mass, however, takes place in such a way that only highly volatile ingredients are evaporated and removed. This process takes place at temperatures up to 100 °C (212 °F), in order to remove only highly volatile acids while avoiding a secondary roasting effect. Thermal damage which has occurred during roasting cannot be reversed in these machines. Normally water is added to the cocoa mass and when this evaporates it takes with it many additional acidic components such as acetic acid. Most processes operate continuously and can easily be controlled, and the degree to which the mass has been treated can be determined by gas chromatography.

The following treatments and processes are at present being operated by commercial chocolate manufacturers: (i) Petzomat treatment; (ii) Convap treatment; (iii) Carle-Montanari treatment; (iv) Luwa and Lehmann treatments (these machines are illustrated in Chapter 5).

Petzomat STC treatment. A Petzomat system consists of one or several spray towers, depending on the required throughput capacity. The cocoa mass in the form of a turbulent-spray thin film falls down the tower as hot air is blown upwards. This counterflow principle ensures that the mass loses its moisture (from added water or a flavour-enhancing material), and the unwanted, highly volatile components, to the air. If sterilization or the removal of bacteriological contamination is required, the Super-Petzomat is used in a further stage of treatment in a pressure reactor. If required, a secondary roasting can also take place in the Petzomat plant.

Individual processing steps for the cocoa mass are:
  (i) Dosing and injecting
 (ii) Pre-degasifying and deacidifying
(iii) Heating
 (iv) Sterilization
  (v) Cooling
 (vi) Roasting
(vii) Cooling.

This equipment has been marketed worldwide. The capacity is 1000–1500 kg/h (1–1.5 tons/h).

The Convap treatment. The principal item of the Convap treatment is the thin-film evaporator itself. Several manufacturers now use this Convap cylinder in their plants. It is very important to mix the added water or flavour development solution with the cocoa mass to the extent that an absolutely homogeneous suspension is produced, which is then fed to the thin-film evaporator. Due to the centrifugal effect of the rotor, the cocoa masse is thrown on to the precisely manufactured flat chrome inside wall and is removed by razor-sharp scraper knives.

The result is a very thin product layer with short thermal diffusion paths. The surface for heat exchange is being permanently regenerated, giving an optimum transmission of heat. This machine can be operated under pressure or vacuum and care should be taken to ensure that beneficial flavour components are not removed together with the undesirable ones. Bauermeister, Alfa-Laval and Frisse all manufacture this type of machine; capacities are normally about 750 kg/h per cylinder.

The Carle-Montanari PDAT treatment. The Carle-Montanari PDAT treatment differs from other thin-film evaporators by its 'batch system'. The cylindrical reactor (autoclave) is equipped with centrally arranged pawl pumps and a stirrer. Pipe assemblies provide for heating and cooling, while inputs for water, sugar solutions, alkali solutions, inert gas as well as for sterile roasting air are available. The additions are volumetrically controlled. The following processes are possible: pasteurizing, removal of acidic components, alkalizing, flavour development and roasting. By adding milk a 'crumb' effect can also be produced. The batch operation allows for individual treatment of each batch, which can be an advantage when the beans used are of variable quality. The capacities available are 500 kg, 1000 kg and 2000 kg (0.5, 1 and 2 tons). The treatment cycle is approximately 1 hour.

References

1. Ziegleder, G. *Deutsche Lebensmittel-Rundschau* (1982) (Sept).
2. Ziegleder, G. *Zucker und Süsswarenwirtschaft* **34** (1981) 105.
3. Dimick, P.S. and Hoskin, J.M. *Can. Inst. Food Sci. Technol. J.* **14** (1981) 269.
4. Ley, D. *Süsswaren* **11** (1981) 423.
5. Bonar, A.R. Rohan, T.A. and Stewart T. *Rev. Int. Choc. (RIC)* **23** (1968) 294.

9 Chocolate flow properties

J. CHEVALLEY

In this chapter the flow properties of chocolate will be limited to those of its liquid state. We will discuss (i) flow behaviour (Bingham and Casson models); (ii) flow measurement; and (iii) factors influencing flow properties.

The following symbols have been used:

A and B	Viscometer constants
D	Shear rate
D_N	Shear rate at the inner cylinder
K_0 and K_1	Constants in the Steiner equation (1)
M	Torque
N	Number of revolutions per minute (rpm)
R	Radius of outer cylinder or sphere
a	Ratio of inner to outer cylinder radii
b	Intercept on the axis $(1 + a)\sqrt{\tau}$ in Casson plot
d_1	Density of a falling body (e.g. sphere)
d_2	Density of a liquid
f	Constant in the Zangger equation (9.8)
g	Gravity constant
h	Height of the inner cylinder
k	Torque constant
n	Viscometer reading
r	Cylinder radius
v	Velocity of a falling body (e.g. sphere)
z	Constant in the Zangger equation (9.8)
η	Viscosity
η_{pl}	Plastic viscosity
η_{CA}	Plastic viscosity according to Casson, or Casson viscosity
τ	Shear stress
τ_0	Yield value in Bingham equations (9.2), (9.3)
τ_{CA}	Yield value according to Casson, or Casson yield value
ω	Angular velocity

9.1 Flow behaviour of chocolate

Molten chocolate is a suspension of particles of sugar, cocoa and/or milk solids in a continuous fat phase. Because of the presence of solid particles in the

142

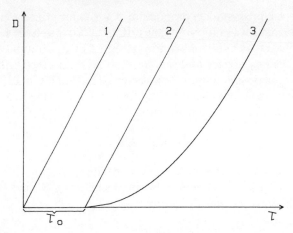

Figure 9.1 Different types of rheogram: (1) Newtonian; (2) Bingham; (3) pseudoplastic (e.g. chocolate).

melted state, it does not behave as a true liquid but exhibits non-Newtonian flow properties. For Newtonian liquids (true liquids), flow begins as soon as a force is applied. This is not the case for Bingham fluids, which are suspensions: with toothpaste, for instance, a minimum force must be applied; below this threshold, no flow occurs. This minimum force is called the yield value.

In a plot of shear rate (D) versus shear stress (τ), three different types of rheogram can be obtained (see Figure 9.1). In the case of Newtonian liquids (curve 1), a straight line is obtained which passes through the origin. Viscosity (η) is the ratio of shear stress (τ) to shear rate (D), and is equal to the reciprocal of the slope (or gradient) of the line.

$$\eta = \tau/D \tag{9.1}$$

For Bingham fluids (curve 2), the straight line does not pass through the origin, but instead intercepts the shear stress axis at its *yield value* (τ_0); the reciprocal of the slope of this line is the *plastic viscosity* (η_{pl}) of the fluid. Its equation is

$$\eta_{pl} = (\tau - \tau_0)/D \tag{9.2}$$

The same equation may also be expressed as

$$\tau = \tau_0 + \eta_{pl}D \tag{9.3}$$

In the case of chocolate (curve 3), there is also a yield value but the rheogram here is curved rather than linear. In plots (2) and (3), if shear stress (τ) is divided by shear rate (D), without taking yield value into account, a derivative called *apparent viscosity* is obtained. Apparent viscosity is dependent upon shear rate.

Amongst the various attempts to express the curvature of the chocolate rheogram in terms of an equation, the most successful was that of Steiner (1),

F

who, in 1958, adapted a model proposed by Casson for printing ink. Steiner's model for chocolate was adopted by the OICC (Office International du Cacao et du Chocolat) in 1973 (2) and is now accepted by most authors.

By taking the *square roots* of shear stress (τ) and shear rate (D), the curve for chocolate viscosity (see Figure 9.1) becomes a straight line. Hence, the rheological behaviour of a chocolate may be expressed by the two parameters, K_0 and K_1, as in the Bingham equation (9.3):

$$\sqrt{\tau} = K_0 + K_1\sqrt{D} \qquad (9.4)$$

Viscosity measurements can be made using rotational viscometers of various different geometries. To allow for this in calculations, Steiner (1), and the OICC (2) take into account the ratio a:

$$a = r/R \qquad (9.5)$$

where r = radius of the inner cylinder
R = radius of the outer cylinder.
In order to obtain correct results, the value of a should be greater than 0.65 (see section 9.2.1); there should be laminar flow of chocolate between the two cylinders and shear stress must be greater than K_0^2/a^2.

The shear rate (D_N) at the inner cylinder surface is given by the relation

$$D_N = \frac{2\omega R^2}{R^2 - r^2} = \frac{2\omega}{1 - a^2} \qquad (9.6)$$

where ω = angular velocity = $2\pi N/60$
and N = revolutions per minute (rpm).
The following relationship, given by Steiner (1), relates shear stress (D_N measured at the inner cylinder) to these other parameters:

$$(1 + a)\sqrt{D_N} = \frac{1}{K_1}\left[(1 + a)\sqrt{\tau} - 2K_0\right] \qquad (9.7)$$

The graphical plot of this equation, as represented by the OICC (2), is shown Figure 9.2. $\eta_{CA} = (1/\text{slope})^2 = (1/K_1)^2$ = plastic viscosity according to Casson (i.e., Casson viscosity); $\tau_{CA} = (b/2)^2 = K_0^2$ = yield value according to Casson (i.e., Casson yield value).

For many years viscosity and yield value were expressed in *poises* and *dynes/cm²* respectively. Since 1978, however, it has been recommended that viscosity should be expressed in *pascal seconds* (1 Pa s = 10 poises) and yield value in *pascals* (1 Pa = 10 dynes/cm²). Shear rate (D) is always expressed in *reciprocal seconds* (s⁻¹).

The calculation of Casson parameters may be simplified by using nomograms (3, 4), or by preparing small computer programs which provide coefficients of the regression line based on experimental points and/or a graph

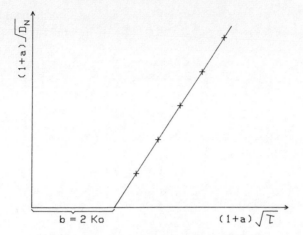

Figure 9.2 Casson rheogram according to OICC.

(5). The Casson equation has not been universally accepted, however. Several equations have been proposed to better describe the flow properties of chocolate (6). Recently, Zangger (7) has proposed a generalized Casson equation with *three* parameters η, f and z:

$$D = (\tau^2 - f^2)^{1/z}/\eta \qquad (9.8)$$

In fact, most of these alternative equations are generally not very practical and, according to Niediek (8), it is justifiable, for routine purposes, to characterize the flow properties of chocolate using the Casson equation.

In our experience, using the Casson equation has always given good agreement with practical results. Viscosity data from dozens of commercial chocolate samples (rheograms with five points in the range $5–50\,s^{-1}$) have been examined; a correlation coefficient r of, on average, 0.9994 (minimum 0.993) was obtained for the regression line of the experimental points plotted as in Figure 9.2. Nevertheless, some authors still have reservations regarding the meaning of the Casson yield value.

9.2 How to measure flow properties

In this section, the focus is mainly upon rotational viscometers, which are the ones most widely used for chocolate applications. Other chocolate visco-meters are discussed in section 9.2.2.

9.2.1 *Rotational viscometers*

Cylindrical viscometers. This is the more common of the two sub-classes of rotational viscometers. Two types exist:

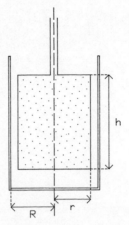

Figure 9.3 Scheme of a coaxial cylinder viscometer.

(i) The cup rotates and the shear stress is measured on the internal coaxial cylinder (e.g. by means of a torsion wire in the MacMichael viscometer)

(ii) The internal coaxial cylinder (usually a plunger or bob) rotates, shear rate being fixed by a geared drive motor (Figure 9.3). Here, the shear stress is measured by means of a torsion spring interposed on the rotating bob. Examples of this type of viscometer include the Brookfield (9), Contraves and Haake (10).

Shear stress (τ) is proportional to the torque (M), which is itself proportional to the viscometer reading (n):

$$\tau = M/2\pi r^2 h = An \tag{9.9}$$

where $l = $ length and $r = $ radius of the internal cylinder.

$$A = \text{constant} = k/2\pi r^2 h \tag{9.10}$$

$$\text{and } k = \text{torque constant} = M/n \tag{9.11}$$

On the other hand, equation (9.6) may also be written:

$$D = BN \tag{9.12}$$

where

$$B = \text{constant} = 4\pi/60(1 - a^2) \tag{9.13}$$

The constants of proportionality A and B are strongly dependent on the geometry of the viscometer. Generally, viscometer manufacturers specify the values of A and B for their instruments as well as any other properties which are relevant. By taking different measuring systems, the range of measurements may be extended.

As has already been noted, the OICC suggests that the ratio a (of inner to outer cylinders) should not be less than 0.65. However, this is not always the case in practice (e.g. for the Brookfield system, $a = 0.62$). It is important that measurements are not made at very low shear rates (where not all of the masse is moving). The OICC also recommends the calibration of viscometers with a Newtonian fluid at a viscosity of not less than 2 Pa s (20 poise) at the calibration temperature (e.g. mineral or silicone oil).

The preparation of the sample plays an important role. This is the OICC procedure (2):

> Take about 100 g chocolate and grate or divide into small pieces (5 g or less). Melt the finely divided sample by placing in a 250 ml glass beaker, heating in a water-bath or incubator at 55 °C, and stirring at intervals until the temperature of the chocolate reaches 50 °C. If mechanical stirring is used, this should not exceed 60 rpm to avoid incorporation of air. Cool the melted chocolate in air to 43 °C with stirring as above. Transfer to the viscometer and bring to 40 °C $\pm$ 0.10 °C while rotating the viscometer at a rate of shear between 5 and 25 s^{-1}. The time required to reach 40° must be determined by prior experiment. If the temperature of the viscometer is controlled by means of a thermostatic bath round the outer cylinder, this must be maintained slightly above 40 °C, e.g. 40.5 °C, in order to keep the chocolate at a mean temperature of 40 °C. When the chocolate has reached 40 °C, measure the viscosity at various speeds within the range 5 to 60 s^{-1} rate of shear, using first ascending then descending order of speeds. At each speed the reading is taken when the viscometer has settled down to a steady scale reading. Calculate the mean values of the readings obtained in both ascending and descending order. Use only those results which lie between 10% and 95% of the total scale.

The calculations are carried out as indicated in section 9.1 (see Figure 9.2).

Today, more and more new viscometers are being equipped with a microcomputer to calculate the Casson parameters and, in some cases, to plot a graph and/or calculate the correlation coefficient of the regression line.

The precision of measurement is not very high. According to the OICC, the coefficient of variation between replicate determinations of Casson parameters should not exceed 10%.

For thick chocolates, thixotropy (decrease in viscosity caused by stirring) can be a source of additional error, and big differences can be found between the curves for ascending and descending orders of speeds. To limit this phenomenon, Seguine (9) recommends pre-shearing the melted chocolate.

According to a recent paper of Pieper (11) concerning interlaboratory viscosity tests organized by the NCA (National Confectioners Association), a better reproducibility is obtained when chocolate is subjected to a known quantity of stress (with the Brookfield, 50 rpm = 17 s^{-1}) for 3 minutes. In this case, the coefficient of variation between the different participants in the viscosity test is about 3–4%.

Concentric rotational viscometers are rather expensive. In addition, although they are metallic, the measuring cups and bobs must be handled with care: If they are distorted (e.g. by being dropped), their physical geometry may be altered so much that the constants A and B are changed significantly.

Cone plate viscometers. An example of the cone plate viscometer used with chocolate is the Ferranti–Shirley (12).

This instrument operates on the principle of measuring the reaction torque due to viscous traction of a cone rotating in the fluid; the apex of the cone just touches the lower plate. It has the advantage of having a uniform rate of shear but, particularly if the chocolate is too coarse, a single particle can easily bridge the gap between cone and plate, leading to quite spurious results (13).

9.2.2 Miscellaneous instruments

The following instruments are inexpensive and make use of time measurements. They are especially useful in factories.

Flow (or efflux) viscometry (cup with drain hole) (14). In this method, the time taken for a known amount of chocolate to drain out through a hole in a cup (or other container) is measured using a stopwatch. This gives a single value which provides an empirical indication of the consistency of chocolate. The Redwood viscometer belongs to this family; it has been applied by Harris (15).

Falling-ball viscometer. The principle of the falling-ball viscomter is the well-known Stokes' law, which relates the viscosity of a Newtonian fluid to the velocity of a falling sphere. If a sphere of radius R and density d_1 falls through a fluid of density d_2 and viscosity η at a constant velocity v, the balance of forces gives the following relationship:

$$\eta = 2(d_1 - d_2)gR/gv \qquad (9.14)$$

where $g =$ the gravitational constant (see (16), p. 272),

   In the case of the Koch falling-ball viscometer (17), a steel ball is suspended on a wire to which are attached two markers. This assembly is lowered into the chocolate, and the interval between the submersion of the markers is timed as the plunger sinks. By using balls of different size and density, it is possible to achieve different shear rates and shear stresses which can be used to construct a multipoint rheogram.

Gardner mobilometer. This instrument has been described in two recent papers (18, 19). It consists of a metal cylinder, which is filled with melted chocolate, and a disc with four holes, which is threaded on to a plunger rod to form a piston. The time in seconds taken by the disc/rod (piston) assembly to fall 10 cm through the chocolate (mobilometer descent time) is measured. The weight of the disc/rod assembly may be modified by adding weights on a tray at the top of the rod.

   By plotting the reciprocal of descent time (proportional to shear rate) versus weight (proportional to shear stress), a rheogram (similar to Figure 9.2) can be obtained. Martin *et al.* (18) have shown that very good correlation exists:

(i) Between mobilometer descent times (at constant weight) and Haake plastic viscosities (see (18), Figure 5)

(ii) Between the intercepts on the weight axis (for the plot of the reciprocal of descent time versus weight) and Haake yield values (see (18), Figure 6).

Although the authors cite Casson, it is not stated in the paper whether the Haake plastic viscosity and yield value correspond to the Casson parameters or not.

On-line viscosity measurements will be treated in Chapter 15, which is dedicated to instrumentation.

9.3 Factors affecting the flow properties of chocolate

In the literature, and from our own observations, the Casson parameter ranges for chocolates are:

$$\eta_{CA} = 1 \text{ to } 20 \text{ Pa s } (10 \text{ to } 200 \text{ poises})$$

$$\tau_{CA} = 10 \text{ to } 200 \text{ Pa } (100 \text{ to } 2000 \text{ dynes/cm}^2)$$

For chocolate coatings, the corresponding values are lower:

$$\eta_{CA} = 0.5 \text{ to } 2.5 \text{ Pa s } (5 \text{ to } 25 \text{ poises})$$

$$\tau_{CA} = 0 \text{ to } 20 \text{ Pa } (0 \text{ to } 200 \text{ dynes/cm}^2)$$

Various factors influence these values: fat content, emulsifier content (e.g. lecithin), moisture content, particle size distribution, temperature, conching time, temper, thixotropy and vibration. Although in reality interdependent, these factors will be discussed separately below.

In a previous review paper (6), the author has cited all references found up to 1975 dealing with the factors which influence chocolate viscosity. Only the most important findings are reported here, taking into account more recent papers in each area, if they exist.

9.3.1 Fat content

It is well known that dilution of a particulate suspension with the liquid which forms its continuous phase (in the case of chocolate: cocoa butter and/or milk fat) reduces its viscosity.

Fincke (20) has shown that both Casson parameters (η_{CA} and τ_{CA}) decrease when fat is added to a lecithin-free chocolate. Generally, however, chocolate contains at least some lecithin. In Figure 9.4, the decrease of η_{CA} and τ_{CA} with increasing fat content is plotted for two milk chocolates (one fine and one relatively coarse) both containing 0.25% lecithin. The figures are taken from previous work from the author's laboratory (21). Note the greater decrease for fine chocolate than for coarse (the reason for this is discussed in section 9.3.4).

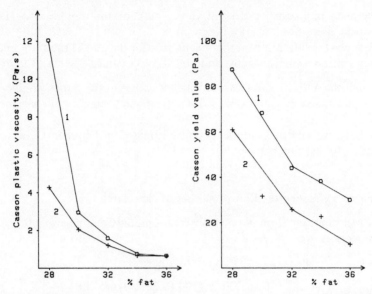

Figure 9.4 Influence of fat content on Casson parameters of two milk chocolates with 0.25% lechithin. (1) Fine chocolate with 5.7% particles $> 20\,\mu$m; (2) moderately coarse chocolate with 16% particles $> 20\,\mu$m.

Harbard (22) adopted a theoretical approach to the study of fat content and the viscosity of chocolate. He assumed that some of the cocoa butter in chocolate is always associated with particle surfaces and so is not free to affect viscosity.

9.3.2 Lecithin and other emulsifiers

Small amounts of suitable surface-active lipids can produce an immediate reduction in viscosity. The action of lecithin as a viscosity-reducing agent was patented by Hanse-Muhle in 1930.

The addition of 0.1–0.3% soya lecithin has the same viscosity reducing effect as over ten times this amount of cocoa butter. This fact is of great commercial importance. If the level of lecithin exceeds 0.3–0.5%, however, thickening of the chocolate occurs. Fincke (20) considered the parameters η_{CA} and τ_{CA} separately. He observed that at a lecithin content greater than 0.3–0.5%, it was only τ_{CA} (i.e. Casson yield value) which increased (see Figure 9.5).

On the other hand, the decrease of the Casson parameters is more important when the fat content is not too high. According to Harris (15) and Minifie (23), the action of lecithin is very weak on cocoa mass and much stronger on sugar/cocoa butter mixtures: i.e. the viscosity reducing effect of lecithin in chocolate is primarily due to its action on sugar particles.

Harris (15) demonstrated experimentally that lecithin binds tightly to sugar:

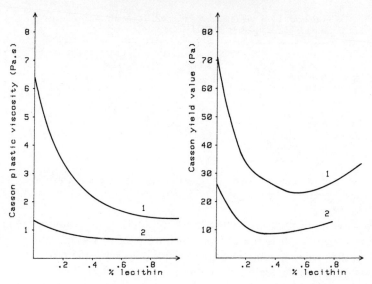

Figure 9.5 Influence of soya lecithin addition on Casson parameters of two dark chocolates (20). (1) 33.5% fat, 1.1% water; (2) 39.5% fat, 0.8% water.

a sugar/cocoa butter mixture was extracted with petroleum ether and the remaining fat-free sugar then mixed with the same proportion of fresh fat; the viscosity of the new mixture was very similar to that of the original, indicating that most of lecithin had been retained on the sugar surface, probably in the form of a monomolecular layer (monolayer). Bartusch (24) studied the percentage surface covered by lecithin: it was found to be, respectively, approximately 50, 67 and 85% for lecithin contents of 0.2, 0.3 and 0.5%.

Soya lecithin is a natural phospholipid; it is the oldest and probably the most commonly encountered surface-active agent for reducing the viscosity of chocolate. Other new surface-active agents have been developed for chocolate: one of the most widely used—especially in Britain—is the synthetic lecithin, YN (obtained from partially hardened rapeseed oil after glycerolysis, phosphorylation and neutralization). YN is more constant in composition and in efficiency than soya lecithin: its flavour is bland and neutral; its efficiency is greater than that of soya lecithin, and at levels above 0.3%, no thickening occurs. When compared with soya lecithin, YN has a stronger effect on τ_{CA} than on η_{CA} (see Table 9.1).

Other surface-active agents include sucrose esters, calcium-stearoyl lactoyl lactate and polyglyceryl polyricinoleate (PGPR), better known as Admul-WOL. Admul-WOL has the ability to strongly reduce or even cancel the yield value of chocolate. This useful property is exploited by the confectioner in such applications as the moulding of Easter egg shells.

The effects of all these agents are summarized in Figure 9.6 and in Table 9.1.

Table 9.1 Flow characteristics of plain chocolate with added surface-active lipids at 50 °C (15)

Addition	Casson plastic viscosity (poise)	Casson yield value (dynes/cm^2)
0.3% soya lecithin	6.1	92
0.3% YN	10.3	30
0.3% sucrose dipalmitate	8.6	166
0.3% PGPR	32.5	25
0.8% PGPR	20.3	0

(Note: sucrose dipalmitate is the least efficient agent in this list).

The stage in the process at which lecithin is added is a very important factor: lecithin added towards the end of conching reduces viscosity by a greater amount than when exactly the same amount is added at the start. The reason for this is that, as the action of lecithin is a surface effect, if it is added to the masse too early, some of it may (by prolonged mixing and grinding) be absorbed into the cocoa particle, thus reducing its efficiency (23). It is also known that exposure to relatively high temperatures for long times reduces lecithin performance.

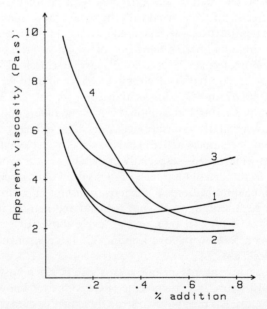

Figure 9.6 Viscosity reduction of dark chocolate by soya lecithin and by synthetic active lipids (15). Apparent viscosity determined at shear rate 15 s^{-1} and 50 °C; initial apparent viscosity before addition: 19.5 Pa s or 195 poises. (1) Soya lecithin; (2) phospholipid YN; (3) sucrose dipalmitate; (4) polyglyceryl polyricinoleate, PGPR.

9.3.3 *Moisture content*

If water is added to chocolate, a marked increase in viscosity occurs. According to Harris (15), the formation of layers of syrup on the surface of sugar particles, which increases the friction between them, could explain this behaviour. This effect of internal friction is reduced by the addition of phospholipids, thus chocolates containing surface-active lipids can tolerate higher levels of moisture (15).

Rasper (25) found that the viscosity increase caused by added water was greater for a fat content of 35% than for 39%. From literature cited by Chevalley (6), it seems certain that a change in apparent viscosity occurs, but it is not clear whether η_{CA} or τ_{CA} plays the more important role.

9.3.4 *Particle size distribution*

As has already been stated, chocolate is a suspension of sugar, cocoa and milk particles in a continuous fat phase. By the end of conching, every particle should be coated by a thin film of fat in order to ensure good lubrication. The presence of a surface-active agent facilitates the formation of this coating.

Particle size distribution also plays a very important role: if the particles are small their specific surface is great and therefore more fat is needed; conversely, if the particles are large, specific surface is small and less fat is needed. According to Rostagno (26), a chocolate is perceived as coarse when more than 20% of particles are larger than 20 μm.

In the laboratory, chocolates with different mean particle sizes were prepared. For the same fat and lecithin contents, η_{CA} and τ_{CA} were plotted against the percentage of particles greater than 20 μm ('% > 20 microns'). The curves shown in Figure 9.7 were obtained.

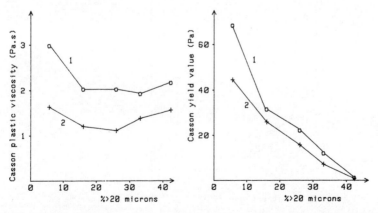

Figure 9.7 Influence of fineness on Casson parameters of two milk chocolates with 0.25% lecithin. (1) 30% fat; (2) 32% fat.

The first conclusion is that the Casson yield value τ_{CA} decreases much more than Casson viscosity η_{CA} when the chocolate becomes coarse. This phenomenon was also noted by Malm (27) and similar curves could have been plotted from his own results. Prentice (13, Chapter 9) gives the following explanation for the phenomenon: when particle size is reduced, then either the number of bonds or the amount of frictional contact between the particles is higher, therefore τ_{CA} is increased. A second conclusion is that η_{CA} seems to pass through a minimum. The increase of η_{CA} at high values of $\% > 20\,\mu m$ can perhaps be explained by the work of Niediek (28) who investigated, separately, sugar/cocoa butter and cocoa/cocoa butter systems. In the first case (sugar/cocoa butter), viscosity increased with increasing specific surface (high specific surface corresponds to a low percentage of particles greater than $20\,\mu m$). In the second case (cocoa/cocoa butter), the reverse was true, the lowest viscosity being obtained for the highest specific surface. As chocolate is a mixture of these components we obtain, in reality, a superimposition of these two phenomena (see Figure 9.7).

In order to explain the rheological properties of cocoa mass, it is convenient to refer to the paper by Kleinert (29). By milling a cocoa mass several times on a five-roll refiner, he clearly showed that viscosity was gradually reduced until after the third pass, while yield value remains constant. The fourth and fifth passes, however, had only marginal effects upon viscosity. For yield value, just the opposite effect was observed. After the fourth pass, and particularly after the fifth, there was a considerable rise in yield value. According to Campbell (30), the reduction of viscosity during the refining of cocoa masses is due to the rupture of fat cells in cocoa particles bigger than $0.0005''$ ($12.5\,\mu m$). Below this diameter, all of the fat cells are already ruptured and the rheological behaviour

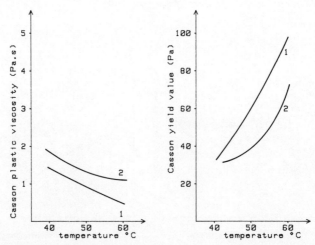

Figure 9.8 Influence of temperature on Casson parameters of two milk chocolates (31). (1) 34% fat, without lecithin; (2) 30% fat, 0.15% lecithin.

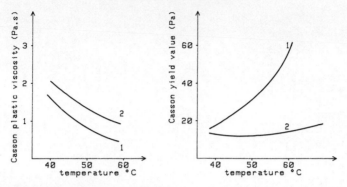

Figure 9.9 Influence of temperature on Casson parameters of two dark chocolates (31). (1) 34% fat, without lecithin; (2) 30% fat, 0.15% lecithin.

of the mass approaches that of sugar. From practical experience in the author's laboratory, it seems that milk particles behave in a similar way to cocoa particles.

9.3.5 Temperature

As temperature increases, two phenomena are observed (31) in milk and dark chocolates (see Figures 9.8, 9.9):

(i) Casson plastic viscosity (η_{CA}) decreases

(ii) Casson yield value (τ_{CA}) may increase, especially for chocolates without lecithin. This thickening effect can be reduced by lecithin addition, especially in the case of dark chocolate.

Harris (15) reported that chocolates containing phospholipid can be heated some 20 °C higher than normal before thickening occurs, and that, if the temperature is not too high, this thickening is reversible.

9.3.6 Conching time

At the start of conching, not all the solid particles are coated by fat and, generally, only a small amount of lecithin has been added (most of it is added towards the end). The moisture content is also slightly higher than at the end of conching. Consequently, the apparent viscosity of an unconched chocolate mass is very high and often difficult to measure.

In the author's laboratory, viscosity changes were measured in a variety of industrial conches. It was established that in all cases, decrease in Casson yield value (τ_{CA}) was the dominant factor and that this decrease was especially important during the first few hours of conching (see Figure 9.10) Casson plastic viscosity (η_{CA}) decreased or even increased according to the type of conche. In any case, the variation of η_{CA} was not so obvious as that of τ_{CA}.

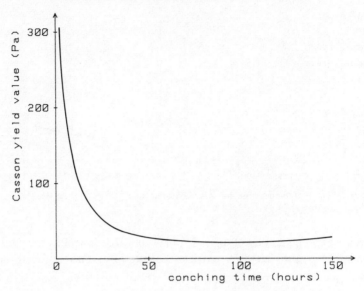

Figure 9.10 Influence of conching time on Casson yield value of a milk chocolate (31).

9.3.7 Temper

Melted chocolate must be tempered before moulding. Temper is the induced partial pre-crystallization of cocoa butter. During tempering, therefore, the amount of solid particles is slightly increased and so too (as one would expect) is viscosity. Moreover, it is in this state that chocolate is used for moulding.

Some authors have found an increase in apparent viscosity which was proportional to the amount of fat crystals in the tempered masse: this could therefore be a way of measuring the degree of temper.

However, according to Rohan and Stewart (32) and from our own experience, for a well-tempered chocolate the amount of seed crystal is relatively low and the difference in apparent viscosity before and after tempering (temperature being held constant, at about 30 °C) is low. When the increase is significant, however, overtempering may have already occurred. In this case, the increase in Casson yield value is responsible for the increase in apparent viscosity.

9.3.8 Thixotropy

Thixotropy is a decrease of apparent viscosity with stirring time at a given rate of shear. It is especially important for thick chocolates. When a chocolate is stirred after a long rest, the shear stress, very high at the beginning, decreases and tends to a steady value after 10 minutes (33). Conversely, when chocolate is intensely agitated then stirred at a lower shear rate, shear stress increases and tends to the same value as that obtained in the ascending curve. It seems that structure is developed when a chocolate is at rest and more or less destroyed according to the rate of shear.

Under the specifications laid down by the OICC the ascending and descending curves cannot be superimposed for thixotropic chocolates if the waiting times at each shear rate are insufficient. The lengthening of waiting time for thixotropic chocolates should reduce the shift between ascending and decending curves and thus improve the measurement.

Fincke and Heinz (34) reported that, in some cases, observed thixotropy may have been an artefact due to slippage on the walls of the cylinders. If the cylinders had a corrugated surface to prevent slippage, the decrease in shear stress disappeared almost completely.

9.3.9 Vibration

The flow properties of chocolate during the moulding and coating operations are influenced also by vibration ('tapping' or 'shaking'). Bartusch (35) showed that apparent viscosity (at a slow shear rate of 1 s^{-1}) decreased with increasing amplitude of vibration and he postulated that the yield value had disappeared. This was confirmed by Rostagno (31) who reported measurements carried out with a Brookfield viscometer fixed on a vibration table. The effect of vibration is very useful for the spreading of chocolate in moulds and in this case the determining factor is the product of vibration amplitude and frequency. Vibration can also be useful in the emptying of chocolate tanks, serving to facilitate the flow of chocolate through the tank outlet.

9.4 Conclusions

In this chapter, we have shown that the flow properties of melted chocolate can be described by two parameters, the Casson plastic viscosity η_{CA} and yield value τ_{CA}, the latter value depending upon the cohesive forces between particles.

Different types of viscometer have been described: rotational (cylinder and cone/plate), flow (or efflux) and falling ball viscometers, and the mobilometer. It has been shown that both Casson parameters (η_{CA} and τ_{CA}) are influenced by fat, lecithin and (probably) water content, but that particle size, conching time, temper, thixotropy and vibration have a more important action on τ_{CA} than on η_{CA}.

Acknowledgement

The author wishes to thank her colleague, Mr M.L. Talbot, for his help in the writing of this chapter in correct English.

158   INDUSTRIAL CHOCOLATE MANUFACTURE AND USE

References

1. Steiner, E.H. *Rev. Int. Choc.* **13** (1958) 290–295.
2. Office International du Cacao et du Chocolat. *Analytical Methods* (E/1973) 10; *Rev. Int. Choc.* (1973) (Sept.) 216–218.
3. Robbins, J.W. *Manuf. Conf.* **59** (1979) (May) 38–44.
4. Solstad, O. *Manuf. Conf.* **63** (1983) (Aug.) 41–42.
5. Marquardt, W. *Zucker und Süsswarenwirtschaft* **33**(12) (1980) 391–396.
6. Chevalley, J. J. *Texture Studies* **6** (1975) 177–196.
7. Zangger, R. *Alimenta* **23** (1984) 13–16.
8. Niediek, E.A. *Zucker und Süsswarenwirtschaft* **33**(4) (1979) 119–125 (German); *Rev. Choc. Conf. Bakery* **5**(3) (1980) 3–10 (English).
9. Seguine, E.S. *Manuf. Conf.* **66** (1986) (Jan.) 49–55.
10. Hogenbirk, G. *Manuf. Conf.* **66** (1986) (Jan.) 56–59.
11. Pieper, W.E. *Manuf. Conf.* **66** (1986) (June) 117–120.
12. Murray, P. *Rev. Choc. Conf. Bakery* **5**(2) (1980) 17–18.
13. Prentice, J.H. *Measurements in the Rheology of Foodstuffs.* Elsevier Appl. Sci. Publ. (1984) Chapter 9.
14. Riedel, H.R. *Conf. Prod.* **46** (1980) (Dec.) 518–519.
15. Harris, T.L. *SCI Monograph* **32** (1968) 108–122.
16. Van Wazer, J.R., Lyons, J.W. Kim, K.Y., and Colwell, R.E. *Viscosity and Flow Measurement.* Interscience (1963).
17. Koch, J. *Manuf. Conf.* **39** (1959) (Oct.) 23–27; *Rev. Int. Choc.* (1959) 330–335.
18. Martin, R.A. and Smullen, J.F., *Manuf. Conf.* **61** (1981) (May), 49–54.
19. Stumpf, D. *Manuf. Conf.* **66** (1986) (Jan.) 60–63.
20. Fincke, A. *Handbuch der Kakaoerzeugnisse.* Springer Verlag, Berlin (1965) 357.
21. Rostagno, W., Chevalley, J. and Viret, D. *First Int. Congr. on Cacao and Chocolate Res.,* Munich (1974) 174–180.
22. Halbard, E.H. *Chem. Ind.* (London) (1956) 491.
23. Minifie, B.W. *Manuf. Conf.* **60** (1980) (April) 47–50.
24. Bartusch, W. *First Int. Congr. on Cacao and Chocolate Res.,* Munich (1974) 153–162.
25. Rasper, V. *Listy Cukrovar* (1962) 38.
26. Rostagno, W. *Manuf. Conf.* **49** (1969) (May) 81–85.
27. Malm, M. *Manuf. Conf.* **47** (1967) (May) 63; *21st P.M.C.A. Conf.,* 50–55.
28. Niediek, E.A. *Gordian* **70** (1970) 244–251, 300–301.
29. Kleinert, J. *Rev. Int. Choc.* (1970) 274–80, 306–323.
30. Campbell, M.O. *Gordian* **73** (1973) (2) 61.
31. Rostagno, W. *5th Eur. Symp. on Food Rheology in Food Processing and Food Quality,* Zurich, October 1973.
32. Rohan, T.A. and Stewart, T. *BFMIRA Tech. Circ.* **337** (1966).
33. Heiss, R. and Bartusch, W. *Fette Seifen Anstrichmittel* (1956) 868.
34. Fincke, A. and Heinz, W. *Rheol. Acta* (1961) 530–538.
35. Bartusch, W. *Fette und Seifen* **63** (1961) 721–729.

10 Chocolate temper

L. HERNQVIST

10.1 Introduction

The crystallization properties of fats in food are very important. Since bought chocolate contains fat mainly in the solid state, knowledge about the possible crystalline states is important in its production, for example in the understanding of tempering.

Failure to produce the correct crystalline form not only results in problems for the manufacturer, for instance sweets may still be sticky on reaching the packing machines, but also gives a product without the gloss, snap and colour normally expected by the customer. Chocolates which have been incorrectly stored are sometimes found to have a white sheen over the surface, or even individual white blobs of fat up to about 1 mm (0.04 inch) in diameter. This is known as chocolate 'bloom' and although harmless, spoils the appearance of the goods, and often results in them being returned by the customer. This is in fact due to the fat melting and then recrystallizing, often in the wrong form.

In order for a product to be called chocolate it must contain cocoa butter. This fat is a mixture of triglycerides, i.e. it has a central glycerol structure to which are attached acid residues of three types. These may be palmitic, oleic or stearic in structure. The physical properties of the cocoa butter depend upon how this structure is built up and the relative proportions of these acids (see Chapter 12 for details of how attempts have been made to copy cocoa butter from other fats). These triglycerides can set in several different polymorphic forms. In some ways this is analogous to the allotropes of sulphur described in most elementary chemistry courses.

This chapter initially describes the techniques which are used to identify the different crystalline states. The transformation between the states is then examined, together with their structure. The latter are so complex for cocoa butter that they have only been determined by analogy with other triglycerides. Finally this is applied to cocoa butter and fat mixtures. The methods used to obtain chocolate in the correct crystalline state are described in Chapter 11.

10.2 Techniques

Triglycerides are known to crystallize in three polymorphic forms, namely α, β' and β. The best technique to use to identify the different forms present in a fat system is x-ray diffraction. The three polymorphic forms are defined according to the following criteria (1):

(i) A form which gives only one strong short-spacing line near 4.15 Å is termed α

(ii) A form showing two strong short-spacing lines near 4.20 and 3.80 or three strong lines near 4.27–3.97 and 3.71 Å, and which also exhibits a doublet in the $720\,cm^{-1}$ region of the infrared abosrption spectrum, is called β'

(iii) A form which does not satisfy criteria 1 or 2 is called β.

When two or more crystal forms of a compound give the same x-ray diffraction pattern they receive the same name, but they should be distinguished by subscripts (e.g. β_1, β_1 etc.) in order of decreasing melting points. Many different nomenclatures exist in the literature for cocoa butter, however. If, for example, the (non-cocoa) triglyceride tribehenin in the α-form is cooled, a chain-packing according to the orthorhombic type is formed. Since this transition is the only reversible one in the case of triglycerides, this form is termed sub-α instead of the numeric subscript β'_3.

Only single-crystal x-ray analysis can give complete information on the crystal structures. Owing to the extreme difficulty of growing the large single crystals required for this analysis, only the structure of the most stable form, the β-form, has been fully derived. A proposed structure of the β'_1-form of the triglyceride triundecanoin has been derived, but is based on incomplete single-crystal data (2,3).

Many other techniques have also been used to study the polymorphism of triglycerides (4), e.g. differential scanning calorimetry, nuclear magnetic resonance, infrared spectroscopy. The disadvantage, however, with most techniques is that the information obtained is rather limited and needs to be supported by other analyses. Differential scanning calorimetry, for instance, can give well-resolved peaks. Since only the net energy for the system is recorded, however, overlapping transitions can occur which cannot be resolved individually. Raman spectroscopy is another technique that can be used in order to characterize lipids in the solid state (5,6).

10.3 Polymorphic transitions of triglycerides

As indicated above, triglycerides are known to crystallize in three polymorphic forms, α, β' and β. Figure 10.1 shows a somewhat more complex picture of the transitions between the states. The two β'-forms are quite similar, as will be discussed later. The sub-α form is not normally present, since very low temperatures are needed for most triglycerides to develop this form. As indicated in the diagram, all polymorphic forms can be obtained directly from the liquid state except the sub-α form. So far only the β' and β-forms have been obtained from solvents (a standard method for preparing crystals).

The transition liquid $\rightarrow \alpha \rightarrow \beta'_2 \rightarrow \beta'_1 \rightarrow \beta$ is the complete route by which

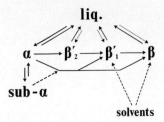

Figure 10.1 Complete picture of the polymorphic transitions of trigycerides. All transitions are irreversible except the $\alpha \rightarrow$ sub-α transition. The dotted lines indicate the possibility of a direct crystallization in other form than the α-form. The possibility of growing crystals from solvent is also indicated (10).

triglycerides reach the optimum packing of molecules. Cocoa butter, however, includes the sub-α form in its polymorphic transition series. As will be discussed below, the transition liquid $\rightarrow \alpha \rightarrow$ sub-$\alpha \rightarrow \alpha \rightarrow \beta'_2 \rightarrow \beta'_1 \rightarrow \beta$ could be the correct way for cocoa butter to reach optimum moleculer packing.

10.4 Liquid state

Based on x-ray scattering technique and Raman spectroscopy, a structure for triglycerides in the liquid state has been proposed (7, 8). Figure 10.2 shows this proposed structure. The hydrocarbon chains are disordered and the proportion of *gauche* conformation is large. As can be seen in the figure, the triglyceride molecules are arranged as 'chairs'; one hydrocarbon chain points in one direction and the other two in the opposite direction. The 'chairs' build up a lamellar unit. The proposed model is dynamic, i.e. both size and orientation of the lamellar units vary with diffusion rates of the molecules. When the temperature is increased, the size of the lamellar units decreases. The size of the lamellar units is hard to determine, although an attempt was made by Larsson in 1972 (7). The approximate size just above the melting point was estimated to be about 200 Å.

10.5 Crystallization

When the temperature is decreased the lamellar units increase in size until finally crystallization takes place. The probability of regions with neighbour-

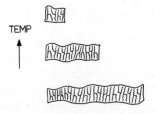

Figure 10.2 Proposed structure of triglycerides in the liquid state (8).

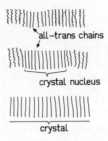

all–trans chains

crystal nucleus

crystal

Figure 10.3 Proposed mechanism of the crystallization of a triglyceride melt (8).

ing hydrocarbon chains arranged in all-*trans* conformation positions also increases. The formation of a crystal nucleus can be regarded as the point in time when the formation of adjacent all-*trans* chains within the lamellar unit dominates over the formation of the *gauche* conformation (see Figure 10.3).

10.6 The α-form

Figure 10.4 shows the proposed structure of the α-form (2). This proposed structure is based on evidence from x-ray diffraction and Raman spectroscopy studies. Since no single crystal study exists of the α-form, the structure can only be illustrated schematically. The chain axes are indicated as lines, since the main part of the hydrocarbon chains is oscillating. The molecules are arranged perpendicular to the methyl end group plane and the hydrocarbon chains are hexagonally (H) close-packed. All molecules show the same kind of 'chair' appearance as in the liquid state, with one hydrocarbon chain pointing in one direction opposite to the others. This 'chair' appearance is common throughout all crystal forms of triglycerides. Two 'chairs' form a dimeric unit which seems to remain unchanged through the polymorphic transitions (2).

From a molecular packing point of view the α-form is unsatisfactory because of the irregular methyl end group region. Due to this irregularity the hydrocarbons would be expected to possess a high degree of mobility (see Figure 10.4). This mobility, together with the hydrocarbon chain oscillation, normally induces a rapid transformation from the α-form to one with better chain packing, e.g. the β'-form.

10.7 The β'-forms

Figure 10.5 shows the main features of the β'_1-form of the triglyceride triundecanoin. The structure has been derived from a single-crystal study (2) which has been recently further extended (3). Measurements of the (*ab*) lattices in the β'- and β forms indicate that the glycerol group region in the β'-form can have the same structure as in the β-form. If the α-form (Figure 10.4) is compared with the β'_1-form (Figure 10.5) it can be seen that the methyl groups

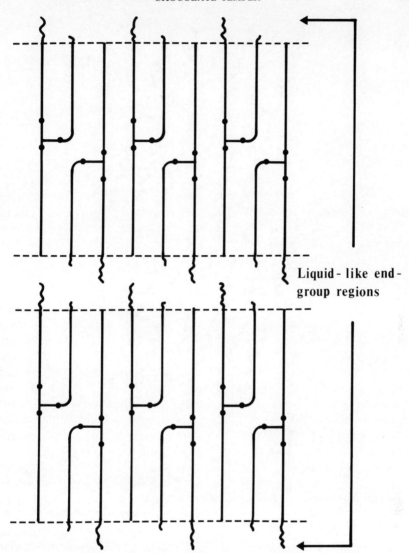

Liquid - like end - group regions

Figure 10.4 Proposed structure of the α-form of triglycerides projected along a short axis. The chain axes are indicated; the main part of the hydrocarbon chains is oscillating and hexagonally close-packed; the methyl end group regions are somewhat more disordered, as in liquid crystals (2).

are better packed in the β'_1-form. This is due to a tilt of the hydrocarbon chains in relation to the methyl end group plane. The hydrocarbon chains are arranged according to the orthorhombic ($O_\perp$)-crystal structure.

The a–c projection of the β'_1-form is closely related to the a–c projection of the β-form (cf. Figure 10.6). The main difference between the β'_1-and β-forms is

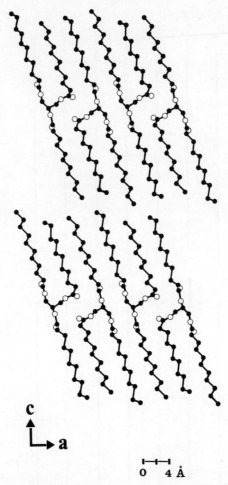

Figure 10.5 Main features of the structure of the *ca*-projection of the β'_1-form of triundecanoin. The glycerol part is conjected to be arranged in the same way as in the β-form of trilaurin (2, 3). ●, CH ($n = 1$ or 2); ○, CO.

seen in the two other projections. In Figure 10.7 all three projections of the α-, β'_1 and β-forms are illustrated, as will be discussed further in connection with the $\beta'_1 \rightarrow \beta$ transition.

The β'_1 form shows two directions of chain tilt in the *cb* plane. This change in tilt is assumed to take place at the methyl end group plane and to give the double chain layers without bending the individual triglyceride molecules (2).

A second β'-form, β'_2, of the triglyceride triundecanoin has been observed (2). It differs from the β'_1-form in its alignment in the different planes. The β'_2-form is tilted only in the *ca* plane but is vertical in the *cb* plane (cf. β'_1, Figure 10.7). The metastable β'_2 has been reported also for tristearin (9).

As mentioned previously, if tribehenin is cooled down, the sub-α-form is developed (10). This form shows the same x-ray diffraction short-spacing data as the orthorhombic structure. Consequently, it can be referred to as β'_3 according to the nomenclature discussed below. However, since the transition $\alpha \rightarrow$ sub-α is reversible (see Figure 10.1) the name 'sub-α' is preferred.

10.8 The β-form

Figure 10.6 shows the crystal structure of the β-form of trilaurin (11). The β-form is in fact the only polymorphic form to have had its structure completely determined. The chains are tilted in relation to the methyl end group planes in

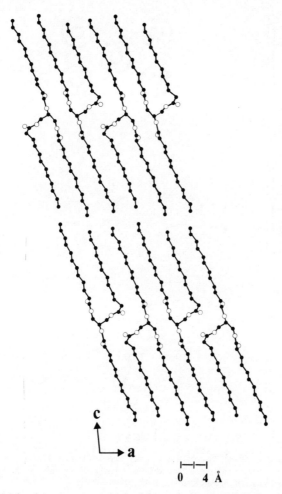

Figure 10.6 Molecular arrangement in the β-form of trilaurin (11) in the *ca*-projection.

the *ca* projection as well as in the *cb* projection. In the *β*-form the hydrocarbon chains are arranged according to the triclinic T‖ crystal structure.

10.9 The β′₁ → β transition

If the three different forms of simple triglycerides, *α*, *β′₁* and *β*, are compared, they all have a quite similar structure. The hydrocarbon chains in all three forms are arranged as 'chairs' with the chains in the 1- and 3-positions pointing in the opposite direction to the chain in the 2-position. The molecular dimer (i.e. two molecules together) can have the same structure in all three forms. The

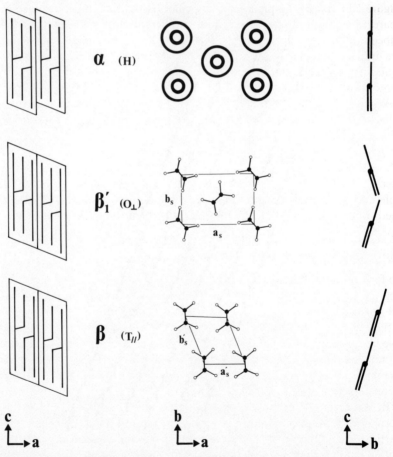

Figure 10.7 Schematic orientation of the proposed triglyceride dimers of the *α*-, *β′₁*- and *β*-forms in the *ca*-projection shown together with respective chain packing subcell and the *cb*-projection where the chain directions are indicated.

main difference is how the dimers are arranged in relation to each other, and in which way the chains are close-packed. The possible ca projections of these dimers are shown in Figure 10.7.

If the β'_1- and β-forms are compared in all three dimensions, the ca projection is similar, but both the ba (the chain packing) and the cb projection differ. The ba projection is illustrated as subcells showing the type of crystal structure. These and other subcells are described in detail in a recent review by Abrahamsson $et\ al.$ (12). From the point of view of the hydrocarbon chains the crystal forms a zigzag structure,with the plane through every chain being perpendicular to that of its nearest neighbours.

One zigzag period of the hydrocarbon chain is seen in the direction of every chain. The β'_1-form shows orthorhombic $(O_\perp)$ subcell. As can be seen in Figure 10.7 the zigzag plane of every chain is perpendicular to those of its nearest neighbours. The β-form show a different close packing, the triclinic subcell $(T_\parallel)$ with all zigzag planes parallel. In the cb plane the β'_1- and β- forms are quite different. In the β'_1 form two adjacent bilayers of triglyceride molecules are tilted in opposite directions, in contrast to the β-form which shows only one tilt direction.

When the $\beta' \rightarrow \beta$ transition occurs, the methyl end group planes of two adjacent bilayers need to slide so that all the bilayers tilt in the same direction. This transition is strongly dependent upon the structure of the methyl end group plane. Consequently some triglyceride molecules give a stable β'_1-form while others are rapidly transformed to the β-form. In agreement with this theory, fat mixtures can be stable in either the β'_1-form or in the β-form. This difference in the polymorphic behaviour is very important when considering the properties required of a particular fat or blend of triglycerides.

10.10 Triple chain length structures

So far only the double chain length type of packing has been discussed, but triglycerides can also be packed in such a way that triple chain length units are formed (Figure 10.8). This can occur in triglycerides containing either saturated or unsaturated fatty acids (13). In the case of saturated triglycerides, one chain in the molecule needs to differ in length by at least four carbon atoms. If one chain in a triglyceride is unsaturated, e.g. 2-oleo-distearin, triple chain length units can be obtained (7) (see Figure 10.8) with all oleic acyl chains in one layer.

In fats the triple chain length structure is quite unusual, since for this to occur all the triglycerides present need to be very similar in form. This is particularly important with regard to the position of the acyl chain. In cocoa butter this happens to be in a good position for the triple chain structure. Cocoa butter consists mainly of symmetrical triglycerides with oleic acid in the 2-position and palmitic or stearic acid in the other positions.

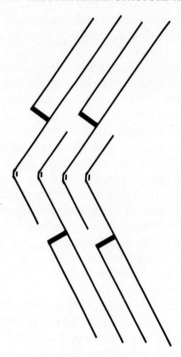

Figure 10.8 Schematic illustration of the proposed structure of 2-oleo-distearin (7). The chain axes, glycerol residues and *cis* double bonds are indicated.

10.11 Cocoa butter

Since cocoa butter shows triple chain length structure the polymorphic behaviour of this fat is different to that of a 'normal' fat. In the literature six polymorphic forms (termed I–VI) have been reported (14, 15). The classification of the various forms in relation to the structures discussed above is, however, rather confusing (14–18).

By using x-ray analysis and comparing the short-spacing of all forms (10) only five forms in fact are found to exist, since two (V and VI) show the same short-spacing, identical to the β-form of 2-oleodistearin. This comparison is illustrated in Table 10.1. Here the short-spacing data of the six forms of cocoa butter, as reported in (15), are compared with the short-spacing data of 2-oleodistearin (sub-α, α and β) and 2-stearodipalmitin (β'_2 and β'_1). The triglyceride composition in cocoa butter is dominated by 2-oleodistearin, 2-oleodipalmitin and 1-palmito-2-oleo-3-stearin, i.e. mono-unsaturated triglycerides with an oleic acyl group in the 2-position. Therefore 2-oleodistearin was used as a reference, while 2-stearo-dipalmitin can be used in order to obtain data on β'_2 and β'_1.

On the basis of this comparison, it appears that the six forms I-VI, are identical to the five forms discussed above, illustrated in Figure 10.1, namely

Table 10.1 Classification of the polymorphic forms of cocoa butter. The short-spacing data are given in Å.

I	sub-α	II	α	III	β₂	IV	β'₁	V	VI	β
3.87(m)	3.78(w)	4.20(vs)	4.25(s)	3.87(w)	3.92(w)	3.75(m)	3.86(s)	3.65(ms)	3.67(s)	3.64(m)
4.17(s)	4.23(s)			4.20(vs)	4.23(diff. s)	3.88(w)	4.06(m)	3.73(ms)	3.84(m)	3.84(w)
						4.13(s)	4.22(s)	3.87(m)	4.01(w)	3.95(s)
						4.32(s)	4.36(m)	3.98(ms)	4.21(vw)	4.05(w)
								4.22(vw)	4.53(vs)	4.29(w)
								4.58(vs)	5.09(vw)	4.52(m)
								5.13(vw)	5.37(m)	4.83(s)
								5.38(m)		5.28(m)

Forms I-VI according to ref. (14).
sub-α; β (2-oleodistearin); β'_2; β'_1 (2-stearo-dipalmitin) (10).
vs = very strong; S = strong, ms = medium strong, m = medium, w = weak, vw = very weak.

sub-α, α, β'_2, β'_1 and β, respectively. Since forms V and VI show the same short-spacing data, they are probably identical.

When liquid cocoa butter is cooled a range of crystal types may be present in any sample. As the β-form is the most stable state in cocoa butter, consequently the β-form has the highest melting point, and the fat will try to eventually take that form. When transitions gradually take place to the higher melting point β-form, heat is released which may in fact melt neighbouring crystals.

10.12 The crystallization of mixtures

A mixture of triglycerides normally co-crystallizes. If the mixture consists of two groups of triglycerides with a large difference in acyl chain length, however, a segregated crystallization could occur. The same could happen if one part of a mixture is dominated by 2-oleo-disaturated triglycerides (e.g. cocoa butter) and the other part of the mixture consists of a whole spectrum of triglycerides with various chain lengths and the unsaturated acyl chains in different positions (i.e. a 'normal' fat).

As a consequence, cocoa butter cannot be mixed with every type of fat. A fat to be mixed with cocoa butter should show the same triglyceride composition, i.e. contain 2-oleo-disaturated triglycerides, if the stable triple chain structure of the β-form is required. If cocoa butter is mixed with a 'normal' fat the 2-oleo-disaturated triglycerides either segregate or co-crystallize with the other triglycerides present. If co-crystallization occurs, a double chain structure of either β'_1 or β (cf. Figures 10.5, 10.6) will be developed. Both alternatives result in altered properties compared to pure cocoa butter.

In a co-crystallized mixture the proportion of unsaturated fat together with differences in chain length induces changes in melting point and sometimes alters polymorphic behaviour. A difference in acyl chain length gives rise to a chain disorder in the methyl end group region. Unsaturations give distortion in the crystal packing, the *trans* unsaturation giving only small distortions compared with the *cis* unsaturation.

When a fat crystallizes, a small fraction can determine the overall crystallization. Even with a normal fat consisting of many different triglycerides, its crystallization behaviour can still be understood in terms of the triglyceride fraction that dominates the polymorphic behaviour.

Sometimes fat mixtures are considered to show 'memory' in the melt, due to crystal forms previously present in the system. A triglyceride mixture containing a few percent of tristearin, however, could contain β-microcrystals in the liquid, which are of micellar size (~ 100 Å) (provided that the temperature is below the β-melting point). This would then act as a seed for subsequent settings, and is the most probable explanation of this type of behaviour.

References

1. Larsson, K. *Acta Chem. Scand.* **20** (1966) 2255–2260.
2. Hernqvist, L. and Larsson, K. *Fette Seifen Anstrichmittel* **84** (1982) 349–354.
3. Hernqvist, L. to be published.
4. Chapman, D. *Chem. Rev.* **62** (1962) 433–456.
5. Lippert, I.L. and Peticolas, W.L. *Proc. Nat. Acad. Sci. USA* **68** (1971) 1572–1576.
6. Larsson, K. *Chem. Phys. Lipids* **10** (1973) 165–176.
7. Larsson, K. *Fette Seifen Anstrichmittel* **74** (1972) 136–142.
8. Hernqvist, L. *Fette Seifen Anstrichmittel* **86** (1984) 297–300.
9. Simpson, T.D. and Hagemann, J.W. *J. Amer. Oil Chem. Soc.* **59** (1982) 169–171.
10. Hernqvist, L. Polymorphism of Fats, Thesis, Universty of Lund (1984).
11. Larsson, K. *Proc. Chem. Soc.* (1963) 87–88.
12. Abrahamsson, S., Dahlen, B., Löfgren, H. and Pascher, I. *Prog. Chem. Fats Other Lipids* **16** (1978) 125–143.
13. S. de Jong, Triacylglycerol Crystal Structures and Fatty Acid Conformations, Thesis, University of Utrecht (1980).
14. Wille, R.L. and Lutton, E.S. *J. Amer. Oil Chem. Soc.* **43** (1966) 491–496.
15. Chapman, G.M., Akehurst, E.E. and Wright, W.B. *J. Amer. Oil Chem. Soc.* **48** (1971) 824–830.
16. Brosio, E., Conti, F., diNola, A. and Sykora, S. *J. Amer. Oil Chem. Soc.* **57** (1980) 78–82.
17. Merker, G.V. and Vaeck, S.V. *Lebensm.-Wiss. u. Technol.* **13** (1980) 314–317.
18. Merker, G.V., Vaeck, S.V. and Dewulf, D. *Lebensm.-Wiss. u. Technol.* **15** (1982) 195–198.

11 Temperers, enrobers, moulding equipment and coolers

R.B. NELSON

Most of the previous chapters in this book have been concerned directly or indirectly with the manufacture of chocolate itself. This chapter looks at the way we use this chocolate to make individual sweets. However it is used, it is essential to ensure that the fat in the chocolate is in the correct crystalline state. The first sections therefore deal with chocolate temper, how it is measured and then the tempering machines used to produce it.

Two distinct forms of sweetmaking exist, enrobing and moulding. In the first the chocolate is poured over the sweet centre, with any excess being removed by shaking or blowing. In the second it is poured into a mould where at least part of the chocolate is allowed to set. The modes of operation of the machines used to carry out these processes are examined in detail, together with some of the pitfalls which can arise during their operation.

Although chocolate will set at room temperatures, it will not do so at the rate required by an industrial manufacturer. Cooling tunnels are therefore normally used. These too must be designed and operated to certain principles for the product to have the best possible appearance.

11.1 Chocolate tempering

This section sets out to give some historical and practical information on chocolate tempering and handling techniques. In addition it outlines some of the pitfalls that can arise in tempering chocolate.

Chocolate tempering, and the handling of chocolate in a correct manner to make it suitable for use in a moulding plant or chocolate enrober, can be said to be basically simple. However, as with many other technologies, a great deal of experience, skill and technique is required in order to obtain the best-quality products.

A chocolate enrober or moulding plant is expected to give certain results:

 (i) To mould confectionery or coat sweets, cakes, biscuits, cookies etc.
 (ii) To give a good-quality finish in terms of colour and gloss
 (iii) To handle chocolate in such a way that the shelf-life is prolonged
 (iv) To give the correct weight of chocolate on the finished product.

In considering these parameters it is of great importance to understand the

172

significant physical characteristics of both the chocolate and the tempering equipment which control the final results. In brief, we need to know some details of:

- (i) The principles of chocolate technology, e.g. fat eutectics
- (ii) The type of pumps available and how to pump chocolate to tempering machines at the required feed rates
- (iii) Tempering machine principles and their control systems
- (iv) Types of tempering machine
- (v) Feeding tempered chocolate to a user plant
- (vi) The user plant, especially coolers and their temperature controls
- (vii) Factory air conditions, such as temperature and humidity.

11.1.1 *Chocolate technology*

Chocolate and similar coatings are produced to many different recipe formulations (Chapter 13), but all contain a mixture of finely milled solids, cocoa, sugar, milk crumb or powder, all suspended and well dispersed in cocoa butter or substitute fat, which at normal processing temperatures is the liquid carrying medium.

When chocolate cools it need to be tempered (Chapter 10), in order to set in a satisfactory manner. To use this tempered chocolate in an enrober or moulding plant it is necessary for the chocolate to have the correct flow properties:

- (i) The chocolate *viscosity* (PV) must be to specification (Chapter 9)
- (ii) The *yield value* (YV) should be correct to suit the application, i.e. be appropriate for enrobing or moulding
- (iii) The *temper viscosity* (TV) should be correct to suit product conditions and the finish required. This will depend on the type of equipment being used.

Temper viscosity is a flow property, which results from tempering the chocolate. The fat content commences to 'crystallize' in the tempering machine in a controlled way so that the conditioned chocolate remains sufficiently liquid to be pumped or delivered to the user plant. Although related to apparent viscosity it is difficult to determine, as any shearing action will tend to change the crystallized state.

Temper viscosity also takes into account that the chocolate following tempering has more than one crystalline state. The viscosity is then determined by the quantity, type and 'maturity' of the fat crystals grown. Cocoa butter and similar fats are capable of setting in several different polymorphic forms, which as they cool and set affect the surface finish, colour, setting times, shelf-life (Chapter 10). A percentage of fat crystals actually become solid in the tempering machine. According to the type of temperer 2–4% may become solid, and hence an increase in viscosity can be seen.

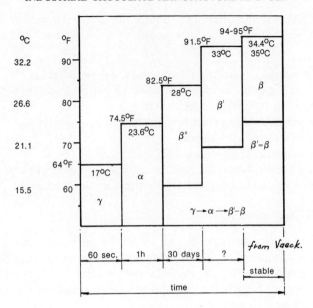

Figure 11.1 The polymorphic forms of cocoa butter.

As a general rule the viscosity increases by a factor of two from the incoming chocolate to that leaving the temperer, although this may be greater if cooling conditions are exceeded; 15–30 Poises (Casson PV) is a typical increase. We now consider chocolate, or more precisely, the cocoa butter, when fully molten, i.e. at a temperature high enough to guarantee complete melting of fat crystals. It is truly free-flowing and is said to be de-tempered at temperatures above 45 °C (113 °F). Above this temperature there are no solid polymorphic forms.

Cocoa butter and many natural fats contain a structure of complex triglycerides with more than one type of fatty acid attached to a molecule of glycerin. The relative proportions of these acids alters the melting and setting characteristics of the fat. Cocoa butter can also set in a number of different crystals known as polymorphic forms. The complex nature of these polymorphs is described in Chapter 9 and in the work of Vaeck (1) (Figure 11.1) who investigated pure cocoa butter crystallization and the behaviour of the different melting point ranges. As is shown in Figure 11.1 there are five recognized crystal types, with a sixth suspected. It has been found that the β-crystals are more stable than the others and in chocolate exhibit better colour, hardness, handling, finish and shelf-life characteristics.

Many different forms of nomenclature exist and are found in the literature. Listed in Table 11.1 are polymorphic forms of cocoa butter according to Vaeck (1).

Table 11.1 The polymorphic forms of cocoa butter

Form	Crystallizes	Melting range	Stability	Contraction
γ	Under 62.5 °F (16.9 °C)	Up to 62.5 °F (16.9 °C)	Unstable	
α	From γ	Up to 75 °F (23.8 °C)	Unstable	0.060 ml/g
β_1	62.5–72.5 °F (16.9–22.5 °C)	59 °F–85 °F (15 °C–29.4 °C)	Semi-stable	0.086 ml/g
β_2	—	Data not available	—	
β	72.5 °F–92.5 °F (22.5 °C–33.6 °C)	68 °F–95 °F (20 °C–35 °C)	Stable	0.097 ml/g

In addition to the melting points and overlapping ranges, another factor, *time*, has an influence; time and temperature are therefore of paramount importance when designing tempering systems.

Incomplete, or bad tempering, results in unstable crystal growth and in poor setting characteristics. These result in a difference in colour, due to reflected light being disoriented by disorganized crystal growth. Apart from bad colour the surface cocoa butter may crystallize and appear as whitish spots or a streaky grey-white finish known as 'fat bloom'. Chocolate prior to coating or moulding, therefore, must be tempered to contain only the highest melting β-form with a high degree of 'maturity'. This gives the best results for quality finish and the best shelf-life.

The coating, or depositing usage *temperature* is dependent on the amount and type of crystal growth due to time, temperature and agitation in a tempering machine. The temperature to which a chocolate can be raised without losing good temper is proportional to the amount of *mature* crystal seed in a particular chocolate, as is the 'contraction' (described later).

Although theoreticians differ about the polymorphic forms of cocoa butter, it is clear that there are degrees of maturity at different temperatures allied to time.

A simple test shows the importance of the tempering *time*. This test is as follows. Normally a rise in temperature occurs on use, where it is usually 2–3 °C warmer at the user point than the outlet of the tempering machine. If we now raise the temperature at the outlet of the tempering machine to the same temperature as the user equipment, we find a decrease in the amount of crystal present with a medium residence-time tempering machine, and if the machine is of a low-residence design, the temper will probably be lost entirely. This proves that time is all-important in conjunction with temperature in tempering.

This temperature rise also shows that further crystallization takes place in the pipeline and associated user equipment, and therefore that optimum temper is not achieved at the tempering machine.

11.1.2 *Advantages of long tempering time*

When a coating has been properly tempered over a longer period of time it contains only the highest melting point crystal formation. It is then possible to raise the using temperature without losing any crystals. As the temperature rises the viscosity falls, making it easier to coat the centre of the product. Thus for any fat content the thinnest using chocolate is obtained at the highest temperature, which in turn requires a long tempering period.

Resulting advantages are:

(i) Good decorative markings, especially on chocolates of high milk fat content that require extra time during tempering
(ii) When bloom-resistant additives such as sorbitan monostearate and polysorbate 60 are supplied in a coating, time is essential to introduce stable nuclei in sufficient quantities to ensure a high coating temperature
(iii) The improved flow at the higher temperature gives better weight control
(iv) A 'mature' chocolate with stable nuclei is not easily affected by adverse conditions, such as a high ambient temperature and varying centre temperatures
(v) A large mass of well-seeded chocolate held at a high temperature, but under the melting point of the stable nuclei, and being well agitated, remains very stable over long periods of time
(vi) The high-temperature coating contains only stable crystals, and with correct cooling, good colour and gloss characteristics will be found
(vii) Better shape, e.g. good bases are a by-product of the thinner chocolate produced.

11.1.3 *Cooling curves*

How do we know if chocolate is correctly tempered and ready to be used? There are two basic methods which indicate the likely end result, one a distinctly practical test, the other more scientific, but serving as a guide only.

(i) Take a sample of a coated centre and place it in a cool packing room at 20 °C (68 °F). If the chocolate sets quickly in still air with a good gloss it is extremely likely that the temper is near the optimum.
(ii) Using a temper meter, take a sample of tempered chocolate, cool and record the cooling curve. This cooling curve is obtained by monitoring the rate of cooling against time. The setting time/temperature relationship will vary in a variety of ways, particularly between dark chocolate and milk types.

The cooling curve can be measured using a system developed by the J.W. Greer Co. who neatly miniaturized the previous time/thermometer sampling by the use of a thermocouple and miniature hard-copy recorder. This is said to give constant cooling conditions for any number of samples. Figure 11.2

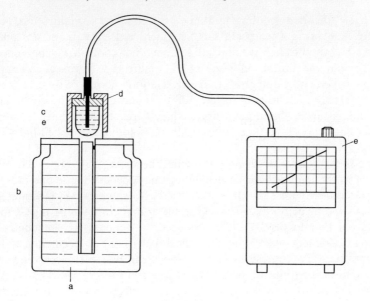

Figure 11.2 Cooling curve apparatus. (*a*) Dewar vessel; (*b*) melting ice at 0 °C (32 °F); (*c*), (*d*) sample tube and thermistor; (*e*) miniature recorder.

illustrates details of this equipment. The 'Temper Meter' consists of (i) a Dewar vessel (or large vacuum flask); (ii) crushed ice and water to give a constant cooling temperature of 0 °C (32 °F), (iii) a sample tube made from highly conducting aluminium or copper; (iv) a probe shroud and insulator; and (v) a thermistor/themocouple probe and miniature recorder. The sample tube is slightly warmed (i.e. 30 °C, 86 °F) and a sample of chocolate is poured so as to fill the sample chamber. The sample tube is immersed in the ice water, the probe and shroud quickly fitted and the recorder switched on. A cooling curve is usually plotted in approximately 5 minutes. Different types of cooling curves are illustrated in Figures 11.3–11.5.

In Figure 11.3 it can be seen that sample cools uniformly for a period then levels off at *A*, maintaining a fixed temperature for a time. This is due to active crystal growth, with the latent heat being released and generating sufficient heat to retard cooling. The sample is in a plastic/solid state at this time. This part of the cooling curve is the 'inflection point'. Once most of the latent heat

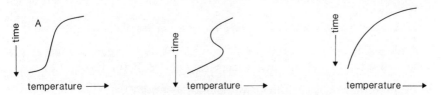

Figures 11.3–11.5 Three types of cooling curve.

has been liberated, cooling continues till complete solidification occurs. Figure 11.3 shows a tempered sample which cools slowly but sets quickly. Figure 11.4 shows a lightly tempered sample which cools slowly, but sets more slowly than the sample in Figure 11.3. Figure 11.5 shows a de-tempered sample which cools quickly, but sets over a long period of time.

The retardation effects of certain additives may also be investigated using the cooling curve. The cooling curve is directly related to 'temper viscosity'. If the cooling curve varies then the viscosity, the colour and the shelf-life will also vary.

The cooling curve serves only as a guide, however, as it varies according to type of tempering machine, i.e. depending on residence time/temperature as is illustrated in Figure 11.6. Here the same curve shape which indicates good temper (*A*) may be produced by less stable or immature tempering (*B* and *C*). The latter two curves are often produced by samples from machines with shorter residence time. Note that the break point is occurring at lower temperatures. To confuse us further, the break will vary according to the chocolate recipe (milk or plain) types of cocoa butter, additional fats and emulsifiers. Both 'maturing' time and eutectic effects can therefore alter what appears to be a good curve.

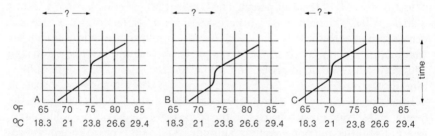

Figure 11.6 Similar cooling curves produced by different tempering machines.

11.2 Tempering machines

There are several ways to temper chocolate, known as 'hand', 'batch' and 'continuous' processes. This chapter is primarily concerned with continuous machinery, although the principles involved are applicable for all methods.

11.2.1 *Pumping chocolate to a tempering machine* (see also Chapter 14)

Many pitfalls may be encountered when chocolate is pumped, usually in metered form, to and through a tempering device. It is important to:

(i) Find the viscosity of the chocolate—Casson PV can vary between 15–150 poises (1.5–15 Pa s)

(ii) Find the required usage rate for the user plant then add approximately 15% more for contingencies
(iii) Calculate the pressure drop to the tempering device
(iv) Check the maximum pressure the tempering machine will stand
(v) Calculate the pressure drop after the tempering machine in the feed pipe to the user plant, bearing in mind the increase in viscosity through the temperer due to temperature changes and crystal formation.

Earlier mention was made of chocolate viscosity (PV) and yield value (YV). We must briefly discuss this area to realize how the distinction is made between enrobing and moulding chocolate, where there are usually two distinct requirements.

Enrobing for high-quality confectionery items nearly always requires a low-viscosity coating with a high yield value, i.e. a Casson PV of 15–20 poises (1.5–2.0 Pa s) and a YV of say 100–300 dynes/cm^2. Both the PV and YV change in value, although not always in direct proportion, during final tempering, which usually takes place in the enrober tank. The low viscosity is required for quality lines, giving precise control to the operator, but the high yield value is necessary for retention of good decorative markings. There are exceptions to this rule; for example, half-coated biscuits are easily scrubbed if there are high peaks, and the pack height can also be significantly adversely affected. In this case a lower yield value may be preferred.

Moulding chocolate can be to the same specifications as for enrobers on occasions, but generally the PV is *higher* and the YV is *lower*, i.e. the PV may vary between 30–120 poises (3–12 pa s), and the YV may be in the range of 30–120 dynes/cm^2. The requirement is a low yield value for easy settlement into moulds to release air bubbles and give a quality finish.

In comparison with enrober viscosities, moulding plants have a slightly broader requirement, starting from fine thin shells to thick 'Easter egg' halves. For both, however, the viscosity must be known before pumping to the tempering machine.

Calculations can be made to size the pipework and pump to suit flow rate and viscosity. Pipes vary from 50–100 mm dia. Standard pipeline friction calculations are not applicable to chocolate and other non-Newtonian masses. The Casson-type equations can be used, however, to predict the pressure drop, pipewall shear stress and motor power for different pipe bores and flow rates (see also Chapter 14). The author uses the Casson equations in preference to others such as Bingham for engineering predictions because the pipeline calculations have been shown to be accurate to the order of 1–5%. This is more than satisfactory when compared with the day-to-day variations expected in chocolate viscosity. A program developed by the author and available from him may be used for non-Newtonian fluids such as chocolate, cocoa mass, batter, ketchups and similar protein suspensions which exhibit a yield value.

Pressures of up to 3.5 kg%cm^2 (50 psi) can normally be used in tempering systems.

11.2.2 *The principle of tempering chocolate*

Up to now we have not indicated how chocolate may be tempered except to mention that it involves cooling sufficiently to form the type of crystal required. The *first stage* in controlled tempering always assumes that the 'metered continuous' infeed to the machine is with chocolate completely free of crystal growth, i.e. at 45 °C (113 °F), although this figure is quite often lower for practical energy conservation reasons, e.g. 41 °C (105 °F).

The *second stage* is to gently cool the warm chocolate through a multistage tempering machine (or alternative), gradually reducing the temperature to 'strike seed' and initiate the first stages of crystal growth—at this initial phase the crystals can grow very fast, and as the viscosity increases the need arises to raise the chocolate temperature to prevent runaway solidification.

The *third stage* reheat happens gradually in two ways. Most heat is applied via heat exchangers. Heat is also generated in a minute way as latent heat is developed. This activity can be roughly determined, as it amounts to approx. 0.9 J/g or 0.4 Btu/lb for cocoa butter. A calculation shows a possible rise of 0.2–0.4 °C. Baker Perkins have utilized this reaction by first detecting the temperature rise, and then by microprocessor computing the control functions. This method is claimed to give control so as to produce the desired cooling curve.

In the *fourth stage*, a retention stage, the design 'time period' promotes crystal maturity, and the temperature may also be adjusted. During progression through the machine, agitation from scraping and mixing blades increases the spread of nuclei in a fine homogeneous structure of small crystals. Continuous temperature control is applied, and in conjunction with the 'time period' the chocolate attains the transition from unstable 'strike seed' temperature condition to a mature, completely stable optimum temper. This, of course, assumes that the user plant is also capable of continuing the 'process'.

There are a few coined phrases like 'Thermo-cyclic', 'Pre-crystallization' and 'Optimum temper' which roll off manufacturers' tongues, but the reader should not be put off from carrying out his own confirmatory calculations on the retention time. The foregoing description has to assume that the tempering machine is capable of carrying out the heat exchange requirement and mixing retention times. Most tempering problems are man-made . Nature endowed the cocoa bean with all the pre-requisites for us to apply properly and produce good results.

11.2.3 *Time for tempering*

The importance of allowing sufficient time for the crystals to mature cannot be over-emphasized.

From experience it has been found that:

(i) 10–12 minutes' residence time is necessary for moulding plants
(ii) 20–360 minutes' residence time is necessary for enrobing.

Differences between moulding plant and enrober plant requirements may be briefly summarized as follows. The *moulding plant* generally needs less fluidity and can make up for a higher viscosity by more intensive shaking and cooling systems. *Enrobers* should have the highest practical coating temperature that can be obtained from the tempering machine and from the enrober tank. This requires a high maturity, i.e. a long residence time. Defects easily show up in enrobed pieces and are less visible in a 'moulding plant' where the product takes its shape and some gloss from the mould.

Chocolate has changed over the years as regards recipe and manufacturing techniques, additives and substitute fats and viscosity-reducing agents. This has an effect on the tempering procedure required. For instance, milk chocolate has a proportion of animal fat (butter) which gives a eutectic effect gainfully on one hand (bloom-resistant) and adversely on the other (lower melting point, softer, slower to temper and set in a cooler). Eutectic effects always seem to lower the melting points and hence also lower the temperature to 'strike seed'. Therefore, more time is necessary to create the mature crystal growth. In the case of the enrober, this time can be in the tempering machine or enrober tank or normally a combination of both. Plain chocolate with a higher cocoa-butter content nucleates easily at temperatures higher than those for milk chocolate. The lack of eutectic activity promotes stability and enables it to be tempered at temperatures higher than for milk, i.e. about 34.5 °C (94 °F) compared with approximately 29.4 °C (86 °F). Thus the practical effect of adulterating cocoa butter (legally) with other fats is that the actual temperature and retention times in the machines must vary according to recipe.

Figure 11.7 is an example of the simplicity of calculations required to give the retention time in a machine specially designed to:

(i) Meter de-tempered chocolate accurately into the machine.
(ii) Efficiently cool the chocolate and keep it fluid during the initial reduction in temperature
(iii) Have a *large* temperature-controlled retention zone to provide the *time* element, to create the optimum temper
(iv) To finally control or adjust temperatures of the product being sent to the user plant.

Temperature controls for tempering. One of the problems associated with many temperature controls is that they are on/off systems giving sawtooth control. A self-learning computer software system that will readjust itself to suit the ongoing changes that occur in heat exchange jackets, due to scaling, is infinitely preferable. In the critical zones in a tempering machine the temperatures should be held to within 0.2 °C (0.4 °F).

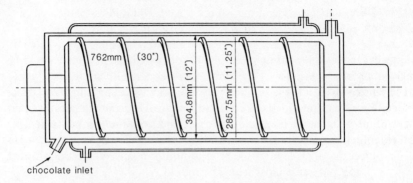

Figure 11.7 Typical retention time calculation in a worm-type tempering machine. Tempering machine tempers 680.58 kg/h or 1500 lbs/h of chocolate.

$$\text{Retention time} = \frac{\text{choc.density} \times 60 \times \text{volume of tempering space}}{\text{hourly rate}}$$

$$\text{Metric: R.T.} = \frac{1281 \times 60 \times (\pi \times 0.1522^2 \times 0.762 - \pi \times 0.1428^2 \times 0.762)}{680.58}$$

$$= 0.749 \text{ mins or } 44.9 \text{ secs.}$$

$$\text{Imp.: R.T.} = \frac{80 \times 60 \times (\pi \times 0.5^2 \times 2.5 - \pi \times 0.4691^2 \times 2.5)}{1500}$$

$$= 0.749 \text{ mins or } 44.9 \text{ secs.}$$

This type of tempering machine produces at the outlet an immature chocolate at 26.6 °C (80 °F) with quick-setting characteristics if used immediately. On a cooling curve recording the inflection point will occur at a lower temperature than on a longer residence time machine. *Note*: The inflection point position is dependent on residence time and is proportional to time within certain limits.

Chocolate kettle tempering. Kettle tempering has been used in batch form for very many years. The kettle is basically a stirred tank whose temperature can be controlled within the appropriate range. It is possible to adapt kettles to work continuously, by metering chocolates in at the base and overflowing out at the top. Additional control is gained by feeding into another kettle, in at the top and metering the chocolate out at the base. The 'time period' is decided by the volume of the kettles and the flow rate. Considerable advantages can be obtained, probably the foremost being the maturity 'time period' which can be from one to two hours, resulting in a high temperature usage chocolate. Further advantages are as follows:

(i)   Simplicity of all mechanical parts
(ii) A simple easily-maintained agitator
(iii) Bearings are mounted outside the product contact area, hence do not have wear or contamination problems associated with pressurized systems

(iv) As there is no pressure in the equipment no seals are required (they are, however, advisable to retain grease)
(v) Kettles are easy to drain and clean (and can be seen to be clean), which is a prime consideration when different colours of chocolate or non-compatible coatings are changed over
(vi) Particulate ingredients can be added (nuts, raisins, crystals of sugar) at a suitable point, provided the exit pump is slow-running with a large swept volume designed for particulate matter (in fact chocolate kettle tempering embodies most of the criteria needed for tempering machines).

Disadvantages are a slightly longer start-up time and a greater floor space requirement, when compared with vertical tempering machines.

11.2.4 Types of continuous tempering machine

Kreuter. We start with the system that has the longest theoretical 'time period', the Kreuter Interval Pre-crystallization procedure, which is illustrated in Figure 11.8. The Interval design consists of a 'batch' stock tank of chocolate that is first pre-cooled and then supercooled till seed formation takes place. An outboard pump recirculates the chocolate, further enhancing the mixing top to bottom in the tank. A stirrer also creates mild disturbance to give good heat exchange at the vessel walls for both cooling and heating to take place. Once the machine strikes seed in the supercooling period, temperatures have to be raised gradually over a period of time. This time period induces mature crystal growth. Up to this point this is the mode of operation of most tempering kettles. In these, however, too long a retention time will result in overtemper. Kreuter on the other hand have made use of the fact that there is a higher temperature that will prevent further solidification, and still retain maximum fluidity with stable mature chocolate resulting (equilibrium). This is possible since the batch system allows for time to create crystal growth. All this sounds to good to be true, and there is of course a trade-off between advantages and disadvantages. Advantages are:

(i) It is possible, with a designed time period, to create the optimum maturity in the chocolate
(ii) This is turn produces a high-temperature coating giving good handling characteristics and longer shelf-life; fat may possibly be removed from the recipe due to lower viscosity
(iii) It is suitable for all types of chocolate
(iv) Since no more crystal growth can take place at the control temperature, a state of equilibrium exists
(v) Energy savings are claimed, since there is no need to reheat excess chocolate feeds that many other systems demand; the user plant takes only what it requires and no return piping is necessary

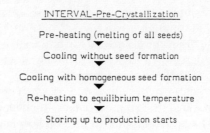

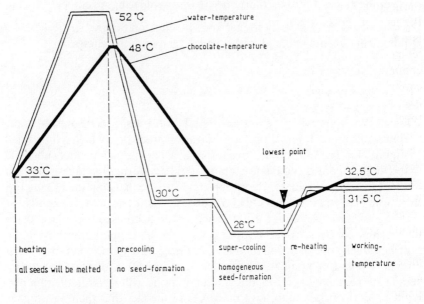

Figure 11.8 Kreuter interval pre-crystallization procedure.

(vi) Prepared tempered chocolate can be stored ready for immediate use, thus reducing start-up times.

The disadvantages of this system are:

(i) The system relies on keeping the seeded chocolate at the pre-determined control temperature within very close limits; this could affect the readiness for usage

(ii) As this is a batch system, a large storage tank would be necessary to suit the high usage rate of (say) a wide enrober plant — one day's usage may require a 16-ton tank, or several smaller tanks, which could create a space problems.

Kreuter also provides the 'K-procedure', a worm tempering device. This is a short-residence time machine with what is claimed to be a fast-revolving worm capable of working at high pressures. The high pressures may be caused by remote location of the tempering machine from the user plant, viscous chocolate, or both. This type of machine falls into the category of the short residence-time period whose relative merits and disadvantages will be discussed later.

Aasted. Aasted tempering machines are world-famous, and probably have the most far-sighted design, which was inherent from inception. Dr Aasted created the first machine in 1946, and since then the principle has not changed. Having the 'feel' for chocolate handling, he also designed a slow-running pump which is used to this day for metering chocolate to the tempering machine. The tempering device is a vertical unit now much copied. It consists of a stack of configurable heat exchange plates scraped continuously to give efficient cooling and mixing. Because the heat exchange plates/discs have a chamber-like space, each one has a retention zone in its own right, and once stacked together they have a sufficient 'time period' to temper the complete range of chocolate recipes with full zonal control able to match the requirements of various coatings. Such is the range of capacities and models available that any flow rate can be matched to the required retention time. The Aasted temperer always handled the difficult high-milk-fat-content English-type chocolate without the problems associated with low-retention-time systems, which speaks highly of its inventor. This tempering device is well suited for moulding plants or enrobers and takes up minimal floor space (Figure 11.9).

The Aasted temperer operates in the following manner. Chocolate is gently metered through the machine. Aasted, recognizing the fact that inadvertent accidents can happen, have fitted a heat-jacketed relief valve to protect the pump and temper from high-pressure damage (pumps can peak out at 14 bar, 200 psi). The chocolate rises up through the control zones, of which there may be as many as seven. The cooling controls strike seed and inititate crystal growth, allowing the remaining zones to be adjusted to suit best conditions. The chocolate leaves the temperer, after being re-heated to the user temperature.

Sollich. Sollich are the company who have probably made the most varieties of tempering machines, ranging from very short residence times through to present-day systems. These include single-stream, double-stream and multi-stream systems. 'Streams' are blends of previously seeded mass, or may represent the new untempered feed into the tempering worm. Sollich have manufactured a range of combined temperer/enrobers and these will be considered in more detail later on. Their separate or outboard temperers include the Solltemper U and the Solltemper MST-V.

The *Solltemper U* (Figure 11.10) is an outboard tempering machine for enrobing and moulding plants. This machine has now been superseded by the MST-V, but is included for its historical value. This machine is an interesting

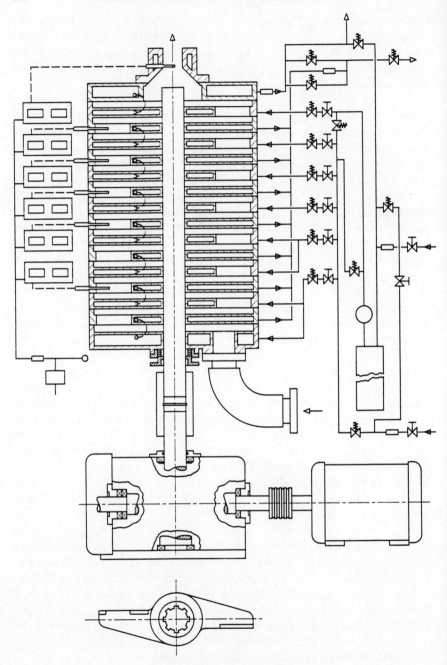

Figure 11.9 The Aasted temperer.

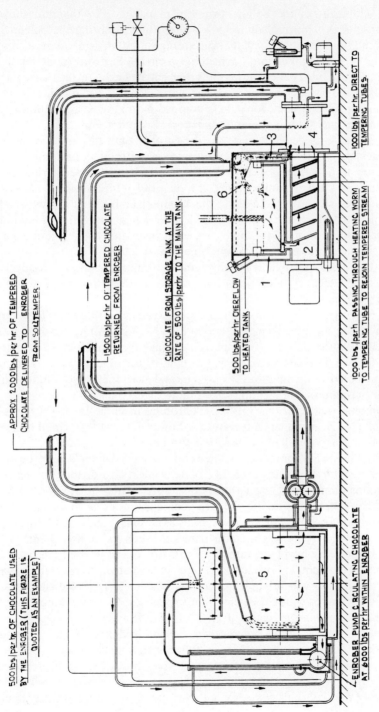

APPROX. 2000 lbs per hr OF TEMPERED CHOCOLATE DELIVERED TO ENROBER FROM SOLITEMPER.

1500 lbs per hr OF TEMPERED CHOCOLATE RETURNED FROM ENROBER

CHOCOLATE FROM STORAGE TANK AT THE RATE OF 500 lbs per hr. TO THE MAIN TANK

500 lbs per hr OVERFLOW TO HEATED TANK

1000 lbs per hr PASSING THROUGH HEATING WORM TO TEMPERING TUBE TO REJOIN TEMPERED STREAM

1000 lbs per hr. DIRECT TO TEMPERING TUBES

500 lbs per hr OF CHOCOLATE USED BY THE ENROBER (THIS FIGURE IS QUOTED AS AN EXAMPLE)

ENROBER PUMP CIRCULATING CHOCOLATE AT 6000 lbs per hr WITHIN ENROBER

Figure 11.10 The Sollich Solltemper U.

development of the Sollich twin-stream patent. The diagram shows the continuous temperer connected to an enrober, utilizing the patent Sollich circulation system. Chocolate from storage is fed on demand to the reservoir tank (1) and is taken into the heating worm (2). Part way along the worm, a stream of seeded chocolate (3) joins the heated chocolate, the two streams are mixed and passed into the tempering worm (4) which contains inner and outer cooling jackets. The chocolate is pumped by the worm continuously to the enrober tank (5) in a quantity in excess of requirements.

The level of chocolate rises in the enrober tank till the overflow level is reached. The excess chocolate not used on the product falls into the return pump and is pumped back (still tempered) to the tempering machine tempered-stock chamber. Since the quantity returning is in excess of the tempered stream being fed into the worm (4), an overflow is created and the remaining chocolate (6) falls into the hot deseeding reservoir (1) to continue the cycle. The overflow principle in the enrober and the tempering machine maintains a constant tempering time period.

This machine provides a freshly tempered chocolate on a low residence time basis. Further crystallization occurs in the relatively low temperatures of the enrober tank.

Recently, Sollich have developed plate type heat exchangers for tempering chocolate, using a system very similar to the well-known Aasted machines. The new machine claims to have more residence time, 'microcrystalline growth' and thermo-cyclic conditioning, through the use of scrapers. This machine is designated 'Solltemper MST-V' and a residence time of five minutes has been the pre-requisite for creating seed (Figure 11.11). This temperer has proved to combine well with the 'Enromat' enrobing machine in producing a good temper. The enrobing temperature is, however, lower than could be achieved with a longer residence time in the enrober tank.

In the MST-V temperer, Metered chocolate is fed in the base of the machine and rises through heat-exchange plate elements, so divided to create three cooling zones, and spring-loaded scraper blades provide efficient heat exchange. The required 'strike seed' temperatures occur in the later stages of the cooling cycle. A further zone reheats as necessary and extends the residence time period. The chocolate now passes through to the user plant.

It is also claimed that water at 16.6 °C (61 °F) can be used effectively as the cooling medium. This presupposes that jackets in temperers are free from scale or furring up which would reduce cooling rates. To avoid this, it is often advisable to provide closed-circuit propylene glycol units feeding temperers at temperatures in the region of 11–13 °C (51–55.4 °F).

There can be quite a difference in 'striking seed' in chocolate when using a cold water in comparison to warmer water temperature. The speed of nucleation is dependent on fast efficient cooling, and large differences occur.

Baker Perkins. This company has over the years produced many temperers

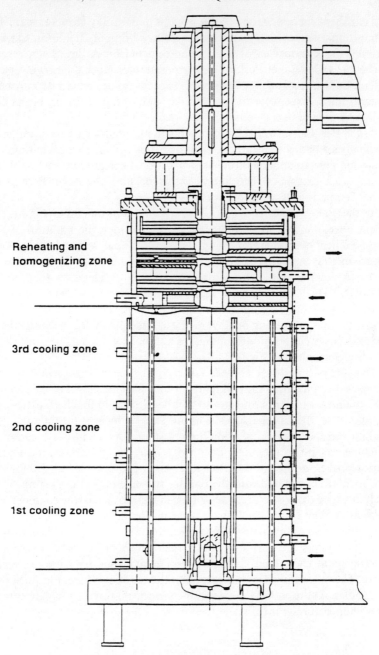

Reheating and
homogenizing zone

3rd cooling zone

2nd cooling zone

1st cooling zone

Figure 11.11 The Sollich Solltemper MST-V.

and enrobers, including the predecessors to present-day enrobers, namely the batch-dipping machines, which simulated hand-dipping methods. Later they supplied the 'standard enrober' under agreement from A. Savy Jeanjean et Cie. In the late 1950s Baker Perkins were licensed to produce their own version of Sollich built-in tempering techniques. These were designed into enrobers as integral units. Except for the present-day units, all had the problems of low residence time systems.

Baker Perkins solved these problems in the mid-1960s by using a large enrober seed bed with a variable retention time between 15–60 minutes. This is enough time, together with efficient agitation, to create the most stable high-temperature coatings. A description of this machine (the Baker Perkins Long Time Tempering System) follows later.

In the Baker Perkins 105 TU ACS tempering machine (Figure 11.12), recent developments in outboard tempering include the addition of a new disc-type heat exchanger for striking seed and a 'real' retention zone, located approximately midway through the tempering machine. In this case 'real' means a definite bulk storage space, specially designed to take and mix chocolate which already has some crystals present, and is ready to be 'matured' into a high-temperature coating. This zone has a retention time of 12–20 minutes depending on flow rate. The time period has been significantly increased over earlier models, and does not include the passage time through the disc elements (see Figures 11.12, 11.13).

The next part of the machine is a disc-type heat exchanger which is capable of precisely controlling the exit temperature. This machine has been designed with the necessary pre-requisites to handle the most difficult eutectics as well as, of course, the easier recipes. The ACS temperer will fill the dual role of feeding enrober plants or moulding systems. As mentioned earlier, this machine recognizes the *latent heat temperature rise* in the retention zone by sensor probes, and can be pre-set to attain a pre-determined cooling curve.

Chocolate is metered into the base of the machine, and rises up through multi-scraped disc cooling elements spaced apart by a narrow chamber, giving efficient cooling to the small mass present. The net result is to form crystals very quickly. The chocolate then passes into the stirred mixing/retention zone, and the crystals are allowed to take their more stable form, giving a high melting point chocolate of minimum increased viscosity with good temper curve. The inflection point indicates that the product is ideal for quick setting in a cooler. At the end of the retention zone the matured chocolate enters the final temperature-control stage then exits to the user plant.

11.2.5 Other methods of tempering

Many other manufacturers produce tempering machines and not all can be mentioned here. Once tempering principles are fully understood, the assess-

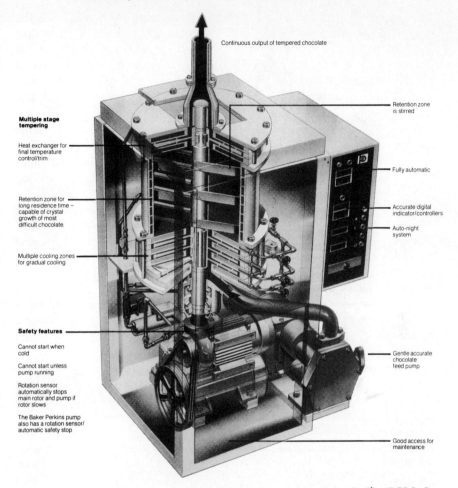

Continuous output of tempered chocolate

Retention zone
is stirred

**Multiple stage
tempering**

Heat exchanger for
final temperature
control/trim

Fully automatic

Retention zone for
long residence time –
capable of crystal
growth of most
difficult chocolate.

Accurate digital
indicator/controllers

Auto-night
system

Multiple cooling zones
for gradual cooling

Safety features

Cannot start when
cold

Cannot start unless
pump running

Gentle accurate
chocolate
feed pump

Rotation sensor
automatically stops
main rotor and pump if
rotor slows

The Baker Perkins pump
also has a rotation sensor/
automatic safety stop

Good access for
maintenance

Figure 11.12 The Baker Perkins 105 TU ACS tempering machine (Baker Perkins BCS Ltd).

ment of individual machines is relatively easy, making selection that much more reliable.

The following list contains some types of tempering systems available.

(i) *Cooling drum.* Multiple or single layouts condition the chocolate before it is trickle-fed to the enrobers. This principle may be used in conjunction with a large seed-bed and can give good results, but it needs experienced operators to maintain optimum conditions.

(ii) *Drip feeding of un-tempered chocolate* utilizes the fact that if the usage rate is low the material already in a tank on the enrober will seed the

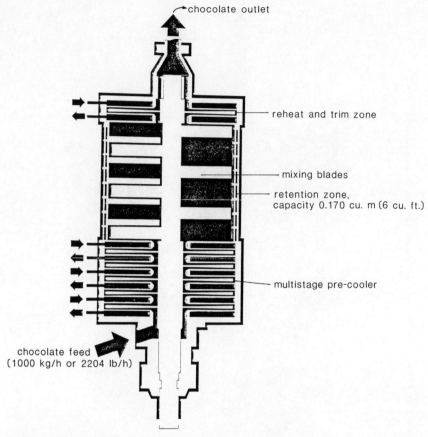

Figure 11.13 The Baker Perkins multi-stage temperer with extra retention zone. The retention time provided by this machine is sufficient to temper most types of chocolate and will produce a mature chocolate at relatively high temperatures when compared with machines of 1 to 5 mins retention time.

Retention time calculation in a retention-zone temperer

$$\text{Retention time} = \frac{\text{Chocolate density} \times 60 \times \text{volume of tempering space}}{\text{hourly flowrate}}$$

$$\text{Metric: R.T.} = \frac{1281 \times 60 \times 0.170}{1000} = 13.06 \, \text{mins}$$

$$\text{Imp.: R.T.} = \frac{80 \times 60 \times 6}{2204} = 13.06 \, \text{mins}$$

incoming chocolate. The idea of trickle feeding is to balance usage. If the usage rate increases more than the rate of crystallization then all the chocolate in the enrober slowly loses temper. Naturally conditions are difficult to keep in equilibrium.

(iii) There have been very many versions of screw-type temperers, some horizontal, some vertical, long narrow worm tubes, of large diameter and short length. One temperer, with multiple screws like a lathe leadscrew, gave variable results.

(iv) One of the most fascinating temperers of former years was the *Lehmann Multi-Roller* horizontally disposed machine (see Figure 11.14). This

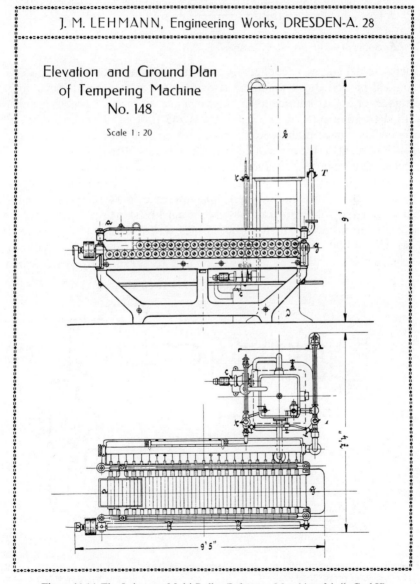

Figure 11.14 The Lehmann Multi-Roller (Lehmann Maschinenfabrik GmbH).

machine consisted of 21 pairs of driven rollers and four infeed rollers. This machine is, to this day, the only device capable of cooling very heavy paste, and delivering it into suitable trays at the end of the machine. At the infeed hopper end, paste falls directly on to rollers which drag the paste or chocolate out of the hopper and into the sequence of hollow cooling rollers, being gradually cooled, as it passes to the end of the machine. This device also had the advantage of zonal reheat between successive pairs of rollers, enabling controlled temperature gradients to be created.

11.2.6 *Bauermeister pressure tempering*

There are difficulties in cooling cocoa butter and cocoa mass to form large moulded blocks. Block moulding is required to aid transport (for a short distance, or across continents and oceans) and storage. The main problem with depositing into block containers is the prolonged cooling time, on average an unacceptable twelve hours, compared with cooling times in moulding plants in the order of thirty to fifty minutes when producing 'normal' blocks. Bauermeister produce the pressure tempering heat exchanger (Figure 11.15), which overcomes this problem. For best chocolate tempering, the author has advocated a relatively long residence time in tempering devices, but when large-volume blocks must be set quickly, short residence times are necessary to produce a high output efficiently. The Bauermeister system is essentially a high-efficiency, scraped cylindrical, horizontally-disposed, heat exchanger, fitted with an adjustable relief valve at the exit. The relief valve may

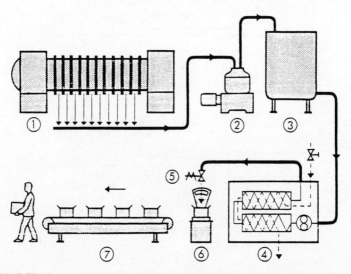

Figure 11.15 The Bauermeister pressurized tempering system (Gebrüder Bauermeister & Co). (1) Cocoa butter press; (2) separator; (3) liquid storage container; (4) pressurized temperer-cooler; (5) magnetic valve; (6) semi-automatic weighing machine; (7) conveyor.

be adjusted to maintain a cylinder pressure of between 1.7–10.5 kg/cm² (25–150 psi).

Using this device, cocoa butter or mass is supercooled down to 23.5 °C (75 °F), where due to pressure and agitation a very fine homogenous crystal growth takes place. Sensible heat is removed along with some latent heat, and since more latent heat is dissipated to the cooling water than in the normal tempering machines, the resultant product emerges from the machine in a creamy, relatively free-flowing form. The pressure release created at the outlet gives a further cooling effect. The result of removing the maximum possible sensible heat and latent heat is to produce a very quick setting and cooling product. Sufficient contraction occurs after one to two hours (depending on the type of cooling or storage), to enable the blocks to be knocked out and wrapped. Tempering in the true meaning is not necessary, and the longer time multistage machines can be disregarded, as they simply would not work without seizing solid at these temperatures.

11.2.7 Thermocyclic or cyclothermic tempering

Last but not least is a tempering principle which is sufficiently different and noteworthy to need a full explanation as to why it is not more popular. Cyclothermic tempering is essentially a system of controlling chocolate through tempering devices which have been separated into four stages. Duck (2) described this method in 1961 (see Figure 11.16). The graph is for cocoa butter and shows the relation between crystal growth over a time period subjected to cooling, heating, cooling and re-heating stages (hence 'cyclothermic'). From Figure 11.16 we see that the cocoa butter has first been heated to 50 °C (122 °F), then reduced in temperature to 29 °C (84.2 °F) over a period of 30 minutes, at the same time being continuously agitated. During this time the viscosity is gradually increasing to a maximum constant value due to

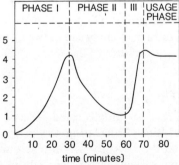

Figure 11.16 Cocoa butter subjected to cyclo-thermic tempering, showing the temper viscosity changes with temperature and time (2). Phase I, 50–29 °C; Phase II, 29–33 °C; Phase III, 33–29 °C; usage phase, 33 °C.

crystal growth (approximately 4% solid seed) and temperature. It is then raised in temperature to 33 °C (91.4 °F) taking 30 minutes. During this period the viscosity falls to reach a new constant minimum value (re-melt of crystals giving approximately 1% solid seed). A further cooling stage reduces the temperature from 33 °C (91.4 °F), to not less than 29 °C (84.2 °F), the theory being that unstable crystals cannot form. This time period is again 30 minutes, during which the viscosity increases to a new constant value (again approximately 4% solid seed). A final 30-minute stage gradually raises the agitated mass to 33 °C (91.4 °F) and this time a change is noted: the viscosity does not fall as seen in the previous stage.

What has happened here? The graph shows that a higher melting point crystalline mass has developed, and since very little melting out has occurred the viscosity has hardly changed. Let us now consider whether this effect is really much different to long-time tempering systems mentioned earlier. We could argue that the lengthy time needed to complete the various stages would have resulted in a high-temperature stable viscosity chocolate in any case. The claim is that the temperature is higher than normal, 33 °C (91.4 °F), although the author has often seen the Greer enrober working at 34 °C (93.4 °F) on plain chocolate. However, let us look at the next derivation of cyclothermic tempering, not just a laboratory experiment. In the previous cocoa butter experiment we saw that we could wait forever to strike seed in chocolate at 29 °C (84.2 °F), so a system of thermo-cyclic tempering described by Kleinert (3) was developed which effectively started seed at 25 °C (77 °F). This action speeded up the crystal development rate, and the remainder of the process follows that propounded by Duck (2), except that a fifth stage was added which is claimed to raise the temperature to 35–36 °C (95–97 °F) for plain chocolate, and 33–34 °C (91–93 °F) for milk-containing chocolate. The only caution here is the work by Vaeck (1) (see Table 11.1), which shows that the maximum temperature at which the stable β-form melts is 35 °C (95 °F), so in theory running chocolate at 36 °C (96.8 °F) will eventually result in a loss of temper.

There is no doubt that cyclothermic tempering will temper chocolate, but the system requires four tempering cylinders with associated stirred retention devices. The cost of this extra machinery can outweigh any chocolate handling advantages.

11.2.8 Feeding to user plant

It is important to beware of the pitfalls listed in section 11.2.1 regarding the pumping of chocolate. When calculating the exit pipeline pressure it is advisable to assume that the viscosity will double during tempering.

Basically two feeding systems have been adopted and more or less standardized on, one for moulding plants, the other for enrobers.

(i) *Ring main*: the moulding plant takes tempered chocolate from the supply as called for, the excess or unused chocolate being returned and reheated (de-tempered) to bulk storage

(ii) Supply to an enrober using the '*Sollich' circulation principle*: a continuous supply of tempered chocolate is fed into the enrober in excess of demand, the feed rises to a predetermined retention level in the machine and the excess is caused to overflow and return for re-heating and storage. The whole process, being continuous and metered, gives an accurate reproducible 'time period' in the enrober tank. If this time is long enough (20–30 minutes) then optimum temper may be achieved.

11.3 Combined temperer/enrober systems

Sollich were an early advocate of this type of system and used their stream principle to operate it.

Streams theoretically assist the seeding rate of chocolate but rarely do so (time period calculations show this), and in fact multiple streams increase the flow rate through the temperer, therefore reducing the tempering retention time. When fed directly to an enrober curtain, streams do have the advantage, however, of meeting the flow rate requirements. For a 820 mm (32″) enrober with two flow curtains, the flow rate needs to be at least 3000–3500 kg/h (3 ton/h). The stream principle is therefore necessary to give sufficient flow rate to maintain a chocolate curtain in the 'Temperstatic' enrober, otherwise the temperer would have to be enormous.

A predecessor to the Temperstatic and Enromat enrobers was an enrobing machine that embodied the U-Temper layout, but instead of a tempered feed to the enrober tank, the chocolate was fed freshly tempered from the tempering worm directly to the enrober flow pan (see Figure 11.17). This limited the flow rate up to the flow pan to that the output of the tempering screw, as there was no independent circulation pump (Figure 11.20). The effect of using the freshly seeded chocolate was that the coating temperature had to be severely reduced to 26.6 °C (80 °F) to develop temper at all. In addition, the retention time was of the order of six to ten seconds, hence little time was available to induce the more mature stable crystals.

As the crystals growth rate was exceedingly fast, the flow pan was jacketed and needed to be continuously scraped to prevent all the chocolate going solid. The chocolate was also setting on the wire belt on which the sweets were being enrobed and not returning by the overflow system. This process was designed so that the fast setting would give faster setting times in the cooler. Although this was found to happen, it was found very difficult to control.

Historically, this method is worth remembering, since contemporary Sollich built-in temperers evolved from the machine. It is possible this principle could be resurrected if product changes with regard to the fat content of chocolate

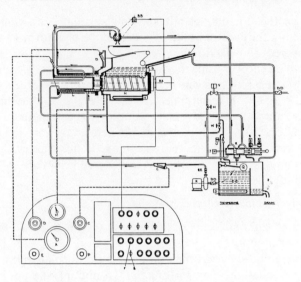

Figure 11.17 Sollich combined temperer/enrober system.

brought about different setting properties which were more appropriate to this scheme.

11.3.1 *Temperstatic enrober (with built-in tempering)*

The early Temperstatic was a three-stream machine. The first stream is hot chocolate de-seeded in a worm heat exchanger, which is then joined by a second stream of partially tempered chocolate. The two combined streams pass into a cooling worm to induce further crystallization, and additional mixing also take place. Part way along the worm, a third partially tempered stream joins the two cooler ones, again uprating the throughput (but not the usage quantity, since this is determined by the first hot stream). Further mixing and re-heating takes place as the coating is pumped to the enrober flowpan.

This machine consists of the following components, illustrated in Figure 11.18.

(*1*) Storage tank for de-tempered chocolate
(*2*) Heating worm to ensure full de-tempering
(*3*) Mixing section for combining stream 1 with stream 2
(*4*) First seeding stream of tempered chocolate
(*5*) Tempering and cooling cylinder
(*6*) Electronic temperature control
(*7*) Seeding with second stream of tempered chocolate
(*8*) Mixing chamber

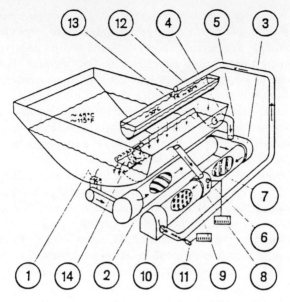

Figure 11.18 The Sollich Temperstatic enrober.

(*9*) Mixing and re-heat cylinder
(*10*) Variable output pawl pump
(*11*) Electronic control temperature for (9)
(*12*) Coating flow pan (scraped and stirred)
(*13*) Residence container for tempered chocolate
(*14*) Level control weir for variable residence times in container (13) also
guides overflow of surplus chocolate into storage tank (1).

The amount of detempered chocolate which is drawn from storage tank (*1*)
through heating worm (*2*) is in excess of the product usage rate of 500 kg/h
(1100 lb/h). Approximately the same amount of tempered chocolate is fed
from the residence tank (*13*) via pipe (*4*) into the mixing section (*3*) and is used
for seeding the un-tempered chocolate. The mixture then passes through the
tempering and cooling cylinder (*5*). A further seeding of tempered chocolate
(the amount of which is adjustable by pawl pump (*10*) is drawn from the
residence container (*13*) into the mixing chamber (*8*). The total quantity of the
chocolate is thoroughly mixed in the mixing and re-heating cylinder (*9*) and is
then brought up to the required processing temperature. The whole then
passes to the flow pan (*12*) via the variable output pump (*10*). Most of the
unused chocolate falls through the wire belt and is returned to the container
(*13*). The amount of chocolate held in container (*13*) and the retention time can
be varied by the adjustable weir (*14*). The surplus tempered chocolate
overflows the weir into storage container (*1*) where it is re-heated and

detempered ready for continued use. The level in storage container (*1*) is maintained by automatic controls topping up the system from bulk storage.

The present-day Temperstatic has been modified by removing the first partially tempered stream, and increasing the flow through the last stream. Recognition of the low retention time has resulted in the tempered chocolate tank being increased in size to lengthen the conditioning time.

This particular novel design allows for a simple installation, with the minimum of pipework, and can be fed direct from bulk storage. Thus a complete installation can be made in two days. The tempered chocolate tank is supported above the de-seeded chocolate tank so that excess tempered chocolate can fall directly into the reservoir for de-seeding. This results in a compact machine with accessories, which were normally outside the machine, now being built-in. A considerable saving is also made in installation services.

The disadvantage with this principle is the 'time period' which does not allow for optimum temper to be produced. At higher belt speeds, chocolate can carry over past the tempered chocolate return chute, which is located under the wire belt, and this can result in tempered chocolate starvation. This system is capable of producing acceptable results with some chocolates with relatively low coating temperatures, but difficulties may arise when some fat mixtures are present which give rise to complicated eutectics. In this case the colour, fluidity and gloss may be unsatisfactory.

11.3.2 *Baker Perkins long time tempering system*

This principle of enrober tempering was designed into a machine to give 'long retention time tempering', a time considerably in excess of any previously known systems available on the market.

In this system (Figure 11.19) the chocolate is circulated by pump *A* from tank *B* to flow pan *C*, the surplus returning to *B*. De-tempered chocolate is drawn from storage tank *J* and metered accurately by pump *E* through the tempering screw *D* to the reservoir tank *B*. The quantity of chocolate in *B* is controlled by the adjustable overflow *F*, the overflow or surplus chocolate is pumped by pump *G* back to storage. The chocolate in tank *B* is held in a tempered condition and acts as a 'seed-bed' to convert freshly tempered chocolate from *D*. The fresh chocolate from source *J* is gently cooled by tempering tube *D* and is mixed with the tempered chocolate in *B*. The quantity in the holding tank *B* is varied according to the type of chocolate being used, by the level control weir *F*. The reservoir is continuously replenished with fresh chocolate at a constant temperature and flow rate, independent of the rate it is being taken away by the product being coated. The only variation is the amount taken away by the product, thus the overflow varies in quantity. This method guarantees a consistent 'time-period', since the supply is designed to be in excess of usage. An automatic by-pass *H* stops the flow from *J* during initial startup, to speed up the readiness time (20 mins). Chocolate is drawn by

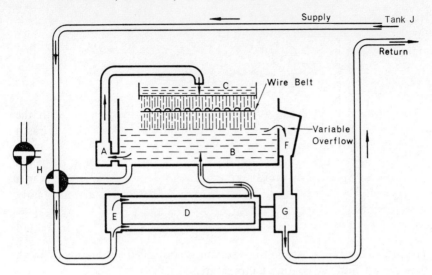

Figure 11.19 The Baker Perkins long-time tempering system.

pump E from B and close-circuited through tempering screw D back to B. As the chocolate comes to temper, a sensor returns valve H to its normal running position when it is drawing chocolate from storage J.

The tube D is a large-capacity single-stream tempering tube. Chocolate is taken away from the machine in two ways: the quantity taken away on the product and the amount returning to source via the overflow F. The combined quantities removed in these ways will equal that being supplied by the tempering tube D. This amount is in excess of that to be taken away by the product to maintain an overflow at F. When the machine is full, it is coating at its maximum speed and capacity. As the chocolate taken away from each source is in the same tempered state, the state of equilibrium is not affected by changes in the quantity of chocolate removed by the product.

In this process, warm de-seeded chocolate is taken in from a bulk storage ring main, and is continuously metered at a designed quantity through the heat exchanger where it is reduced in temperature to 28 °C (83 °F) by cooling water at approximately 11 °C (51 °F).

This temperature introduces fast-growing nuclei of tiny crystals which are most likely to be immature β-crystals. The continuously agitated stream of freshly seeded chocolate enters the enrober reservoir with a constant and fine crystal growth (these conditions remain constant due to precise continuously metered chocolate and good temperature control). So the mass entering the coater reservoir contains a fast-growing pattern of crystals, and to control the development of these crystals the whole masse is raised in temperature to 31–33 °C (88–92 °F) over a controlled 'time period'. Beaters provide an efficient mixing action and assist the conversion to mature and stable β-crystal forms.

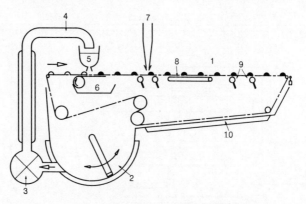

Figure 11.20 Components of an enrober.

The rate of crystal growth in the reservoir is controlled automatically by the time period, and the coating temperature.

The length of the 'time period' can be determined by the type of chocolate being tempered and the coating temperature desired. The time period is adjustable from 15–60 minutes (long time tempering). The state of equilibrium is obtained by accurate flow rates, temperature, time and agitation.

The coating temperature is decided by the amount of temper (seed, or grain) since the coating temperature is proportional to the amount of, and type of crystal growth in the coating. It must be remembered that different coatings may need varying seeding times, therefore coating temperatures will vary, especially with eutectics. However, this principle produces a high coating temperature with the many processing advantages described earlier.

11.4 Enrobers

Figure 11.20, showing the basic components of an enrober, may help in the following descriptions. An enrober consists of a driven wire grid conveyor belt (*1*) on which to transport centres to be coated. This passes over a stirred reservoir tank (*2*) which hold the tempered chocolate. A chocolate pump (*3*) circulates the chocolate up a 'riser pipe' (*4*) and into the top flow pan (*5*). The flow pan creates one or two curtains of chocolate through which the uncoated centres pass. Positioned below the top flow pan and beneath the grid is a bottoming trough (*6*) (also called the surge roller trough). This trough retains chocolate falling from the curtain and feeds it on to a flat plate, thus forming a bed of chocolate which is moving with the wire grid. The centres entering the enrober are semi-suspended by this bed and become coated on the underside.

An air nozzle (*7*) is located after the chocolate curtain together with a grid shaker frame (*8*). A number of grid licking rolls (*9*) are shown located after the grid shaker; these remove excess chocolate from the grid and centre. The chocolate is drained from the roll scrapers and is returned to the reservoir tank, by the grid belt running along a heated extension trough (*10*). The coated sweets leave the enrober passing over the anti-tailing rod, which, as its name suggests, controls the tails or residues left by the liquid chocolate as it separates from the wire grid and transfers on to the cooler belt or plaque.

11.4.1 *Types of enrober*

Until the early 1900s chocolates were coated by hand, or by automated batch techniques emulating hand-dipping methods. The traditional hand-dipping of chocolate with a batch tempering procedure results in an attractive finish difficult to copy. The major problems with hand or batch methods were low production rates and a high labour content. Thus the necessity for automation arose, and the enrober was born.

The first enrober is credited to Magniez (1901) and was produced by A. Savy Jeanjean et Cie. This machine, the 'Standard Enrober', was also supplied by the agents Baker Perkins and manufactured by the National Equipment Co. of America. Such was the advanced nature of the machine that many were sold throughout the world. In a sense this machine was revolutionary for its time, and it became the basis for all future designs of enrober with the principles little changed to this day (Figure 11.21). The designer recognized that there could be two parts to the enrober coating area, one the flooded area where the coating took place and a second, weight control zone. So he provided two wire mesh conveyor belts, one in the wet area and the other in the weight control area, thus aiding the removal of excess chocolate, leaving only slight carry-over chocolate from the product to be shaken off. The wire mesh conveyor belt is a miraculous achievement for its time. This belt is unique in its construction and needs considerable accuracy in forming the loops and bends in a high-quality wire. The separate wires link together simply, forming a strong flexible continuous length, and considerable skill was required to join the length into an endless circuit. The requirement was to keep the joined section from being distorted and making sure the join is flat (this still applies today).

The Savy enrober had many features, some of which are lacking on present-day enrobers, mainly because of cost-cutting exercises and possibly different requirements. The early machines did not run particularly fast, and coolers had not reached the efficiency of modern units, so there was relatively more time available for good coating to take place. Nevertheless, the following features are important:

(i) All of the many rollers were driven and scraped (a necessary feature for good weight control, puts less force on the belt thus extending its life)
(ii) There were no fixed or hinged bars for chocolate build-up to take place — all revolving parts were scraped
(iii) Chocolate was elevated from a stirred heated chocolate tank by scraped ring feed to the flowpan, i.e. no pump

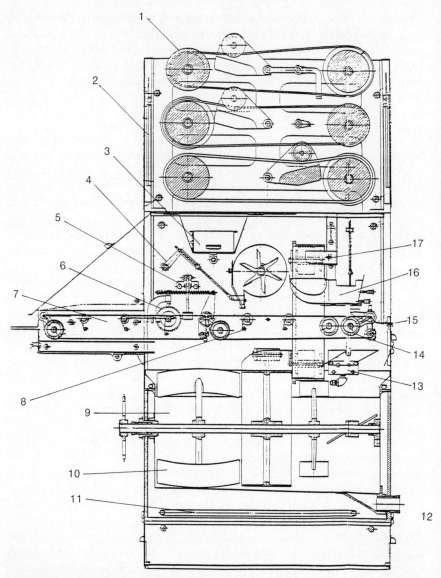

Figure 11.21 The 'Savy' enrober. (1) Variable-speed cone pulley drive; (2) window mechanism; (3) gas hood heater; (4) turbo fan angle adjustment; (5) shaker adjustment; (6) shaker and awl; (7) extension belt; (8) scraped rollers; (9) chocolate reservoir; (10) stirrer; (11) heater; (12) drain point for chocolate; (13) decorator gear pump; (14) twin surge roller, (15) in feed belt; (16) flow pan; (17) pick-up wheel.

(iv) The blower to remove excess chocolate was of the turbo layout, the rotor in the blower being the same width as the belt

(v) The wire belt drive had variable speed

(vi) Gas gave quick heating, though we would be hesitant to use this medium today. Weighted windows gave quick access and heat retention

(vii) A sideways driven movement of the delivery belt produced zigzag decorative markings from the piping system

(viii) An in-built gear pump for the decorator was complete with essential filter system and flow adjustment

(ix) The Kilgren design of chocolate decorator (see Figure 11.22) was used

(x) A pre-bottomer was included to put some chocolate on the underside of the sweet.

It can be seen that there has been little improvement on this design, and the Savy enrobers in some companies still produce excellent work, though there must be only a few left operational. The tempering for this machine was processed in batches in chocolate kettles, and experienced operators produced work as good as that of present-day automatic machinery. Modern enrobers have developed in all areas of confectionery, and the biscuit and bakery industry processes have not missed out on the machine developments, each section of the industry providing a widening scope for enrober manufacturers to innovate and develop new ideas.

Enrobers are provided for the smallest producer to the largest, and this gives a wide variety of different designs to meet requirements. We have belt widths from 125 mm to 1820 mm (5 in–68 in), although obviously only the largest manufacturers can utilize the wider machines. The author recollects designing the first 1730 mm (68 in) enrober in 1967 for a company noted for its efficiency of countline production. These wide machines are now producing enormous outputs with less labour in production and in maintenance. For instance, the flow rate up to the flow curtain equipment was metered at 31 750 kg/h (70 000 lb/h) giving enough chocolate to provide four curtains and a good bottoming supply. The author also responsible for designing the 'Turboflow' cooler which was easily stretched out to suit the wide machines, and whose air cooling cushion provided the necessary support for the heavy load and gave much extended belt life.

What we do specify when we wish to purchase an enrober and how do we decide which machine to purchase when there are so many available?

Usually there is some experience within a company with existing machinery that leaves little to be desired, or on the other hand with bad equipment that needs rectifying. The latter may give some ideas on how to request a design improvement. If a chocolate that is causing difficulties on current machines, a test should be carried out on another version, even if it means transporting the chocolate thousands of miles to a new supplier.

There is of course controversy over which is the best or the right type of

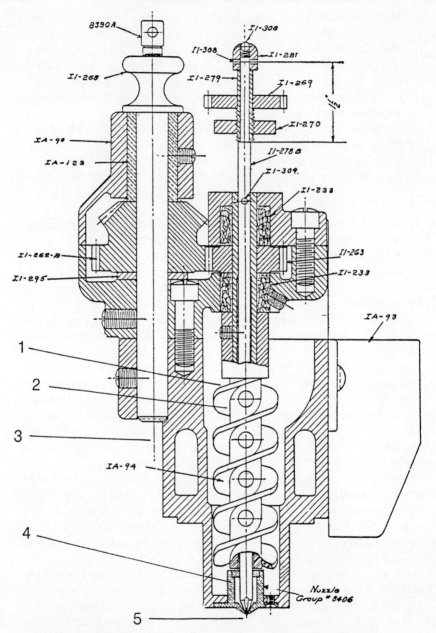

Figure 11.22 Kilgren chocolate decorator. (1) Chocolate reservoir; (2) chocolate feed screw; (3) heated water jacket; (4) nozzle assembly; (5) pricking and sealing pin.

enrober for the job in hand. Important considerations may be the price, or whether the tempering should be outboard or in-built, or indeed whether tempering is needed at all. Narrowing down the selection is usually easy when the requirements are fairly specific, as the following examples show.

(i) Quality assortments requiring the best possible appearance, a long shelf-life, and almost certainly automatically piped decorations. Tempering may be inboard or outboard.

(ii) Countline coating with good finish required but capable of high speed operation. Usually outboard tempering is recommended.

(iii) Biscuits with full and half coatings are usually coated at high speed direct from ovens, via ambient coolers or forced convection coolers, depending on the type of biscuits. The enrober may need to operate with a fat blend compound coating which does not need tempering, or with a recipe which requires complete tempering.

Special enrobers may be designed for half coating only; these are usually cheaper when committed to dedicated lines. A turnover device is fitted to invert the rows of biscuits, so that the base coating becomes uppermost. Thus the coating retains the grid marks on cooling and presents a more decorative finish (see Figure 11.23). For deep shoulder dipping (a combination of bottom coating, sides and end coating) at heights between 5–18 mm (0.25–0.625 in) a depressing roll or wire mesh retaining conveyor will be needed. These methods prevent the product from floating on the bed of chocolate, and thus allow the coating to reach the highest level. The depth of shoulder coating is controlled by the depth of bottoming surge created; this is in turn dependent on the surge roll speed and the viscosity. For instance it is virtually impossible to create a deep coating level with a low-viscosity chocolate.

(iv) Tandem enrobers, close coupled or spaced apart, may be used for coating irregular products covered with cereals or nuts. With these difficult products, more high pressure blowers may be found to be necessary to fill the crevices with chocolate and give complete coating. Weight control also is difficult to maintain within close limits.

(v) Cake coating, full coating or shoulder dipping enrobers with the capability of accepting large cakes, swiss rolls, etc., usually have a piping decorator built in or adjacent to the enrober itself. Decorating may alternatively take place on a separate belt with a different-coloured chocolate being piped for additional effect. Striping flow pans are also used for different effects, e.g. coating one or more sections of a biscuit.

(vi) Ice cream enrobing is similar in some ways to enrobing confectionery, but is more specialized to suit the problems arising when coating a freezing product. Machines have to run much faster to prevent too much solidification in the enrober coating zone, whilst still maintaining coating weight control, and fluidity during the passage through the machine. Belt

H

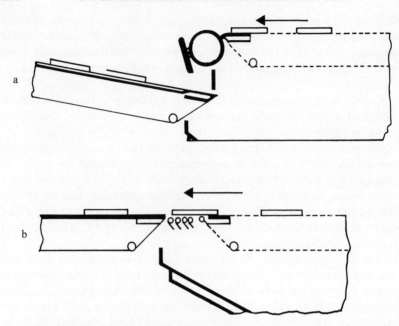

Figure 11.23 (*a*) Biscuit turnover device. This device is fitted at the end of the enrober grid in place of the detailing shaft. Biscuits are bottom-coated using the bottom flow pan only. At the discharge from the coater the biscuits pass over a special roller. This is driven at a speed to match the grid speed. The chocolate-coated surface adheres to the roller and carries the biscuit around with the roller through 180° where it falls off. A decorative pattern is left on the liquid chocolate by the enrober grid. The upper surface (in the enrober) is clear of chocolate. Finished biscuits are then cooled in a cooler. (*b*) Bottom scraping device. In some biscuit coating applications a very thin bottom coat of chocolate is required. Standard licking rolls as mentioned previously can remove quite a lot of chocolate. Even more chocolate may be removed by fitting a bottom scraping device in place of the de-tailing shaft. This device usually consists of up to four small scraped rollers driven at the grid speed. Chocolate is picked up from the base of the product by the rollers and returned by the scrapers to the enrober hopper.

speeds can be in excess of 30 m/min (95 ft/min). Hygiene is most important when dealing with frozen food, and stainless steel is the main material of construction as it allows for hose-down cleaning and dismantling. Pure chocolate is not used for coating ice cream: usually a high fat temperable coating is used (50–55%).

Let us take a closer look at a high quality assortment enrober capable of handling milk or plain chocolates. The milk chocolate is high in animal fats, and requires time in tempering. The importance of the time needed to temper chocolate in a tempering system and how to apply this to an enrober was described earlier. If one decides to purchase an enrober with a small capacity holding tank, one must purchase a tempering machine with at least 15 to 20 minutes' retention time. This may be run at a high coating temperature and will produce a product with a good gloss, snap, and maximum shelf-life. This method virtually guarantees good results, provided that there is sufficient flexibility of control of the system. If one purchases an enrober with a large capacity reservoir then a tempering machine with a lower retention time can be used. The author still, however, advocates operating with a maximum retention time in a tempering system as this gives additional latitude under adverse conditions. In choosing a tempering system for the enrober, it is important to first work out the flow rate for usage requirements, and therefore, avoiding putting a massive flow rate through the tempering machine, when it is not necessary, as the latter only wastes cooling water energy and reduces the chocolate retention time. Some enrober manufacturers believe this high flow rate helps them control the coating temperature at the enrober. This is of course correct, but the enrober controls become ineffective and flexibility is lost. This makes it very difficult for the enrober operator to change the chocolate fluidity at the enrober itself.

Having determined which process is required it is necessary to look at the design of the enrober itself. Good quality assortments usually need two enrobers set apart with a water cooling table between them. With two enrobers the coating thickness can be built up for better results. The water-cooled table facility allows time for partial setting of the chocolate between the enrobers. The water-cooled 'table' is a conveyor with a cooled base located under the transport belt. The belt has good heat exchange properties and aids the setting of the chocolates. The partial setting of the base allows chocolate to be stripped off the belt after 2–3 minutes, i.e. prior to entering the second enrober. Thus the base thickness can be built up to give a thicker overall coating and better sealing for products with liquid centres. Having obtained two enrobers, the first machine may also be used for bottom coating only, with the second machine then providing the final coating. On occasions the first machine may be required to run a completely different chocolate or one of a much lower viscosity to cover an irregular centres such as nut or cereal combinations.

The enrober specifications could be as follows.

(i) Pipes and tanks should be completely jacketed with operator-controlled heating and cooling facilities.

(ii) Pipes conveying chocolate should be fully jacketed outside the product zone. It is false economy to leave them unjacketed, since this gives rise to blockages which waste time and damage sensitive equipment, thus outweighing cost savings.

(iii) Chocolate pumps should be slow-running with large swept volumes to enable some particulate matter to be pumped without damaging the pump or the product. The latter can occur for example when nuts are present in the chocolate. If these are crushed in the pump the nut oil is released into the fat phase and may prevent the chocolate tempering. The pump should be fully jacketed and provided with drainage facility for easy maintenance.

(iv) The enrober reservoir should have a waste-material collecting sump to trap waste and wire belt fractures. This protects the pump and allows easy removal. For highly contaminated material it is advisable to take a line from the reservoir to a sieve and return the chocolate for de-seeding. The reservoir should have an effective stirrer to give an efficient mixing action, and should be of easy clean design. The stirrer should provide good heat exchange, which is aided by low clearance at the reservoir jacket walls.

(v) Wire belt support rollers should be grooved and scraped in the weight control areas (under the blower section and after the shaker section). Scrapers should preferably be adjustable and heated. If it is felt necessary an adjustable 'weight control roll' may be fitted after the shaker section. This roller is adjustable up and down under the belt so that it can wipe chocolate off the product in a controlled manner. The chocolate picked up by the roll is scraped off and allowed to drain back to the reservoir tank by falling on to the return run of the wire belt. Belt support rollers should be supported in sealed bearing housings with a separate seal. A run trap should be provided to prevent chocolate reaching the seal (this trap is usually cut out of the scraper bar). Belt support rollers should be driven by a timing belt to prevent the oily maintenance residue one usually finds inside enrober covers, since they do not require lubricating. In the decorating area, or extension of the enrober, rollers should be 120 mm (4.5–5 in) apart to give good control on the belt for precision marking. This prevents vibration from the shaker being transmitted back and affecting the decorations.

(vi) Wire belt drives should be either single or duplex gear systems. The tensioning arrangement should give minimal tension, for over-tensioning greatly reduces the belt life. Badly aligned drive gears also cause considerable wear.

(vii) The decorating and shaking section should be long enough to provide sufficient time on the shaker for weight control to take place. This area should also be covered and heated. The shaker should not transmit vibrations to the end of the machine, neither should it grip the belt or allow chocolate to built up and create extra cleaning problems. It is possible to specify the fitting of a rear bottoming bath which allows a better bottom coverage on large-area products or to cover up faults created by starch moulding. This bottoming bath is fitted just after the shaker and requires the addition of weight control rolls to remove the excess chocolate.

(viii) For the highest-quality tailing control, a 3 mm (0.125 in) dia. rod is generally thought to be best. The rod should be adjustable up and down, also adjustable to and from the belt for scraping requirements. The advantage of running the anti-tailing rod close to the belt is that it allows the wire belt to take away the excess chocolate picked up by the spinning rod. The driven rod should usually run with the direction of product flow to help small goods over the transfer point into the cooler. The drive speed is not very high and a variable speed control is preferable. A range of speeds of 500–1500 rpm is generally used.

(ix) The blower or fan section is a problem area for enrobing, as it generally makes a lot of noise due to air quantity and pressure. However, with quality enrobing, the shaker is now frequently preferred to the fan for weight control. Where it is necessary to severely reduce the weight of chocolate on a centre, a more powerful high water gauge fan with a low air quantity can be used to advantage e.g. 250 mm wg (10″ wg) with 420 m³ h (250 cfm). This type of fan requires anti-splashing guards to help prevent the windows being coated with droplets of chocolate carried away by the air stream.

(x) The surge roller area is important to provide a sufficiently deep bath of chocolate, for bottom coating and shoulder dipping. This area should be fully adjustable so the flow may be cut off or altered, and drainage should be possible with access for cleaning. Usually the surge roller runs at three times belt speed and is 100 mm (4 in) in diameter. These are modifications that can be carried out to the surge roller pickup plate to create a deeper bed of bottoming chocolate for shoulder dipping.

(xi) The top half or hood of the enrober usually contains the suspension equipment for the blower systems, flow pan, heating and ventilating fans. The hood should be sufficiently high above the product to give easy access for cleaning and visual inspection of the process. Windows should be of non-splinterable material and be removable for cleaning.

(xii) If the chocolate recipe is regularly changed, it is often advantageous to install a removable reservoir tank for easy drainage and cleaning out. Some manufacturers will supply these if specified.

(xiii) Piping or decorating equipment can be chosen according to specific requirements. If you are using tank chocolate for decorating, you may fit a suitable pump to feed tempered chocolate to the piping system. A filter or twin filter is advisable for continuity of production since most blockage breakdowns are caused by particles of waste or agglomerates. If the chocolate is pumped from an outside temperer, changes in temper can occur by the time the chocolate reaches the decorating equipment. It is advisable to use this system, however, if a special chocolate is used for the decorator. When decorating equipment is used, the environment must be kept warm.

Having described the principles of the enrobers themselves, it is appropriate to look at some of the best ways to operate them.

11.4.2 Hints and tips on enrobing

(i) Coat at the highest coating temperature that the temper in the system will allow. This gives the maximum chance of obtaining the best fluidity and results. If warm biscuits or centres are feeding into the enrober plant, however, the heat will have to be counteracted by the heaviest temper that the system will provide; usually only long retention time tempering will resist temper loss. Excessive heat will destroy the best in tempering, so there are temperature limits to which tempered chocolate can be exposed: 35–37 °C (95–100 °F) this is also dependent on the type and recipe of the chocolate used. It is also possible to coat slightly warmer biscuits if they are only half coated, at 40 °C (105 °F). It is undesirable to feed very cold products as they will tend to affect even the best viscosity control. It is recommended that the closer the product temperature is to the coating temperature, the fewer the problems will be in the enrober and in the cooler.

(ii) Lower the flow pan as close as possible to the product. This helps to prevent air entrapment which will eventually show as air bubbles in the final product.

(iii) Keep the machine clean by adjusting scrapers and drainage devices internally.

(iv) Make sure the wire belt is only lightly tensioned. If it requires heavy tension there is a fault in the system.

(v) Keep the anti-tailer rod straight and well maintained. The adjustment of this device is critical to quality.

(vi) Change the decorating tube filters regularly.

(vii) If the wire belt continually stretches and breaks before the expected life-time, consider a heavier gauge belt. These are available in a graduated range of diameters, though quality assortment coating demands a fine gauge. Heavy gauge belts are associated with high-speed heavily laden machines.

(viii) Monitor the temperature controls regularly and check the chocolate viscosity.

(ix) For weight control accuracy on half-coated biscuits, align the biscuits to feed into the enrober between the wire belt interlacing sections. If this is not possible and some biscuits fall on the border, order a belt to suit your spacing requirement. This is important to achieve consistent weight control, and a consistent pattern on turned-over biscuits. To achieve weight control accuracy, not only must all aspects of tempering and machine control be stable, but so must the size of the biscuit or sweet centre. One millimetre can make quite a significant difference, and it is often worth calculating the percentage change in pickup with size.

(x) Weight control in most cases is carried out mainly by the blower, shaker and licking rolls. Each must be accurate:
  (a) A uniform air velocity distribution across the width of the machine
  (b) Shaker amplitude and frequency across the width
  (c) Licking rollers and scrapers evenly adjusted to provide constant, and adjustable conditions.

(xi) Select the chocolate characteristics to suit the product: the Casson plastic viscosity and yield value are important factors.

It is not unknown to come across chocolate that cannot be tempered in a particular system. If this problem should be encountered, test by the 'hand method', i.e. take a spatula and hand temper on a cold slab. This is the quickest way to sense the crystallization rate. Experience has indicated that if chocolate is un-temperable, it is probably for one of the following reasons:

(i) The addition of too large a quantity of lecithin or similar surface-active agent
(ii) The addition of too large a quantity of the wrong type of lecithin e.g. some mustard seed lecithins can reduce crystallization speeds
(iii) The addition of too much butter oil (cow fat)
(iv) The addition of any other eutectic additive, e.g. other vegetable fat which in quantity will render tempering impossible.

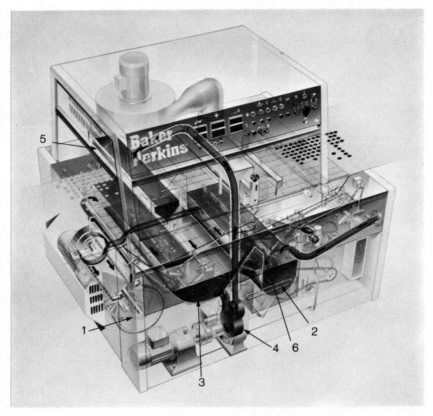

Figure 11.24 A Baker Perkins enrober (Baker Perkins BCS Ltd).

11.4.3 *Enrober operation*

A typical present-day enrober by Baker Perkins is shown in Figure 11.24. This machine is fitted with in-board tempering independently driven (*1*). It contains a built-in hot chocolate reservoir (*2*) alongside the tempered chocolate tank (*3*). An independent pump circulates tempered chocolate to the flow pans, a useful point when flow rates have to be altered. This machine requires the minimum of work and services to install.

Chocolate is drawn in from hot tank (*2*) which is automatically kept full from bulk storage, and is metered continuously through the tempering tube. The flowrate is maintained in excess of usage. The freshly tempered chocolate feeds into the tempered chocolate reaction tank (*3*) for fully maturity to develop. Due to time, temperature, and agitation the chocolate quickly attains optimum conditions. Chocolate pump (*4*) independently circulates the tempered chocolate to the flow pan (*5*). Tempered chocolate equilibrium is maintained by the overflow (*6*), where the excess chocolate falls into the hot tank (*2*) for de-seeding to take place.

A similar principle can also be applied using an out-board temperer connected to a conventional enrober configuration.

11.4.4 *Additions to enrobers for decorating*

In the hand-dipping days, assortments were marked by hand, using hand tools such as seen in Figure 11.25. The well-known piping bag enabled difficult designs to be made, which are not easily repeated by machine automation. Currently we have some quite ingenious machines producing a wide range of design. The Decormatic (Figure 11.26) and the Woody Stringer

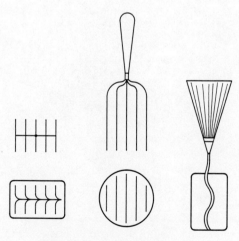

Figure 11.25 Hand decorating tools.

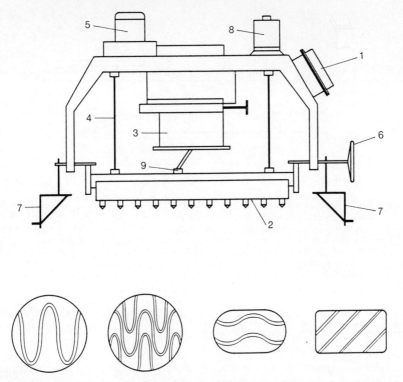

Figure 11.26 (Top) The Sollich Decormatic: (1) control unit; (2) nozzle tube; (3) stroke amplitude adjustment; (4) suspension rods; (5) variable speed motor drive; (6) centralizing adjustment; (7) enrober sideframes; (8) nozzle cleaner solenoid; (9) pattern adjustments. (Bottom) Some typical decorative markings. The Decormatic enables decoration to be applied in which additional chocolate is deposited onto the goods through individual nozzles. A small tempering machine provided for the decorator only tempers the decorating material continuously and conveys it through the nozzle tube. The nozzle tube can carry out a variation of movements, e.g. circular or oscillating movements. In addition the length of stroke or the diameter of the circle can be varied, as can the stroke frequency and the rpm. Various nozzle tubes with optional nozzle arrangements up to a minimum pitch of 4.5% can be supplied for the Decormatic. A trailing wire device can be also supplied. All in all, an almost unlimited number of decorations or patterns can be produced. Because of the independent tempering machine, milk chocolate decorations can be applied onto dark chocolate and vice versa. The nozzle tube is fitted with a patented cleaning device with periodically cleans each nozzle with a special air-operated 'needle' during use.

(Figure 11.27) are two examples of decorating machines, both of which utilize pumped chocolate. (Tempering chocolate for decorating machines follows the same requirements as those needed for enrobers.) Other methods are driven markers of various designs giving a variety of patterns by rotating devices in contact with the wet chocolates.

Decorating usually now takes place in the enrober environment, but it may also take place using the same or a different chocolate, on the infeed to the cooler (the decorating table). During the days of hand-marking teams of girls

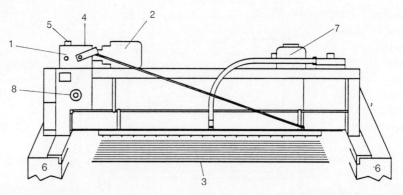

Figure 11.27 The Woody Stringer. (1) Two-speed gearbox; (2) motor driving stringer; (3) piping tube; (4) front crane; (5) rear crane; (6) enrober sides; (7) built-in chocolate filter; (8) motor speed control. The machine is similar to other piping machines. A piping tube (3) is supported freely so that it can move both longitudinally and laterally across the enrober grid and the sweets. The motor (2) drives a gearbox (1) which has two output shafts on which cranks 4 and 5 are fitted. The front crank (4) controls the amount of lateral movement of the nozzle tube (adjustable from zero to approximately 20%). A second adjustable crank (5) is fitted to the other output shaft of the gearbox. The operating point (phase angle) of this crank in relation to the front crank (4) is also adjustable. The rear crank (5) may also be run at twice the speed of the front crank. The speed of the motor can also be varied to adjust the frequency. Two forms of design can be made; single-loop and double-loop.

sat here on opposite sides of the conveyor using piping bags. There were many problems to overcome, one of course being the continual replenishment of tempered chocolate, making hand-marking in fact a very skilful operation.

11.5 Moulding plants

Moulding plants are a method of producing a precision-sized product. As the name suggests, moulds are the primary part of the plant. They can be attached by carrier frames to the circuit-carrying chains. Another method is the loose mould system, whereby the extension pins in the carrier chains fit into specially designed slots in the mould. This method has the added attraction of allowing the mould to be automatically removed from the plant for changing or cleaning. Cleaning is a very necessary requirement, as moulds quickly pick up excess chocolate on the leading and trailing edges of the mould surrounding the sweet. This can cause severe infestation problems when some of this material drops off and falls into the main coolers.

Moulds were initially made of metal, with stamped-out impressions representing a female replica of the required shape of sweet. These impressions were fitted into stamped-out flat sheets with a datum for location and final assembly. The two were sweated together using food-quality pure tin solder, and the necessary very high-quality polished finish was provided by the 'Platinol' layered impression. Metal moulds were very heavy and have now been replaced in most plants by injection-moulded plastic moulds. The latter have the advantage of good demoulding characteristics, and their lightness results in lower drive loads and less noise. New methods of demoulding have been developed, which involve twisting the mould in the demoulding zone. This has almost totally reduced the noise produced by the original high-speed noisy vibrators and hammer devices. Using this method the working environment has greatly improved and the efficiency of the system has increased.

11.5.1 Chocolate contraction or expansion in moulding plants

When chocolate is un-tempered or only partially tempered, or even if mistakes are made in tempering, or (say) the mould heating is too fierce, chocolate will not de-mould. Why is this so? A graph which illustrates the cooling curve and contraction of un-tempered chocolate is shown in Figure 11.28. The cooling curve of un-tempered chocolate is seen to cool quickly, but set slowly, and this is indicated by the position of the inflexion point which occurs at 4 °C (39.2 °F). The time taken to reach this 'latent heat' release point is very prolonged (50 minutes). This graph was produced using cooling water at 0 °C (32 °F). Therefore, the cooling setting point is not due to inefficient cooling, but is an indication of the much lower crystallization temperature and the slower

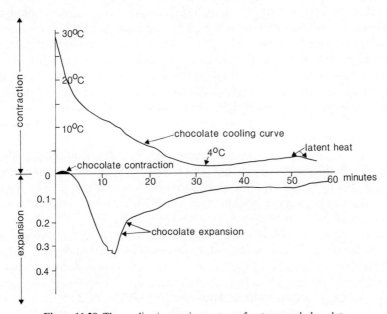

Figure 11.28 The cooling/expansion curves of untempered chocolate.

crystallization rate of un-tempered chocolate. Further examination of the graph indicates that un-tempered chocolate does not contract, but instead has a massive expansion in the early stages. It is not till prolonged fierce cooling takes place that eventual, but minimal contraction occurs. This is why un-tempered chocolate does not de-mould. Tempering for moulding plants is in principle similar to tempering for enrobers, the main difference being that it is usual to temper for a shorter time (10–12 minutes). The depositing temperature is lower, approximately 30 °C (86 °F). These temperatures produce very acceptable results. Cooling principles follow those for tunnel coolers (see section 11.6), but the temperatures can be lower, since the mould acts as a partial insulator. As mentioned earlier, chocolate for moulding plants can generally be of a higher viscosity than for enrobing, but there are occasions when thin shell mouldings are required, and therefore a much lower viscosity chocolate is necessary.

Moulding plants are supplied for solid bars with metering equipment for adding nuts, raisins and many other chocolate-compatible ingredients. Shell moulding plants operate by inverting the chocolate-flooded mould before it has set. This leaves a skin or shell (cup) which on cooling allows for a further deposit to be made in the shell. Liquid toffee may be deposited, or fondant, liqueurs, jellies and biscuits may be placed in the shell by special machinery.

11.5.2 Split moulds

Special split moulds are available for making 'Easter egg' shells; the moulding takes place in a manner similar to the standard shell production by filling the mould cavity full with chocolate, then, after the necessary time, inverting the mould and shaking out the excess chocolate. Shell thicknesses are difficult to control precisely, since the final weight depends on constant conditions and accurate temper viscosity. An in-line viscometer is most useful in pinpointing shell moulding deficiencies. The layout and cooling requirements of moulding plants have generally resulted in multi-tier cooling units (see section 11.6).

(Note: There is another method of producing accurate shell mouldings: this is the spinning method. A split mould is clamped together in a rotary arm through which a metered amount of tempered chocolate is injected, and the mould is then rotated, centrifugally creating a shell in the mould. After a suitable cooling period the mould is opened and the completed shell is ejected.)

Moulding plants have become very complex, and may contain several depositors of different designs to suit the application. These must be spaced apart according to cooling and timing requirements. As present-day plants become larger and faster, mechanically-driven cam-operated depositors have been unable to stand up to the pressure, and drives have been replaced by hydraulically driven systems providing the necessary 'muscle' to overcome the work rate. Depositors may be of the piston type, which are more versatile and more accurate than the rotary gear type and are easily configured to suit complex nozzle layouts. Incidentally, the pressure on the nozzle plate on a plant 1.2 m (48 in) wide can be greater than $18 \, kg/cm^2$ (250 psi).

The rotary gear type depositors are simpler in design and smaller than the comparative piston designs, and are quite sufficient to fill moulds where a biscuit or other centre will be inserted before the final backing off facility.

11.6. Tunnel coolers

In order to cool and set chocolate, it must first be correctly tempered and be in the best possible condition for setting. Optimum temper is advisable, especially when cooling times are required to be as short as possible. Cooling is a combination of conduction, convection and radiation. Coolers for chocolate have taken many shapes and used numerous principles, incorporating many traditional views and practices, not all of which were correct. It is also interesting to consider how to cool enrobed chocolate.

Before good-quality cooling belts and cooling machines were available, assortments were placed on plaques, which were then taken on multilayer trolleys to a cooled room where they were gently cooled over a period of time, then removed and de-plaqued. These methods were labour-intensive, but sufficient cooling time was available to completely set and produce the finest gloss. Bases were completely set and a full gloss was revealed on de-plaquing.

In all coolers, once a centre is enrobed, advanced crystallization must take place without further treatment as the chocolate is setting. It only takes a slightly lower temperature to complete the transition from the plastic to the solid, since tempered chocolate is already partially solid. Despite the fact that chocolate will set quickly in open cool air, we do need coolers. These coolers provide a standard environment and can give all-round efficient cooling at production manufacturing rates when their settings are correct.

To cool and solidify chocolate properly it must first be allowed to cool gently, in either radiant or gently moving air conditions. It is recognized that chocolate leaving the enrober must not be subjected to fierce cooling, as this has the effect of drawing the cocoa butter up to the surface of the product, quickly resulting in fat bloom.

The second stage of cooling may be by forced cooling at mild temperatures (13 °C, 55 °F), or by convection/radiation. Zonal cooling is found to be advantageous when cooling loads are high. This means that the coldest part of the cooler can be situated to counteract the latent heat release point in the chocolate cooling cycle. Note that the specific heat is smaller than the latent heat, so large quantities of energy are suddenly released after the initial cooling phase, hence the cooling curve measurement. Temperatures in a jet turbulence cooler may then as low as 10 °C (50 °F), giving very effective cooling with jets at 28 mm. wg (1.2″ wg) 1300 m/min (4300 ft/min). It is inappropriate in this chapter to go into details of cooler design, or to refer to psychrometric charts relating to dew points. To determine the minimum safe temperature of

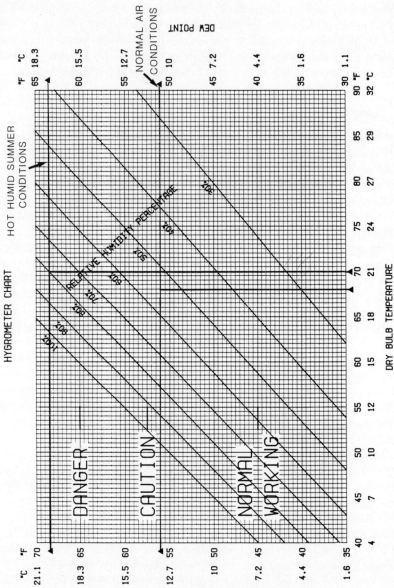

Figure 11.29 Hygrometer chart (from a Lesme original).

products leaving a cooling tunnel or moulding plant cooler, Figure 11.29 can be used, however, as an alternative to calculations. Assuming that the packing room conditions are 16 °C (68 °F) at 55% RH take a vertical point from the dry bulb base line at 16 °C (68 °F) up to the 55% RH line as shown in the example in Figure 11.29. Draw a horizontal line at the dry bulb and RH intersection point to both sides of the graph. Then read the dewpoint scale on the right-hand side of the graph which is reading approximately 10.5 °C (51 °F). This reading is the dewpoint and is the condensation-forming temperature of any part or product in contact with air at 16 °C (68 °F) and at 55% RH.

The advised minimum safe temperature of goods leaving the cooler is read off the left-hand scale; this temperature is 13.36 °C (56 °F). Therefore, providing the room humidity does not rise appreciably above the control limits, there should be no moisture deposited on products entering the packing room atmosphere. A word of caution is necessary: moisture may appear on the surface of products actually inside a cooler of the radiant plate type. This can happen if the radiant plate is below the dewpoint of 10.5 °C (51 °F), even though dehumidifier systems are fitted. For instance in a hot, 'humid' summer of 20 °C (70 °F) and 80% RH, and when the packing-room air conditioner cannot cope with the load, the radiant cooling plates condense moisture which falls as droplets on to the product. The result is that the product suffers a dulling off on the surface (the moisture being almost imperceptible).

This problem does not exist in turbulent air coolers since the cooling heat exchanger is nearly always located underneath the product zone where all condensed moisture is trapped and released to drain. Calculation examples illustrate how to calculate air conditions in a cooler using the dewpoint formula (see Table 11.2). If the air temperature in a cooler is below the dewpoint, moisture will precipitate in the cooler and on the product. This results in a dull finish and the likelihood of sugar bloom occurring.

A booklet is available from the author for calculating cooler refrigeration requirements, insulation losses and flow rates and pressures.

Should we cool by radiation or contact, or by air stream (convection)? There is a controversy over which method is the best for cooling chocolate, since both methods work. Engineers can, however, calculate the percentage cooling by each principle, and, surprisingly, radiation cooling removes only approximately 7% of the cooling load (radiation is only effective when with massive temperature differences exist, in this case, between the product and the plate temperature: see Table 11.3). It is recognized that conduction or contact cooling is the most efficient, because it has almost intimate contact with the product. Calculation shows that the base area of a product provides a large area for contact cooling which plays a major role in the overall cooling. It can, on some low-profile products, take 50% of the cooling load.

Forced convection cooling is the next most efficient form of cooling, since the top and sides of the product are in the air stream. This is analogous to the weatherman's 'chill factor', i.e. the quicker removal of heat due to wind speed.

Table 11.2 Dewpoint

(i) *To calculate the air off coil temperature at dewpoint to a design RH*
Design requirement is 45% RH with a dry bulb temperature of 13 °C (55 °F) after re-heat.
The SVP of water at 13 °C (55 °F) = 11.2 mm of mercury. The actual vapour pressure in air of
45% RH is 45/100 × 11.2 = 5.04 mm (dewpoint).
The figure of 5.04 mm is the SVP of water at 1.5 °C (34.5 °F). Thus if air is first vapour
saturated at 1.5 °C (34.5 °F), and then warmed to 13 °C (55 °F), it will be at the correct
design temperature and humidity, therefore the air off the coil = 1.5 °C (34.5 °F).
Saturated vapour pressure of water in mm of mercury

Temp. ° F	° C	SVP	Temp. ° F	° C	SVP	Temp. ° F	° C	SVP
32.0	0	4.6	44.6	7	7.5	57.2	14	12.0
33.8	1	4.9	46.4	8	8.0	59	15	12.8
35.6	2	5.3	48.2	9	8.6	60.8	16	13.6
37.4	3	5.7	50	10	9.2	62.6	17	14.5
39.2	4	6.1	51.8	11	9.8	64.4	18	15.5
41.0	5	6.5	53.6	12	10.5	66.2	19	16.5
42.8	6	7.0	55.4	13	11.2	68	20	17.5

(ii) *To find the % RH from dewpoint and the air temperature*
Example: air temperature is 17 °C (62.6 °F) (SVPT) and the dewpoint is 10 °C (50 °F)

$$RH = \frac{SVP\,(Temperature)}{SVP\,(Dewpoint)} \qquad \begin{array}{l} SVP\ at\ 17\,°C = 14.5\,mm \\ SVP\ at\ 10\,°C = \ 9.2\,mm. \end{array}$$

Therefore relative humidity $= \dfrac{9.2}{14.5} = 0.634$ or 63%

Dewpoint = (RH) × (SVP Temp.)
   = 0.634 × 14.5 = 9.2 mm = 10 °C (50 °F)
Check: air at 17 °C (62.6 °F) and 63% RH
The vapour pressure of air at 63% RH and at 17 °C = 63.4/100 × 14.5 = 9.195 mm. (from
table)
Therefore the dewpoint = 9.195 mm = 10 °C.   (from table)

(iii) *Infiltration*
Leaks in and out of a cooler can cause a serious 'latent' load on evaporator coils. This effect is
termed 'latent' due to latent heat being absorbed, condensing moisture on evaporator coils in
a cooler. This is because the cooling coils are below the dewpoint, thus condensing moisture
from surrounding air. The magnitude of this condensation depends on the amount of the leak
and the relative humidity, i.e. say at 60% RH and a temperature of 21.1 °C (70 °F) there will be
4.8 grains per cu. ft. Note − 1 lb of moisture = 7000 grains. In calculations we normally allow
for 10% of total air circulation to be infiltration (this is a significant amount.)

Fast-moving air and even slow-moving convection currents are more efficient
than radiation cooling, allowing operation at higher temperatures, and
condensation problems can be completely avoided.

When cooling chocolate the process cannot be hurried too fast, otherwise
poor results will be obtained. What then should the cooling time be? The
cooling time is limited by the temper, and the type and thickness of the
chocolate coating, and is also determined by the base quality desired.
Assuming the temper is at the optimum, then the time will vary only according
to the type of chocolate. Recipes in some countries are so varied that trials may

Table 11.3 Comparison between effectiveness of radiation cooling and convective air cooling at 0 °C (32 °F)

Temperature of object		Heat radiated to absorber @ 0 °C (32 °F)		Heat transferred to cooling air 240 m/min (800 fpm) at 0 °C (32 °F)	
C°	F°	W/m²	BTU/ft²/h	W/m²	BTU/ft²/h
(a) 1648	3000	788500	250000	34063	10800
(b) 1093	2000	195548	62000	22708	7200
(c) 537	1000	22078	7000	11354	3600
(d) 315	600	6781	2150	6781	2150
(e) 121	250	1009	320	2523	800
(f) 25	78	*126	*40	*630	*200

Radiation cooling is extremely effective at large ΔT (temperature differences): (a) shows a difference of 23 times the efficiency of moving air. However, under (d) with the object at 300 °C the efficiencies are equal, and still ΔT is large.
Once ΔT drops below the 270 °C level, 'convection' air cooling becomes more efficient.
Line (f) is a realistic comparison in a chocolate cooler and clearly indicates that moving air at the above rates is approximately five times more efficient. Present-day jet coolers with air velocities 1234 m/min–1520 m/min (4000–5000 fpm) are even more effective than 'radiation' cooling.

be required. Contact cooling, although excellent, can be deceptive—only surface cooling may have occurred and contact cooling, and a problem occurs on removing the product from the belt. Even though the surface cooling gives an apparent set and glossy appearance early in the cooling cycle, the base condition is unseen and is slow to harden off fully. Inferior base finishes with gloss and dull patchy areas result when using cooling times of 6–10 minutes when 20–30 minutes are required. Standards have, however, relaxed generally in the confectionery industry, and the base quality has been reduced (the base finish determines the overall cooling time: if the base is set, then the whole product is set).

Two types of cooler are shown in Figures 11.30 and 11.31. These are the Gainsborough Craftsman, Baker Perkins cooler and the Sollich contact radiation cooler. The Sollich cooler requires a de-humidifier system since the radiation plates can be below the dewpoint, especially in humid summers. This system has a fan in the cooler, so it also becomes a forced convection cooler. There remain, however, limitations on the plate temperatures. It is interesting to note that radiation coolers are never used on non-coated confectionery because they are not efficient enough in practice.

11.7 Multi-tier coolers

The design of a tier cooler arose from the need to cool and pack chocolate in a limited space. For instance, a three-tier cooler for any given length gives approximately six times that length, so cooling time increases sixfold. In early

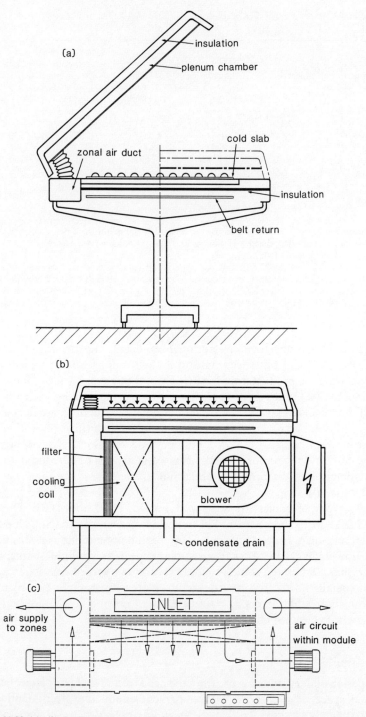

Figure 11.30 (*a*), (*b*) Sections through the Gainsborough Cooler; (*c*) Gainsborough Cooler, showing air circuit.

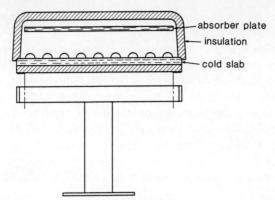

Figure 11.31 The Sollich cooler.

designs, chocolates on plaques were carried by hand to ambient cooling units, which were then located on metal trays suspended from the carrier chain circuit. The chain circuit runs continuously but may be stopped at will. To enable the trays to stay horizontal, special turning-point mechanisms assist the trays into the next track run of chain. This machine was the forerunner to the in-line process multi-tier cooler now in use. This allows a plant to be very compact, yet provide 140 m (460 ft) of cooling, a vast saving in factory space. The main difference between tunnel coolers and multi-tier coolers is that chocolates feeding to a tier cooler are placed on discrete plaques of metallized or similar foil, whereas with tunnel coolers chocolates feed directly on to a continuously moving endless belt which is a simple conveyor system. Further complexities with the multi-tier cooler are:

(i) The plaques have to be attached to the moving tray circuit, giving a timing and location problem.

(ii) Since the tray chain circuit and plaque circuit were separate for feeding and de-plaquing reasons, a special rotary or reciprocating 'knife-edge' is required at the transfer between the enrober and tier cooler. This allows the plaque carrier rod to pass up through the transfer gap before the knife-edge returns to its transfer position. Since the knife-edge has to retract, no sweet transfer can take place during this period, therefore some output is lost, nearly 10% in some cases, depending on how well a plaque can be filled.

(iii) Chocolates therefore have to be run through the enrober in batches with the space between them coinciding with the plaque feed at the transfer point. This has resulted in several ingenious batching and timing devices.

(iv) Once sweets are transferred onto the moving plaque, the plaque then has to become attached to the moving tray circuit, another difficult timing problem to overcome, and one which is never completely free of supervisory maintenance.

(v) The advantage of a tier cooler, its zigzag circuit, also posed another problem, that of keeping the tray with its plaque-laden load of coated sweets horizontal, with the minimum of mechanical vibration and disturbance. Similarly, all tiered-type coolers with moulds or trays require special turning points or star wheels which are necessary to pick up on tray guidance rollers as the tray is conveyed into the next tier. This method of guidance keeps the trays horizontal, but always represents a jamming area with questionable reliability due to metal fabrication inaccuracies and wear and further mechanical distortion once a jam has occurred.

(vi) Once cooling has taken place, the plaque has then to be stripped off the tray, timing mechanisms pull the plaque by its carrier rod past a retractable knife-edge (the reverse of the feeding procedure), the knife-edge then returns to the transfer point, and transfer takes place on to a separate conveyor belt. As the plaque is pulled off its carrying tray, chain lugs carry the plaque back to the feed end to continue the cycle.

(vii) Cooling in a tier cooler can be zoned, but care has to be taken to prevent empty plaques being moved by countercurrents of air and therefore causing jams.

It can be realized, therefore, why multi-tier coolers are less popular than the much more simple tunnel coolers, which have far fewer maintenance problems. The cleaning and hygiene regime is also hard to maintain in a tier cooler.

References

1. Vaeck, S.V. 'Cacao and fat bloom'. *14th PMCA Production Conf.*, Pennsylvania (1960).
2. Duck, W. 'The measurement of unstable fat in finished chocolate'. *18th PMCA Production Conf.*, Pennsylvania (1964).
3. Kleinert, J. *CCB Rev. for Choc. Confect. and Bakery*, (1980) (March) 19–24.

12 Vegetable fats

G.G. JEWELL

12.1 Introduction

The first use of vegetable fats in chocolate occurred around the turn of the century. The Danish company Aarhus Oliefabrik recorded in 1897 the first delivery of a lauric fat, stearine, to a Danish chocolate manufacturer. During the same period the British company Loders and Nucoline was also supplying 'butter substitutes' based on lauric stearines. The chocolate manufacturers were active in looking for alternative fats with similar properties to cocoa butter and had utilized such materials as Borneo Tallow and Illipe when it was available.

What prompted the search for suitable vegetable fats? The explanation is related to the high cost and variability in supply and quality of cocoa and hence cocoa butter. The possibility of 'substituting or extending' the expensive cocoa butter with a cheaper vegetable fat offered considerable financial savings.

Although the detailed triglyceride chemistry was not appreciated at this time, it was recognized that one of the principal physico-chemical attributes of cocoa butter was its sharp melting point at body temperature (e.g. in the range 33–36 °C, 91–97 °F). Hence early studies concentrated on finding fats with a similar melting performance. The most common vegetable fats at the turn of the century were coconut and palm kernel. It was found that when such oils were shipped from the tropics into Europe in so-called 'Ceylon pipes'—which were really large wooden barrels—the oils 'set up' in such a way that the higher melting fractions solidified into relatively large crystals.

These crystals were separated by wrapping the fat in woollen 'bags' which were subjected to hydraulic pressure. The liquid 'oleine' was forced out from between the crystals and filtered away, leaving behind the high melting crystals or stearines. This is the process which was practised by European fat suppliers to provide the early stearine or hard butters to the confectionery industry. Although the stearines did generally have a melting point in the range 35 °C (95 °F), they also suffered from two major drawbacks. Firstly, it was found that when these 'lauric-based' fats (rich in the C12 fatty acid lauric acid) were mixed with cocoa butter they formed eutectic mixtures which have a lower melting point and hence give a markedly softer chocolate. The second problem was that in the presence of the enzyme lipase (which could come from such diverse sources as cocoa powder, milk powder, recipe ingredients such as desiccated

227

coconut or even microbial contamination) the lauric fats are hydrolysed to liberate fatty acids, which impart soapy off-flavours to the product. For these reasons the early lauric stearines were generally associated with 'cheap' chocolate, since the recipe required minimal cocoa butter (to avoid softening) and low-fat cocoa powder instead of mass.

The next stage in the development of better vegetable fats occurs following the advances in fat hydrogenation techniques after World War. I. Many different types of hydrogenated coconut and palm kernel oil were produced, and although the hardness was increased this was also associated with the production of some very high-melting stearines (44 °C, 111 °F) which would not melt at blood heat and so left a waxy/chewy residue in the mouth. Nevertheless, these hydrogenated fats have found application in some types of chocolate and coatings and other confectionery applications, e.g. caramels and toffee.

The final breakthrough to being able to produce a range of highly compatible vegetable fats required two developments which both took place during the 1950s. The first was the greater understanding of glyceride chemistry, which permitted the identification of the unique glycerides of cocoa butter which dictate its desirable physical behaviour. The second was the scale-up of solvent fractionation techniques from a laboratory to full manufacturing process.

The significant increase in the understanding of the chemical composition of oils and fats arose from work in both universities and industrial laboratories in Europe and the USA. 1957 saw the publication by workers from Unilever (1) and from the USA (2) concerning the symmetrical mono-oleoglycerides of cocoa butter. With the recognition that 2-position mono-oleoglycerides were the dominant class of glycerides in cocoa butter came the appreciation that for

Table 12.1 Natural vegetable fats containing POP, POS and SOS

Generic name of plant source	Common name of source	Fatty acid composition (wt%)			Glyceride composition		
		Palmitic (C16)	Stearic (C18)	Oleic (C18:1)	POP	POS	SOS
Burseraceae							
Dacryodes rostrata	Java almond	11	40	44	✓	✓	✓✓
Dipterocarpaceae							
Shorea species.	Illipe butter	21	39	38	✓	✓✓	✓✓
Vaterica indica	Dhupa butter	10	39	48	✓	✓	✓✓
Guttiferae							
Allan blackia sp.	Kagne butter	3	53	44		✓	✓✓
Garcinia indica	Kokum butter	3	56	39		✓	✓✓
Pentadesma butyracea	Kenya butter	5	46	48		✓	✓✓
Palmae							
Elaeis species.	Oil palm	42	6	40	✓✓	✓	
Sapotaceae							
Butyrospermum parkii	Shea butter	7	39	49		✓	✓✓
Mimusops njave	Njave butter	4	36	58		✓	✓✓
Palaquium oleosum	Siak butter	6	54	40		✓	✓✓

a vegetable fat to be compatible it would also require 2-oleo-monoglycerides. This in turn explained the failure of many of the early lauric stearines and hydrogenated stearines, since they did not contain the requisite 2-oleoglycerides and were thus incompatible. A search for oils which contained the right glyceride types produced a list (Table 12.1) which revealed several drawbacks. Firstly, it is a fairly short list, but more importantly, with the exception of the oil palm, none of the remainder are plantation crops, they are merely indigenous tropical species and therefore offer little hope of significant regular commercial exploitation. The oil palm was relatively abundant, but the use of straight palm oil with cocoa butter was again found to lead to eutectics and significant softening of the chocolate.

Again, in 1957 Loders and Nucoline developed a pilot-scale (20 tons/week) solvent fractionation process based on acetone. Using this approach it was possible to remove the so-called mid-fraction from palm oil and blend it with a source of 2-oleoglyceride fat (from Table 12.1) and produce a vegetable fat which had a chemical composition containing the prerequisite 2-oleo-monoglyceride, and in turn a physico-chemicial behaviour very close to cocoa butter. Unilever called the product they developed Coberine, and the material became commercially available from 1960, receiving wide acclaim from chocolate manufacturers. A more detailed description of these historical aspects of vegetable fats may be found in (3) and (4).

The next 25 years have seen further activity in the development of types of vegetable fats. The driving force over this period has been the significant increase in our understanding of relationship between chemical composition and physical properties, particularly in the areas of polymorphism and tempering, and also improvements in physical and chemical fractionation techniques. The rest of this chapter will consider the various classes of vegetable fats now available and discuss aspects which control their compatibility.

12.2 Legislation

The inclusion of up to 5% of a vegetable fat in chocolate is currently permitted only in the UK, Eire, Denmark, and Japan. There are no detailed specifications enforced by government for the type of vegetable fat, other than it is suitable for human consumption. The type of fat used (at a maximum of 5%) is left totally to the discretion of the manufacturer in relationship to the type of product manufactured. Elsewhere in the world a product may not be called chocolate if it contains a vegetable fat. This has led to the adoption of 'chocolate-flavoured' types of description for products. The EEC has been debating for many years whether or not to permit vegetable fats in chocolate. In order to enforce such a relaxation of current EEC Regulations requires a strict definition of the types of fat permitted or excluded and also of analytical procedures to ensure the legitimate usage of any such materials. After a period of much confusion over nomenclature (5, 6), there have emerged two distinct classes of vegetable fats—CBEs (Cocoa Butter Equivalents) and CBRs (Cocoa Butter Replacers). Some manufacturers choose to subdivide their range of products within each category (7).

From detailed discussions within the EEC came both a tentative definition of a CBE and also analytical procedures to ensure verification (8). The EEC definition of a CBE is a vegetable fat which complies with the following criteria:

(i)   Level of triglycerides type SOS $\geq 65\%$ (S = saturated fatty acid, O = oleic acid)
(ii) Fractions of the 2 positions of triglycerides occupied by unsaturated fatty acids $\geq 85\%$
(iii) Total content of unsaturated fatty acids $\leq 45\%$
(iv) Unsaturated fatty acids with 2 or more double bonds $\leq 5\%$
(v)   Level of lauric acids $\leq 1\%$
(vi) Level of *trans* fatty acids $\leq 2\%$.

These criteria are totally related to chemical composition with no physical specification, and as such will preclude all of the lauric or hydrogenated fats that could and are currently being used by the UK industry.

12.3 The CBE class

After the introduction of Coberine, several manufacturers offered similar products, which were in turn followed by extended product ranges claimed to be more tailormade to a particular customer's requirements. Why should there have been the scope to develop and offer such a wide range of CBEs? The explanation lay in the composition of the rest of the fats phase of the chocolate. As previously discussed, chocolate contains 30–35% fat; in a dark or plain chocolate the cocoa butter will probably account for 95% of this fat with the remainder being fats added for their anti-bloom properties (e.g. butter oil). In milk chocolate, however, up to 30% of the fat could be butter oil with the balance being the cocoa butter. UK legislation permits up to 5% of the chocolate to be vegetable oil, which since the fat is approximately 33% of chocolate, means that up to 15% of the total fat phase would be vegetable oil, this accordingly reduces the cocoa butter content of dark chocolate to 80% and for milk chocolate to 55%. As described in Chapters 2 and 13, cocoa butters vary in composition and physical properties depending upon geographical origin and time of harvest. Hence as different combinations occur in individual manufacturer's recipes, such as soft Brazilian cocoa butter in low milk fat milk chocolate, or a hard West African cocoa butter in high milk fat chocolate, it was possible to develop a more or less suitable vegetable fat for a particular application.

Both chocolate manufacturers and fat suppliers put a lot of effort into developing methods to understand the physical behaviour of these complicated mixtures of fats. Numerous techniques have been developed and proposed for predicting fat phase behaviour and thus ensuring a satisfactory performance of the vegetable fat in the recipe. Early results were based on cooling-curve performance (Chapter 11) and measurement of dilation (i.e. contraction and expansion with changes in temperature) whereas more recent methods have involved the use of NMR (nuclear magnetic resonance) to measure how solid/hard the fat phase is, coupled with x-ray diffraction, DSC (differential scanning calorimetry), and electron microscopy to study crystallization and polymorphism (Chapter 10).

The key to achieving good performance in terms of the fat phase in a chocolate is that, when subjected to a tempering regime (see Chapter 11) the fat phase readily solidifies to give the correct quantity of fat in the correct crystallographic form. The tempered chocolate is then formed (i.e. moulded, enrobed, hollow goods etc.) and then cooled to give the finished piece. Again during the final cooling regime the tempered chocolate must solidify to the correct level of solids of the correct crystal form.

It is well established that cocoa butter displays a complex polymorphism with up to six distinct crystallographic types. The tempering process is normally recognized as achieving 2–5% seed of the β-crystal type which subsequently solidifies as further β-types, to give a stable glossy product with good snap and shelf-life. The work from Unilever previously described correctly attributed cocoa butter's physical performance to the dominance of the 2-mono-oleoglyceride type and recognized that a compatible vegetable fat would require a similar glyceride type which would also require tempering. What these early workers did not fully appreciate was that relatively low levels of other glycerides or partial glycerides could markedly influence the physical behaviour of the complex fat mixture. These interfering species could come from the vegetable fat itself, milk fat or even an added emulsifier.

Rossell (9) reported a method for describing triglyceride interactions based on isodilatation, a process whereby dilatation results from either classical dilatation or SFI (Solid Fat Index) results from NMR are plotted as a series of contour lines, and sharp peaks or troughs in the blend composition v. solids curve denote regions of low solids and hence incompatibility. Paulicka (10) used a combination of conventional phase diagrams coupled with x-ray polymorphic behaviour to follow mixed fat compatibility. He confirmed the requirement that a CBE should exhibit β-type crystallization for compatibility with cocoa butter, since β′ - tending fats usually lead to solid disruptions in the presence of cocoa butter which frequently resulted in bloom formation.

Further studies on the behaviour of both CBE and CBR fats were undertaken (7, 11, 12, 13). Jewell (12) found that no one single technique was capable of adequately predicting performance of a vegetable fat in a recipe. Furthermore, a study of the vegetable fat alone provided little insight into how it would perform in the complex mixture of fats in a typical recipe. The preferred techniques from these studies were NMR to study level of solids, a tempering test to compare ease of tempering and x-ray diffraction for confirmation of polymorphic form. These results showed that even in many β-tending CBE systems the level of solids could vary significantly, which could lead to an unacceptably soft chocolate. As previously described, a typical CBE consists of the triglyceride POP (usually obtained from palm oil by either acetone or other solvent fractionation) and this is then blended with a fat or mixture of fats rich in POS and SOS. Typical sources of SOS have been fractionated shea oil, while POS and SOS can be obtained from *Shorea robusta* (Sal) Illipe etc. In very simple terms, the higher the content of SOS the harder the fat will be; conversely the higher the content of POP (the cheapest component), the softer the fat will be. However, very high contents of either POS or SOS can lead to tempering problems whereby the POS/SOS preferentially crystallize as the seed component and may lead to a high viscosity in the tempered chocolate (this can be particularly troublesome in enrobing operations), and problems in subsequent setting on cooling. High levels of POP lead to a fat which has a β' crystallization tendency and this causes problems of softness and tempering.

It is clear that the level of solid fat and the rate at which it is achieved is a critical parameter for chocolate manufacture, so there has been much debate on the type of preconditioning the sample should receive before measurement (14, 15).

In conclusion, CBEs have been successfully applied in UK, Denmark, Eire and Japan for over 25 years, and indeed, a partial list of CBEs offered by the major European suppliers is given in Table 12.2. The proposed EEC definition of a CBE which is based solely on chemical composition will exclude some fats which have been previously found to exhibit perfectly acceptable physical properties.

Table 12.2 Selected CBEs from European suppliers

Name	Supplier
Coberine	Loders and Nucoline
Choclin	
Illexao range	Aarhus Olie
Superit	Friwessa
Akomax E	Karlshamns

12.4 The CBR class

This class may be divided into two broad groups, the lauric types (based on coconut and palm kernel oil) and the non-lauric (usually based on cottonseed or soya bean oil). Both groups are usually fractionated and the non-lauric are frequently hydrogenated as well. Further sub-definitions have been discussed by Hogenbirk (7), Wilson (15) and Paulicka (16).

Under current UK legislation up to 5% of such CBR fats could be used and the product could still be called chocolate. Under the proposed EEC definition of a CBE, CBRs could not be included in a product called chocolate. Traditionally, in the UK most chocolate manufacturers have found that including 5% of a CBR in a typical UK chocolate formulation would lead to problems with tempering, product softness and possibly bloom formation. The CBRs have found their place in producing good-quality lower-cost coatings using specially developed recipes.

Virtually all of the CBRs are so-called non-tempering fats, since they solidify from the melt directly into a stable β' form. This feature avoids the expense and complications of tempering facilities required by chocolate and CBE-containing chocolate.

The European market developed the lauric fats, both fractionated and hydrogenated and fractionated products being available. The USA market developed on domestic soya and cottonseed oil, usually with hydrogenation, which leads to high levels of *trans* acid isomers, e.g. elaidic instead of oleic acid. Both markets now offer both lauric and non-lauric types. The fractionated palm kernel fats exhibit very low compatibility with cocoa butter (4%) and thus all recipes need to be used on low-fat cocoa powder (7). The fractionated non-laurics are claimed to be compatible at up to 95% cocoa butter, whilst the fractionated high-*trans* acid-hydrogenated fats are said to be compatible at 20% cocoa butter (16). When used in a correctly balanced recipe, the CBRs of either type will give a product with excellent mouldability, melt, snap and contraction.

As mentioned in the introductory section, all lauric-based fats run the risk of developing soapy flavours if adequate attention is not paid to good housekeeping. Further, since there are considerable compatibility problems between CBRs and cocoa butter, great care must be taken to avoid cross-contamination through storage tanks, pipework etc.

However, with the appropriate attention to detail and the wide range of

Table 12.3 A selection of CBR fats

Name	Supplier
Shokao	Aarhus Olie
Cebes range	
Kaomel	Durkee
Satina	
Ivora Noua	Friwessa
Wesco range	
Ako Nord range	Karlshamns
Ako Ext range	
Hycoa	Loders and Nucoline

good-quality products available (Table 12.3) very acceptable products can be produced from CBRs. Indeed in some cases only an expert tasting panel can distinguish a CBR product from a true chocolate.

A further complication has arisen with the introduction of the term 'supercoating'. Supercoated products contain a CBE, usually at levels of 50% or greater substitution for cocoa butter. The high level of inclusion means they cannot be regarded as chocolate, while the fact they are based on CBEs tends to given an advantage in terms of mouthfeel and flavour release, but the disadvantage of requiring a full tempering operation. The supercoatings are gaining popularity for enrobing operations where the products tend to be identified by their brand name rather than a specific association with being chocolate.

12.5 Quality control

Vegetable fats, like all food ingredients, should have an adequate specification to ensure that the product received is of the agreed quality and hence will perform correctly.

The specification for vegetable fats is usually a combination of chemical, physical, microbiological and organoleptic parameters. For most CBEs a precise chemical composition in terms of triglyceride types is not specified, since the fats are usually blends of several oils. However, limits of particular fatty acids are usually specified, and indeed if the EEC definition is accepted a more rigorous composition will be required. The fatty acid and glyceride types are readily checked by GLC or HPLC (16).

More important parameters will be those which indicate adequate pretreatments (e.g. refining, deodorizing etc.), moisture level, free fatty acids and peroxide levels. Also, atomic absorption can be used to check for levels of trace inorganics. The important physical tests will be for melting behaviour, solid fat index and cooling or solidification curves. The microbiology aspect will specify absence of pathogenic species such as *Salmonella*. The organoleptic aspect will specify a bland product, with a good colour and no foreign odour.

In the case of CBRs which have been hydrogenated, an iodine value is specified, and in some cases the level of *trans* acid isomers is declared.

12.6 The future

There are two big questions for the future. Firstly, will the EEC permit the use of CBEs in chocolate, and if this happened would the USA follow? If the answer is yes to either market then there will be a significant increase in the quantities of CBE produced. This in turn identifies the second major issue, which is the availability of the exotic oils to produce CBEs at both present or any increased level of demand. It is not expected that there will be any shortage of the base raw material, i.e. palm, oil mid-fraction. However, the make-up oils are subject to wild fluctuations in availability and hence price. There was significant shortage of shea nuts in the 1985/86 crop, with a consequent increase in the cost of CBEs.

The Illipe forests are being logged out in Borneo as new agricultural schemes are developed. The Sal crop in India is relatively abundant in the tropical forests, but until access is improved only a small percentage of the nuts can be gathered before the monsoon causes them to germinate. There would appear to be no conventional short-term solution to the supply of exotic oils. However, certain companies are looking to a biotechnological source of exotic oils. Two routes are being explored, an enzyme interesterification process, and a microbial lipid route. Both processes have been proven technically possible at the small scale; it remains to be seen whether they can be turned into an economic reality.

However, it is not yet clear whether or not the EEC will approve CBEs. If Europe does not it is highly unlikely that the USA will, so the supply and demand equation may be balanced by traditional means in the foreseeable future.

Acknowledgements

I am extremely grateful to the following manufacturers of vegetable fats who kindly supplied information to me. Aarhus Olie, Denmark; Durkee Industrial, USA; Friwessa, The Netherlands; Karlshamns, Sweden; Loders and Nucoline, UK.

References

1. Chapman, D., Crossley, A. and Davies, A.C. *J. Chem. Soc.* (1957), Part II, 1502–1509.
2. Lutton, E.S. *J Amer. Oil Chem. Soc.* **34**(10) (1957) 521–522.
3. Whetherall, R.L. *Tech. Circ.* **166** (1959) BFMIRA, Leatherhead, UK.
4. Anon. 'Selected glycerides by continuous solvent fractional crystallization'. *Industrial Chemist* (1963) (August) 401–405.
5. Wolf, A.J. *29th PMCA Production Conf.* Pennsylvania (1975) 82–87.
6. Meara, M.L. *Tech. Circ.* **697** (1979) BFMIRA, Leatherhead, UK.
7. Hogenbirk, G. *Manuf. Conf.* (1984) (June) 59–64.
8. Padley, F.B. and Timms, R.E. *J. Amer. Oil Chem. Soc.* **57**(9) (1980) 286–293.
9. Rossell, J.B. *Chem. and Ind.* (1973) 832–35.
10. Paulicka, F.R. *Chem. and Ind.* (1973) (Sept) 835–839.
11. Gordon, M.H., Padley, F.B., and Timms, R.E. *Fette Seifen Anstrichmittel* **81** (1979) 116–121.
12. Jewell, G.G. *35th PMCA Production Conf.* Pennsylvania (1981) 63–66.
13. Jewell, G.G. and Bradford, L. *Manuf. Conf.* (1981) (Jan.) 26–30.
14. Shukla, U.K.S. *Fette Seifen Anstrichmittel* **85** (1983) 467–471.
15. Wilson, L.L. *Manuf. Conf.* (1985) (Oct.) 49–55.
16. Shukla, U.K.S., Schiotz Nielsen, W., and Batsberg, W. *Fette Seifen Anstrichmittel* **85** (1983) 274–278.

13 Recipes

K. JACKSON

3.1 Tastes in different countries

When Columbus sailed to the New World, the flavour of 'cocoa' in one form or another was already established, if one could accepted a mixture of burnt cocoa nibs, water and other spices. In man's search for 'different tastes', this must have been one of the worst discovered. However, the addition of sugar to the mixture by the Spanish was the turning point of acceptance of this product and a direction for the future in all countries.

High fat levels in the drink (cocoa butter), even with the sugar, was also a deterrent to an acceptable product. It was not until the 1828 invention of the cocoa press, by Van Houten in Holland, that a more acceptable product was forthcoming.

Van Houten's introduction of the cocoa press made quantities of cocoa butter available, so that 'eating chocolate' could be made as an alternative to 'drinking chocolate'. After 1828 the first moulded 'dark chocolate' tablets (bars) were made. J.S. Fry and Sons of Bristol, England (now part of Cadbury's) is generally considered to be the first, in 1847, to produce large quantities.

Milk chocolate was a much later invention. Literature attributes it to Daniel Peters of Vevey, Geneva, Switzerland in 1876 (1) and links it with two aspects of the milk industry in Switzerland: (i) a surplus of milk in the country, and (ii) the development of methods for the preservation of milk by the Nestlé Co. Milk crumb originated around 1918–1920 in Britain, for the same reasons as the Peters/Nestlé invention, i.e. a surplus of milk in the country and also a need to produce a flavour change (2). It can be said that the world of chocolate as we now know it started from these simple beginnings.

13.1.1 Milk chocolate

Defining tastes in different countries obviously has to be given in broad strokes, as to be too defined will only lead to controversy. In general, milk chocolate follows the following lines.

(i) *Europe:* Generally a chocolate where a good milk flavour predominates and a mild but clean cocoa flavour is required (i.e. a fresh milk flavour as opposed to a cooked or heated milk flavour).
(ii) *Great Britain* (UK): The predominant flavour is the traditional rich, milk caramelized 'crumb' flavoured chocolate.

236

(iii) *North America:* Follows a stronger, cleaner cocoa flavour than Europe, with an earthy background from the milk content and a somewhat sweeter taste than European chocolate. Processed milk contributes the familiar 'barnyard' or cheesy flavour in the US chocolate.

(iv) *British Commonwealth countries:* Generally follow the UK type of flavours because the major manufacturers follow their parent companies. This influence so predominates that local companies have changed to UK type chocolate. Imports obviously create or set the tone.

(v) *Other countries:* Obviously major manufacturers or parent companies have set the tone in the other countries too, with the introduction of imports and then, perhaps, of manufacturing centres.

13.1.2 Dark chocolate

Here the differences between countries are much smaller.

 (i) *Europe:* A bitter chocolate with good rounded cocoa flavour and fruity background notes, derived from the careful blending of 'flavour' cocoas. along with the so-called 'filler' beans or 'bulk cocoas'.

(ii) *Great Britain:* As Europe, but a little sweeter in flavour.

(iii) *North America:* Good cocoa flavour, somewhat lacking in the fruity notes found in European chocolates, due to the use of South American beans and different roasting techniques.

(iv) *British Commonwealth:* Follows generally the UK/European types because of the origin of the companies and their imports.

(v) *Other countries:* Generally follow parent countries.

An overall view would be that economics (using the closest sources of beans) dictated which beans were used originally in various countries, and because 'tradition' is still extremely strong in the industry, there has been and still is a reluctance to change and experiment. Colonization also had a great effect on the types of and quality of beans used: Britain and France obtained their source materials from thier own colonies. Finally, availability, quality and marketing all have had their part to play in the flavours of chocolate.

The quality and quantitiy of the types of cocoa beans used can sometimes be affected by factors outside the industry. For example, the change in Nigeria from an agricultural to an oil-based economy led to a decline in cocoa production in that country. Conversely in Malaysia the reduced demand for natural rubber resulted in a massive increase in cocoa production. These and other changes have had an effect on the availability of cocoa and hence had an influence on the 'flavour' of the products.

Marketing has also played a major role in the development of the types of chocolate currently available. We have only to look at the major manufacturers and suppliers of chocolate and chocolate products to see how successful they have been in the marketing of their products by use of the printed word and other visual media, radio and television, both in the countries of origin as well as in the countries to which they export. Well-known names include Hershey, Wilbur, Nabisco, General Foods, Rowntree Mackintosh, Cadbury, Mars, Terry's, Van Houten, DeZaan, Suchard-Tobler, Lindt & Sprüngli, etc.

We must also acknowledge the influence of regulatory authorities and consumer groups, which today influence what 'chocolate' and 'chocolate products' may and may not contain. This influence is mainly 'after the fact', because many products have been manufactured for a good number of years. It is generally the industry and government working in co-operation that sets and determines the standard, which may in turn affect the flavours of the products.

In the consumer's eyes, taste has to be synonymous with texture. As a teacher and lecturer the author is often asked, 'Which is the best chocolate in the world?' and 'Why is European chocolate different from English and North American chocolates?' Student's own replies to these questions provide some insight into how they perceive the differences. One great overwhelming difference is texture: 'European chocolate is perceived as much smoother than either English or North American chocolate.' This reply leads one to believe that the 'total' properties of chocolate have to be considered, not merely taste and colour, etc. In the USA and Britain, chocolates are refined to on average between 20–30 μm (8–12 × 10^{-4} in), and European chocolate to between 15–22 μm (6–9 × 10^{-4} in). This is not to say that all chocolate fits this profile.

Chocolate refined to over 35 μm (1.4 × 10^{-3} in) can generally be detected as 'gritty' and 'sandy' and is not very well accepted. Exceptions could be chocolate for filled bars or enrobed bars where the centres influence the 'texture' more than the chocolate. Conversely, chocolate and coatings under 15 μm (6 × 10^{-4} in) are considered 'greasy' and 'clinging' and are not generally acceptable as a good eating chocolate.

13.2 How to get different tastes

To answer this question could perhaps be linkened to 'dancing on quicksand', but common sense should prevail, based on several factors such as (i) the relevant market; (ii) the product one wishes to sell; and (iii) the cost of the chocolate or 'compound' one wants to make. For the purposes of this chapter, compounds or products other than those generally recognized or accepted as chocolate are now included. *Chocolate* is defined here as follows:

  (i) *Dark*: Sugar, cocoa solids, cocoa butter, anti-blooming agents, lecithin (emulsifiers), and flavours
 (ii) *Milk*: Sugar, milk solids, milk fat, cocoa solids, cocoa butter, lecithin (emulsifiers), and flavours.

Compounds are defined as above, but the fat portions are not cocoa butter, or not all cocoa butter (depending upon local legislation). In the author's view, there is a general trend in the industry towards two distinct product groups: (i) chocolate; and (ii) non-chocolate. Both groups are very important to the economics of the industry. Factors affecting taste (flavours and textures) are common to both.

Considering basic materials then, we should study each one and its effect on flavour.

13.2.1 *Sugar*

Sugar is generally assumed to be an inert ingredient with regard to the subtleties of flavour, contributing only to sweetness, but it should not be so easily dismissed in chocolate making. A change of 1-2% in the sugar content has a great effect on costs; sometimes economic factors increase this to 5%, at which stage large flavour changes become very evident.

In dark chocolates and compounds, sugar is added for flavour purposes to offset the bitterness of cocoa solids, and can also have an effect on processing techniques. The effect of sugar in milk chocolate, particularly crumb-based chocolates, and to a lesser degree in milk-powder-based chocolate and compounds has a great influence on the finished flavour. Milk crumb technology is based upon making a condensed milk, adding cocoa mass (liquor) and then drying the mixture under vacuum. Although most of the processing is done continuously under vacuum, milk protein and sugar in intimate contact and in the presence of heat undergo a chemical change (Maillard reaction) and the masse takes on a mild caramelized flavour.

This reaction can take place in the conching process above 40 °C (104 °F) (Chapters 7, 8), and is dependent upon the temperature and time of conching, and to some extent, the moisture content which can affect the reaction. Although the effect on caramelized flavour is much less than in the crumb-making process, it is nevertheless an option which is used.

13.2.2 *Milk*

As milk chocolate is the most popular of all the types of chocolate, and its name highlights one of the ingredients as 'milk', it is not surprising that the consumer has a preconceived idea as to what they are expecting *before* they taste the product. This preconceived idea means that the chocolate is either accepted or rejected after the first taste. The question of the best milk chocolate, and the best kind of milk to use in a milk chocolate, has been the source of many debates in the industry over the years.

Liquid milk, from which all milk products are derived, has a flavour with which we are all familiar. However, almost any processing applied to milk alters its flavour. In all products made from milk, the flavour obtained may arise from several factors. These should be considered first before using, as they

J

can have an appreciable effect on the character of the finished chocolate.

Liquid milk rapidly takes on sour notes, especially if left unrefrigerated. Speed of processing and hygiene are paramount factors in milk processing, and sterilization is also a major contributor to flavour change.

In every case milk and milk products used in chocolate manufacture have to be dehydrated before use, and doing so requires that heat be applied in the process. Applying heat to milk during any of the dehydration processes will affect the finished flavour of the milk. One of the debates in the industry over the years has concerned the differences between the flavour of roller-dried powders and spray-dried powders. If one takes 'run-of-the-mill' dried powders irrespective of processes used, then flavours will and can vary from dairy to dairy or process to process, due to the differences in types and times of process and the temperatures at which the milk was processed. It is important therefore, that all chocolate manufacturers have either approved suppliers and flavours, or their own processing plants.

Perhaps three distinct milk flavours can be picked out in milk chocolate, following these broad guidelines: (i) fresh milk; (ii) matured milk; (iii) cooked milk.

Fresh milk takes advantage of capturing the flavour of fresh milk and keeping it as 'fresh' as possible by rapid low-temperature vacuum processing and spray drying, but it can still have 'cooked milk' notes due to the drying process.

Matured milk can be enzymatically or microbiologically changed, or ripened. The flavour has been referred to as 'barnyard', 'cheesy' etc. Matured milk has been used in the USA for many years. The type of enzymes or microbiological process used give different-flavoured milk powders. This should be taken into consideration when the powder is either manufactured in-house or purchased in order to guarantee continuity.

Cooked milk denotes the flavours obtained when fresh milk is cooked in the presence of sugar to create condensed milk. This can be used in the chocolate-making process as such, and the whole dried out during processing or more commonly during manufacture of 'crumb'. White crumb can be made *without the addition of cocoa mass*. More usually, milk chocolate crumb is manufactured, which incorporates cocoa mass in the process.

These 'crumbs', especially the latter, are greatly used in and manufactured in Ireland, Britain and Commonwealth countries, and are the basis of the 'caramelized' flavour in the chocolate of these countries. The percentage sugar addition and types of cocoa mass used have a very great effect on the finished flavour of crumbs. It can be seen that variations on the above can be accommodated, as no doubt they are in some companies.

There are also many claims made that each product and hence process, produces the best chocolate, not only from a flavour point of view, but also from a processing point of view, i.e. roller-dried milk, being more flaked than spray-dried, is less absorbent to fats, and therefore requires less cocoa butter

Table 13.1 Sources and characteristics of cocoa beans

Major sources	Characteristics
Brazil	Acidic in nature with 'smoke bacon' notes
Nigeria	Generally good all-round beans but lack distinctive flavour
Ghana	Similar to Nigeria but flavour considered in some circles a little better than Nigeria
New Guinea	A fine-grade aroma bean but quality varies between estates
Trinidad	A good-quality bean with some particularly fine flavour—has varied over the years
Venezuela	Some good-quality flavours (Criollo), but crops of lower grade have been known
Ivory Coast	A variable quality crop, weaker in flavour than Nigeria or Ghana
Malaysia	Very acidic in nature with weak cocoa flavour
Cameroun	Different flavours to Nigeria and Ghana–is variable and smokiness is prevalent
Sierra Leone	Is generally under-fermented and has green notes because of this.

(fat) in processing. Amorphous sugar is present in fresh crumb to a greater extent than in roughly ground powdered sugar. When the two types of chocolate ingredients are refined, different proportions of fine particles are produced, thus requiring differing amounts of cocoa butter. Nevertheless, the best form of milk product to use to make the best milk chocolate is still open to debate.

13.2.3 Cocoa beans

This constituent has the biggest potential influence on the flavour of all chocolate products. This may seem simplistic, but considering the botanically different beans, the countries of origin, the growing conditions and cultivation, and the differences in handling and fermentation processes before the beans ever arrive at the factory (plant) door, one can easily see the point that is being made. Considering the countries of origin and the generally accepted flavours and aromas from each (Table 13.1, 13.2), we can begin to get an idea of the complexity of the problem (see Chapter 2 and ref. (3)).

As mentioned previously, economics in the countries of origin, supply and

Table 13.2 Bulk and flavour cocoas

Some base or bulk cocoas	Some flavour and aroma cocoas
Accra (Ghana)	Ceylon
Bahia (Brazil)	Arriba (Ecuador)
Cameroun	Java
Nigeria	Caracas (Venezuela)
Grenada	New Hebrides
San Thome	Samoa

demand, wars and strikes all have taken their toll on the quality, flavour and aroma of cocoa beans over the years. Some countries, for economic reasons, have moved away from the exportation of beans to the exportation of liquor (mass), and hybrid beans which create high yields when processed into liquor have been planted. The flavour of these beans (liquor/mass) is not necessarily the one required for chocolate manufacture, but may be acceptable for cocoa powder processing.

It is clear from these facts that chocolate flavours and aromas have changed, however subtly, over the years and, except for the surviving 'art' of the blending and roasting techniques of chocolate artisan, the flavours of today are not those of chocolates of past years.

Unfortunately, science has not yet succeeded in producing successful synthetic chocolate flavours, as in many other food products, because of the complexity of defining flavour and aroma. However the industry has not altogether lost its artisans: the best flavour and aroma machine (man) is still alive and well.

The main botanical types of beans are Forastero and Criollo, which differ in flavour and aroma. These should be considered in arriving at the finished flavour. *Forastero* is considered the general filler-type bean, giving the characteristic strong cocoa flavour, but inclined to be bitter. *Criollo* is the variety supplying the fine-grade flavours. They may be variable but are generally light-coloured, and contribute a mild and nutty flavour not found elsewhere. Hybrids considered scientically as Forastero types, but exhibiting a different flavour, are Nacional and Arriba beans, which contribute distinctive aromatic floral notes—see Chapter 2 and ref. (4). As mentioned, use of specific

Table 13.3 Examples of cocoa mass (liquor) blends

Milk chocolate	%		Arriba	25	
			Trinidad	25	
African	65	Mild roast	Surinam (Guiana)	25	
Bahia	35	Full roast	Caracas	25	
			Dark chocolate		
African	30	Medium roast		%	
Estates Trinidad	30	Medium roast	African	50	Full roast
La Guayra Caracas	40	Medium roast	San Thome	10	Medium roast
			Bahia	40	High roast
Caracas	50				
Guayaquil (Ecuador)	50		African	50	Medium roast
			Bahia	25	Full roast
Caracas	50		Caracas	25	Mild roast
Bahia	50				
			African	40	Mild roast
Ceylon	48		San Thome	10	Full roast
Java	24				
Caracas	28		Estates Trinidad	20	Full roast
			Seasons Arriba	30	Mild roast

beans from different countries affects the characteristics of chocolate in the countries of use for example, West African beans are used in Britain, whereas Brazilian beans are used in the USA.

Some interesting uses and blends of beans from different countries are recorded in the cocoa mass (liquor) blends in various publications, (Table 13.3).

As can be seen bean/mass blending is very important and totally reliant upon several factors: (i) bean quality; (ii) operator expertise in blending beans and roasting; (iii) the equipment available to the operator.

This brings us to the next phase of flavour and aroma development in cocoa beans which is the real key to chocolate flavour, namely roasting. Elsewhere in the book (in particular Chapters 5, 7) the differences in types of roasters and their chemical effect on the beans have been discussed. In essence, a roaster is a method of:

(i) Using heat on beans in order to affect a flavour change by the removal of water

(ii) Development of flavouring components which enhance the taste and aroma; changes in texture also occur so that the shells can be removed (bean roasting only)

(iii) Changing the colour of the beans (nibs), which has an effect on the appearance of the finished chocolate.

Research makes it obvious that there are different schools of thought on the best methods of roasting, the best temperatures, etc. Kleinert (5) describes three segments to the process: drying, hot air treatment and roasting, and as stated, it is impossible to define these segments by temperature alone.

Another misconceived idea is that the temperature of the air is the temperature of the bean. Kleinert (5) makes reference to work by Heiss, Mohr and Gorling and also by Dimair, Acker and Lange, which demonstrates considerable differences between air temperature and bean temperature. Other parameters in this equation are the time the beans are in the roaster and the moisture content of the beans. This emphasizes the role of the operator and the art of chocolate making.

A very general range of air temperatures for the three parts of the roasting process is as follows:

Drying   100–110 °C (212–230 °F)
Heat treatment   110–150 °C (230–308 °F)
Roasting   125 °C–200 °C (257–392 °F)

It may be said that the finer (Criollo)-type flavour beans should be processed at the lower temperature ranges, and the general filler beans at the higher temperature. Once again, this is only a very general observation because a great deal of flavour development (or better yet, modification) takes place in the conching process. Therefore, in the blending of beans or mass, the degree of roast has to be taken into consideration along with all the other variables.

A more modern approach to flavour development, in order to overcome some of the inadequacies of bean roasting, is thin-film cocoa mass roasting, which is described in Chapters 5 and 8. Suffice to say it is a relatively new process, but has great potential to 'even out' the flavour of roasting, and hence the flavour of finished chocolate. Again, time and temperature of roasting and the origin and quality of beans from which the mass is produced still have an effect on the finished flavour.

A modification of this method is the cocoa mass treatment plant for the treatment of 'origin liquor' (or mass) produced in the country of origin and shipped in block or chip form to the chocolate manufacturer. The treatment modifies the 'roasted' flavour so that a reduction in conching time, or elimination of conching, can take place. A by-product of the process is sterilization of the mass.

A third alternative is the roasting of nibs. This again is said to 'even out' the roasting, and all the anomalies of bean origins are said to be eliminated, as is also claimed by mass roasting and treatment.

Both nib and mass roasting systems require some type of heating/drying of beans, in order to facilitate the removal of the shell from the nib before further processing is carried out. This in itself may give rise to some flavour changes.

Both systems require the artisan's skill, as with standard bean roasting systems, because Nature does not supply the same raw materials twice. Although this new technology is beginning to open up possibilities for the reduction and/or elimination of conching, its effectiveness remains to be seen in practice.

If we find that roasting is complicated in the development of flavour and aroma, then conching must be considered both mysterious *and* complicated. In research on conching, many claims are made as to what really happens in the process. The only positive evidence (obtained by taste testing and to a lesser extent by chemical analysis) is that it modifies the flavour by removing the undesirable volatile acidic components remaining after previous processing (Chapter 7). It also homogenizes the mass, effecting the viscosity and removing moisture. Changes in particle shape are sometimes claimed, as well as reduction in agglomerates.

Flavour changes or modification are affected by the following:

  (i) Type of conche
 (ii) Temperature of conching (product and jackets)
(iii) Time of conching
(iv) Exposure to air
 (v) Blend of ingredients in conche.

Ley (6) describes two tendencies in conching: low-temperature conching at 45–55 °C (115–130 °F), called 'cold conching', and conching at relatively high temperatures of 70–80 °C (160–175 °F), called 'hot conching'. Both can be applied to dark and milk chocolate. Lees and Jackson (3) give similar temperatures: milk chocolate, 45–60 °C (113–140 °F); dark chocolate, 55–85 °C (131–185 °F) for British tastes and up to 100 °C (212 °F)* for European tastes. Chocolate intended for high-temperature conching should be given low-temperature roasts, and mixes containing spray-dried milk should be conched at up to 5 °C (9 °F) higher than those containing roller-dried milk. No doubt claims for temperatures outside these ranges can be made. Ley (6) also points out that conching is still a highly cherished, well-kept secret, and that each manufacturer through experience has developed a product which satisfies customers' flavour requirements.

The duration of conching is very variable, ranging in literature and papers from 1 hour to seven days. The latter applies to untreated cocoa mass, so that all flavour modification is carried out in the conche; the former, on the other hand, relies on fully treated mass and controlled ingredient additions, and whether the method can be claimed as conching as we know it is still open to question. The average period of conching generally runs between 8 hours (one working shift for expediency's sake) and three days (72 hours), the shorter times for milk chocolate and the longer times for dark chocolate. The deciding factor is the percentage of cocoa mass in the chocolate, and the treatment it has undergone before conching.

13.2.4 Flavours

This section deals with the flavour additives which can be or have been added to chocolate. As discussed earlier, most 'standards' for chocolate have been set in co-operation between companies and government agencies dealing with food, and therefore most flavour additions have been used for many years.

The most common flavour addition is vanilla, in the form of the natural vanilla bean or the now more commonly-used artificial vanilla flavour, vanillin, either in the methyl or ethyl form (the general level of addition lies between 0.06% and 0.09% of the batch for milk chocolate, and can be up to 0.1% for dark and bittersweet chocolate). Salt is perhaps the second most common addition.

Both of these are added to enhance the flavour rather than mask it, the vanilla or vanillin creating creamy notes, whereas salt accents the clean crisp notes. Many spices such as nutmeg, cinnamon, etc., have been added, in order to bring out or enhance certain notes. Here again, the artisan's skill has considerable scope.

Attempts over the years to replace cocoa butter, cocoa mass and cocoa powder in chocolate by creating 'compounds' which taste the same as chocolate have led to a flurry of activity by flavour chemists to produce 'chocolate' flavours. None of these have been totally successful, for one reason

---

*This figure given by Lees and Jackson is higher than would now normally be used, but does illustrate the wide range of conditions which have been investigated

or another, but in recent years, flavour profiles have been obtained during the conching of chocolate and on cocoa butters (7, 8) and show the complexity of the situation we are dealing with. To capture these flavours in a bottle is obviousely a considerable challenge.

This does not mean that work should not continue. However, one mistake can easily be made: due to the removal of cocoa butter from a formulation, not only will the setting and texture characteristics change, but also its flavour characteristics. Bearing in mind that added cocoa butter alone can represent between 5–25% of total weight, depending upon the type of chocolate, immediately this 'flavour' addition is removed, this proportion has to be made up in some other way. This involves either adding more cocoa mass or cocoa powder and/or changing the degree of roasting, etc.

Some of the simpler additions to chocolate to modify the flavour are, of course, sugar and dairy products which we tend to accept as being 'normal' additives. However, without these additions and their quality control, the flavours we know today would be missing.

Any flavour additions must be acceptable in the country of use, and flavours which are considered adulterations should be checked before the chocolate is manufactured or shipped to other countries.

13.3 Dietetic and low-calorie chocolates

Very little has been written on this subject; sugars are discussed elsewhere in this book (Chapter 4) and so here, I will comment only on their use.

As the name implies, dietetic and low-calorie chocolate is used by those persons who are (i) calorie-conscious, or more importantly, (ii) who are diabetic and must limit their intake of sugar, primarily sucrose, but also dextrose, and other carbohydrates.

Some good-tasting products have been made (a great benefit to those who cannot eat the normal type of chocolate) and are sold in several countries. The restricted demand tends to limit their manufacture.

The main ingredients in dietetic chocolate, along with the customary cocoa mass, cocoa butter, milk powders, etc. are mannitol, maltitol, xylitol, fructose and sorbitol. All have been used alone or in combination, and sorbitol (1) has been used in conjuction with saccharin and a proportion of nut meats to produce an acceptable dietetic chocolate. Here the nut meats help to reduce the cost and provide 'bulk' in the batch, because the price of sorbitol is several times the price of sugar (1).

Excellent dietetic milk chocolate has been made in Germany, and two recipes using other sugars are as follows (courtesy of Ch. Kruger, Finnsugar GmbH, Hamburg).

	No. 1	No. 2
Cocoa mass	10%	10%
Cocoa butter	25%	25%

Skimmed milk powder	13.5%	20.5%
Butter fat	4.5%	4.5%
Maltitol	39.6%	—
Xylitol	7.0%	—
Fructose	—	39.6%
Lecithin	0.4%	0.4%
Vanillin	0.02%	0.02%

The biggest problem, as has been suggested, is to find a product which has the bulk of sugar yet has little or no caloric value. The price to the eventual consumer also has to be considered. The industry should try to fill this void in order to attempt to address the current world of 'slim', 'lite' and 'low-calorie' challenges from other products. When current fashion changes the industry must be ready to address any changes that may be thrown its way. Perhaps now is the time for such a challenge.

Minifie (1) discusses the addition of products such as carboxymethyl cellulose, which when ingested creates bulk in the stomach on the absorption of water, thus reducing appetite, pointing out that incorporation of this product into normal chocolate production creates no difficulties.

No matter how one looks at the situation, any reduction in fat increases the carbohydrate content, and vice versa. Therefore, the only possible change to be made is a low-calorie filler, which has little or no side effect such as reduction in appetite, loss of 'good' flavour, etc.

Any claim that increased air in the product reduces calories is spurious, because this depends on the weight of the product ingested and not the volume. Chocolate, aerated or not, has the same number of calories per gram.

Dietetic chocolate is generally sold in the form of a solid piece, but can be used in conjunction with dietetic centres to create enrobed and moulded centre-filled products—in fact, a whole product range could be devoted to dietetic and diabetic 'chocolate' lovers.

13.4 Comparison of chocolate for moulding, enrobing, panning and chocolate drops

In this section several factors are taken into account with regard to the flavour and texture of chocolate (milk, dark or bittersweet) according to its uses: (i) solid eating chocolate; (ii) moulded chocolate with centre filling and/or additions; (iii) enrobed products; (iv) panned products; and (v) chocolate drops.

(i) *Solid eating chocolate.* Either milk or dark chocolate in this form is the real test of the artisan's work. It must bring to the eater the ultimate in chocolate flavour and texture and fulfil any claims made by advertising on its wrapper or by its appearance. Whether the piece is 'thin' or, in the latest format, 'thick', the important consideration is the fat content. The thinner and more defined the piece (mouldwise), the higher the fat content (up to 35–36%) and the smoother (15–20 μm, 6–8 × 10^{-4} in) the chocolate has to be to give the highest impression of quality. 'Thick' or 'chunky' bars are generally considered to deliver good flavour but not necessarily smoothness, nor be too rich in fat content (30–32% fat and 25–35 μm, 10–14 × 10^{-4} in). Lower fat contents tend to distract from good eating.

(ii) *Moulded chocolate with centre filling and/or additions.* Here one has to consider not only the viscosity required to create a 'shell' and get definition from the mould, but strength of the chocolate (fat content) when considering the ratio of chocolate to centre material. Again, the newer system of 'centre in shell', based upon 'one-shot' depositors, demands very similar viscosity of chocolate and centre. Therefore, control of fat content and particle size in order to achieve a correct viscosity is important. Current technology in 'one-shot' depositing gives 60% chocolate and 40% centre, and thus this becomes one of the controlling factors.

Another consideration is the texture of the centre. In this category are included additions such as nuts, fruit, cereal and wafers, etc. When they are relatively rough, is it necessary to carry out extra work to ensure that the texture of the chocolate is smooth (15–20 μm, 6–8 × 10^{-4} in) or should we consider flavour only? This also holds true for centre-in-shell products, most of which are fondant- or caramel-based, and some other products. We may think of the example of alcohol-based liquor-filled chocolates with coarse sugar crystals in the crusted liquor centres. This contrasts strongly with the chocolate shell which is very smooth and 'continental' in texture. One also has to consider fat content of the chocolate, which greatly affects 'overall eat' of the product.

(iii) *Enrobed products.* A very wide range of products covered by chocolate constitutes this category. All are, however, restricted by certain limitations due to the centre itself and/or the equipment used. For example, the viscosity of chocolate required in order to allow it to flow over the product and then be partially blown off for weight control purposes is one; the smoothness of the centre being covered is another. A nut-coated centre generally needs two coats of chocolate to fill in the cavities, whereas a smooth starch-deposited cream or jelly, with no definition, requires perhaps only a single coat. Another criterion is the 'total eat' as defined by the manufacturer.

(iv) *Panned products.* These include nuts, jellies, caramels, fruits, etc. Here again, the texture of the product controls the amount of chocolate added (have you tried to fill the wrinkles in a raisin?) and the 'total eat' of the product. Equipment should be taken into consideration, in as much that the chocolate may be added by ladle or spray gun. Each method demands a different viscosity (which is fat content-related) of chocolate.

(v) *Chocolate drops*. The criterion here is one of low fat content (25–28% in general) in order to make the defined shape. The chocolate used is almost invariably dark chocolate, because of its total contribution to flavour follow-through in the cookies or cakes, etc. Milk chocolate does not deliver the chocolate flavour required in the finished product, and the milk components tend to overheat, giving objectionable tastes. Size of pieces is also very important to the user, because invariably cookies are deposited or rotary-moulded. The cookie shape and equipment in this case defines the size of the piece.

All of the categories above can be rationalized, and it is the author's experience that fat content from 28% up to 36% can be used for any product or method one wishes to choose. The deciding factor is then what 'total eat' the manufacturer wishes to create, and the limitations of the equipment he uses. The only exception to this is 'spinning' (where the chocolate is in a rotating mould) where a high-fat-content chocolate (low in viscosity) is essential in order to create hollow goods with no surface air bubbles and mould defects. This also gives the products strength. However, equipment and techniques have changed the situation and shell moulding is pre-empting spinning as a technique.

The following ranges give some idea of fat content in each of the categories. These must be taken only as very general guidelines, however, and not hard and fast rules.

Solid eating chocolate: 29–36% (dark or milk chocolate)

Moulded chocolate with centre filling: 28% up to 35% (dark or milk chocolate)

Enrobed product: 29% to 36% (dark or milk chocolate)

Panned product: 28–32% (dark or milk chocolate)

Chocolate drops: 25–29% (generally dark chocolate).

It should be noted that:

(i) Where there are pre-set standards for fat contents in different chocolate then, of course, these must be adhered to

(ii) The fat contents are based upon 'tempered chocolate', and the degree of temper (temper viscosity) has a great effect on end usage (Chapter 11)

(iii) Equipment limitations may also influence the fat content of a product—equipment and cooling tunnels of 20–40 years ago were not nearly so efficient as today's more modern machines, nor were the old metal mould as efficient as the latest plastic ones.

Centre-filled products, and products including ingredients by addition, must show some of the flavour characteristics of the centre and/or addition. Chocolate then becomes an integral part of the overall flavour and texture and plays a rather supporting role. Chocolate chips or drops must be strong in flavour to overcome the flavour and sweetness of the cookie, cake or ice cream in which they are used; the coldness of ice cream particularly retards the release of flavour. Clearly the *art* of chocolate making once again comes to the fore, so that a chocolate can be tailormade to the product or products.

13.5 Compounds

In the previous sections we have been defining the use of chocolate as such, but there is really no reason why 'compounds' cannot be used where applicable. Indeed, in many instances, they are used today to great advantage.

Compounds are used in some sections of the industry in cookies, cakes, icings and ice creams, etc., to bring out more of a chocolate flavour, generally using cocoa powder or high-roast liquors. Also, because 'tempering' may not be necessary, there is frequently a decided mechanical advantage over 'chocolate' to some users.

There is also a price advantage over chocolate, as generally cocoa butter is more expensive than the other fats used in compounds. Manufacture of compounds is generally cheaper and requires less energy, since less processing is required compared with chocolate. Frequently they do not require as much refining and/or conching as they are often less sophisticated in flavour and texture. The last statement is not a criticism of compound, because all the subtleties and nuances of chocolate making can be applied to compound manufacture. When new products are being developed, one must simply compare the possible advantages or disadvantages of using chocolate or compound.

All the major fat suppliers that make specialty fats for compounds provide recipes and methods of manufacture. These should be adhered to in general, because some of the fats can only tolerate low-fat (9–11% cocoa butter) cocoa powder, whereas others can tolerate cocoa mass. Whole milk powders and/or milk fats in many instances cannot be tolerated. Therefore further restrictions apply, as the use of skimmed milk powder tends to limit some of the 'milk' flavours which can be achieved when using cocoa butter.

As with chocolate, the art of compounds-making is required in order to tailor-make a product for a specific end use. Manufactures have not only to consider the taste of the compound, but also the texture with regard to plasticity or hardness in the finished product. When the finished product is at the point of consumption, very hard compounds on a soft sponge cake-type centre, for instance, could splinter and crack off when being cut or eaten. Therefore, a softer fat is required in order for it to become an integral part of the finished product. Ice cream coating is another good example. Shelf-life is generally considered to be lower when products contain compounds rather than chocolate, because they lack the natural anti-oxidant properties of cocoa mass. In many instances- lipase-free milk powders and cocoa powders must be used in order to avoid the risk of soapy rancidity found when using most compound fats.

There are, of course, non-lauric fats which are not susceptible to soapy rancidity. These fats tend to be compatible also with cocoa butter; therefore cocoa mass can be used, producing coatings with far better flavour characteristics than with cocoa powder only. The cost of these fats tends to be between that of cocoa butter and general compound fats, and this, together with the fact that cocoa mass is often used with them, tends to raise the cost of the product.

13.6 Coatings

13.6.1 *Wafer, biscuit and cake coatings*

Here the considerations must be viscosity, flavour, and texture (both fineness and plasticity). There is a tendency towards very thin coverings on some wafer, biscuit and cake products and also towards cheapness. This is because all these products are light in weight for volume, and any heavy covering of chocolate or compound rapidly puts up the cost. This in turn results in the use of 'hard fats' which tend to 'crack off'. Flavour has to be compatible with the centre of the product, so careful blending of ingredients has to be considered. Due to the shorter shelf-life of this type of product, the cheaper ranges of fats can often be used.

In some cases, straight cocoa mass moulded into blocks or chips is sold for household use in cakes, biscuits, icings and chocolate sauces. Where rich chocolate flavour and smooth texture are required, high-quality dark or bittersweet chocolates are used. An example is the 'Schwarzwalder Kirsch Torte', not complete without the above together with copious amounts of Kirsch.

Fineness of the protect can in most instances be coarser than that of good eating chocolate—30–50 μm ($12–14 \times 10^{-3}$ in) can be considered. Much less refining time and energy is required, which again represents savings in the cost of manufacture.

13.6.2 *Ice cream coatings and inclusions*

Again, any fat may be used in chocolate-flavoured covering for ice cream bars and novelties. Because of the hardness and plasticity of frozen cocoa butter coatings when eaten, soft lower melting point fats are preferred, such as 24 °C (76 °F) coconut oil and cottonseed oil. One of the characteristics of an ice cream coating is that, because it sets up within seconds of application, little or none of the coating drains off. Control over viscosity is essential therefore, so that costs do not soar owing to over-use of the coating.

Coatings must be formulated with very strong chocolate flavours in order to compensate for the reduced flavour release at low temperatures, and also for the fact that between 50–70% of the coating is fat. The flavour may be accentuated by use of stronger-flavoured, high-roast cocoa beans (Bahia and West African). A cheaper class of coating is made using cocoa powder to replace the cocoa mass, with a corresponding decrease in strength of chocolate flavour.

Inclusions in ice cream are generally in the form of chocolate chips of various sizes, or slivers of chocolate or compounds. Fat contents can greatly be reduced in chocolate chips (25–28%), but flavour strength must be paramount to clearly identify the chocolate quality to the consumer.

13.6.3 *White coatings and pastels*

'White chocolate', as it is commonly known, is made using all the conventional ingredients of chocolate except for cocoa solids. Therefore, the predominant flavour is one of varying strengths and flavours of milk powders.

In the past, cocoa butter was used as the fat medium, and is still used where white, milk and dark chocolates are layered together or blended (stratified) to create interesting specialty confections. However, only highly deodorized cocoa butters are acceptable flavourwise. If poor cocoa butters are used, the flavour of the milk powders is overpowered by cocoa butter flavours. Therefore, when other hard fats and CBEs became available, cocoa butter was replaced, the flavour of milk powder predominated, and a whole array of white coatings and pastels become practical, from a flavour point of view. This in turn led to flavoured and coloured coatings such as lemon and orange, with powdered acids added in order to accentuate the flavours. Peppermint, coffee and butterscotch are also popular flavours. White coatings and pastels can be manufactured on conventional chocolate-making equipment, omitting the conching process or using the conche as an homogenizer. Specific pastel manufacturing equipment is now available, but equipment should at all times be kept separate from conventional chocolate-making equipment, (i) to avoid cross-contamination of colours and flavours, and (ii) to avoid cross-contamination of fats, causing eutectic problems (Chapters 10, 12).

White coatings and pastels are susceptible to oxidative rancidity due to lack of protection afforded by the natural antioxidants found in cocoa butter and cocoa mass. In addition, like all compounds, they are open to lipolytic randicity, and all milk powders used should be lipase-free in order to avoid this problem.

13.7 Recipes

Typical recipes follow which include milk chocolate, dark chocolate, milk crumb chocolate and compounds. The wide range of possibilities found in researching various recipes and formulae has been reduced by giving a 'norm', and highs and lows around these norms based upon nib or cocoa mass content for the different types of chocolate. Some formulae from books by E.J. Clyne (1955) (9), C. Trevor Williams (1950) (10), and H.R. Jensen (1931) (11) are included as reminders of past usage, and of how few changes have been made in formulation. Rather more changes have been made in equipment, however, in order to reduce cocoa butter (fat) content and reduce processing time.

When comparing formulae for North American, British and European chocolate, the author could find no definitive differences in percentage composition terms. This suggests that the major regional differences (apart from 'crumb' chocolate) are in milk flavours, degree of bean roasting or cocoa mass treatment and blending (which can be very different), and fineness of the finished chocolate. Even if there were definitive differences between individual manufacturers' recipes for chocolates which could be printed here, it would be inappropriate to do so—it is important to maintain the art and mystique of chocolate making, in order that all companies should be able to survive by maintaining their individual characteristics.

The section on compounds quotes from a Unilever Manual (12). This interesting article is used as an example only, and not as an endorsement for Crocklaan Fats or Unilever products above any other products. The fat content of these compounds can, of course, be varied to suit any application. The percentage composition of different types of fats and their tolerances can be ascertained from the recipes which follow.

13.7.1 Milk chocolate

Analytical determinations made by the Jensen (11) of a number of well-known brands of milk chocolate are shown in Table 13.4. Nos. 2, 3 and 6 were popular Swiss makes, and No. 4 was German. As these are all high-grade chocolates, stearin fats were assumed to be absent in calculating the milk fat. Beyond that, some error is possible from the use of average fat constants.

The ideal balance for true milk chocolate flavour lies between 2 and 6, with

Table 13.4 Milk chocolate composition (parts by weight) (moulded bars)

	Cocoa nib	Sugar	Non-fat milk solids	Milk fat	Total fat	Milkiness index*
1	40	40.5	12	4.5	29	1.4
2	29	37	13	8	36	2.0
3	26	39	16	5.5	33	2.3
4	25	37	17	6	33.5	2.0
5	20	45	12	4.5	33	3.1
6	19	36	17	6	32	3.1
7	17	42	15.5	8.5	34.5	3.9
8	15	44	17	4.5	32	4.4
9	11	45	19	6.5	32	6.4
10	8.5	42	25	3.5	30	8.3

*The "milkiness" index is computed from $\dfrac{\text{Total milk solids} + \text{sugar}}{\text{Cocoa nib}}$

Table 13.5 Milk chocolates (4)

	Milk chocolate High %	Milk chocolate Average %	Low %	Typical milk crumb %	Enzyme-treated milk chocolate %
Cocoa mass	14.00	11.78	10	12.00	12.00
Whole milk powder*	23.60	19.08	18		10.92
Whole milk (roller)					
Whole milk (spray)					
Enzyme-treated (WMP)					3.00
Sugar	46.00	48.73	55		52.23
Added cocoa butter	16.00	19.98	16.50	18.00	21.35
Lecithin	0.40	0.35	0.38		0.30
				Milk crumb**	
Salt				70.00	0.06
Vanillin		0.08	0.06		0.06
Total	100.00	100.00	100.00	100.00	100.00
Total fat	30.00	31.50	27.00	31.32	32.00

*Whole milk powder is sometimes substituted with skimmed milk powder and butter oil
**Milk crumb (4) F.C.M. Solids 37% Sugar 54% Cocoa mass solids 8% Water 1% *Total 100% Total fat 13.32%*

an optimum of about 3, unless high milk fat is used. Outside this range, either the cocoa or the milk flavour is smothered. The use of very mild cocoas allows a certain amount of latitude, but the final adjustment of perfect milk chocolate is an art.

Formulae for milk chocolate taken from L. Russell Cook (4) and Clyne (9) are given in Tables 13.5 and 13.6.

Table 13.6 Milk chocolate (9)

	Dutch milk chocolate (1955) %	British milk chocolate (1955) %
Cocoa mass	15.70	15.84
Milk powder	16.22	13.28
Whole milk (roller)		
Whole milk (spray)		
Enzyme-treated (WMP)	Skimmed milk powder 7.57	
Sugar	30.83	45.28
Added cocoa butter	29.20*	25.36
Lecithin	Cinnamon 0.16	
Salt	0.16	0.20
Vanillin	Vanilla pods 0.16	0.04
Total	100.00	100.00
Total fat	42.06	37.50

*Extra fat added to counteract extra milk powder (9)

13.7.2 *Dark chocolate*

Formulations are shown in Tables 13.7 and 13.8 (10).

13.7.3 *Coating and couverture chocolate*

Formulations for coating are shown in Tables 13.9–13.11. The composition of a couverture chocolate is shown in Table 13.12.

Table 13.7 Dark chocolates

	Bittersweet			Dark		
Ingredient	High %	Average %	Low %	High %	Average %	Low %
Blended beans/cocoa mass	70	60.65	55.00	43.00	39.62	35.00
Sugar	29.8	36.25	42.04	48.00	48.08	50.37
Added cocoa butter		2.60	1.70	7.83	11.75	14.15
Lecithin	0.30	0.30	0.30	0.30	0.35	0.30
Salt				0.06	0.06	0.06
Vanillin		0.20		0.04	0.14	0.12
Vanilla beans						
Total	100.00	100.00	100.00	100.00	100.00	100.00
Total fat	37.10	35.35	31.40	31.20	33.10	33.10

Note: Fat contents can vary for specific uses, e.g.
      Moulding 32
      Enrobing 34.5
      Piping 29.0
      Spinning 38.0

Table 13.8 Dark chocolates, 1950 (10)

Ingredient	Dark chocolate
Blended beans/cocoa mass	43.18
Sugar	43.40
Added cocoa butter	13.04
Lecithin	0.30
Salt	
Vanillin	
Vanilla beans	
Total	100.00
Total fats	36.35

Table 13.9 Typical CBE coatings

	Dark		Milk		White	
	A	B	A	B	A	B
Cocoa mass	40	20	10	10	—	—
Cocoa powder		10	—	—	—	—
F.C. milk powder	—	—	20	—	20	—
Skimmed milk powder	—	—	—	15	5	20
CBE	10	20	22	27	27	32
Sugar	50	50	48	48	48	48
Total	100	100	100	100	100	100
Total fat	32	32	32.5	32	32	32

In all recipes 0.4% of lecithin is used.
Note: The above are only for use with CBEs which are compatible with both cocoa butter and milk fats.

Table 13.10 Typical lauric fat coatings (for covering centres, cookies, panned goods)

	Dark	Milk		White	
		A	B	A	B
Cocoa powder (10–12%)	14	5	5	—	—
F.C.M. powder	—	10	0	20	—
S.M. powder	7	10	17.5	5	20
Lauric CBR	31	29	31.5	27	32
Sugar	48	46	46	48	48
Total	100	100	100	100	100
Total fat	32	32	32	32	32

In all recipes 0.4% of lecithin is used.

Table 13.11 Typical non-lauric fat coating (for cakes, wafer and crispy centres)

	Dark		Milk		White	
	A	B	A	B	A	B
Cocoa mass	10	—	10	—	—	—
Cocoa powder	15	20	—	5	—	—
F.C. milk powder	—	—	6	—	20	—
S.M. powder	—	—	12	17	5	20
Non-lauric CBR	28	33	28	34	30	35
Sugar	47	47	44	44	45	45
Total	100	100	100	100	100	100
Total fat	35	35	35	34.5	35	35

Table 13.12 Couverture chocolates composition (11)

Cocoa nib	Sugar	Added fat	Total fat	Protein	Nib/sugar index
41	46	13	36	5.5	0.89
38	46.5	16	35.5	5.35	0.82
37	51	11.5	32	5.18	0.72
37	45.5	17.5	37	5.25	0.81
34.5	48	18	36.5	5.0	0.72
32	46.5	21	38	4.44	0.69

Note: Any nib/sugar index higher than 0.7 demands a good blend well conched.

Although both full and mild flavoured types of couverture chocolate are used, the variation in the 'index' is seldom very large, owing to the necessity of balancing the chocolate with delicate flavoured centres. Liqueur-flavoured centres permit the fullest-flavoured cover, and such have a great popularity on the Continent. Nib blends used in good covering usually contain a proportion of pale Criollo cocoas.

13.8 Conclusion

It is hoped that his chapter explains to readers how different flavours and textures can be created in order to obtain a 'total eat', and the variations there are in the industry. As chocolate is amongst the must popular and most widely used flavours in the world, we must carefully protect and nurture it. The recipes, however varied, all contribute to this aim.

Acknowledgement

The author wishes to thank his wife Margaret, children Nikki and Chris, and all his compatriots in the industry for their help and support over the years.

References

1. Minifie, B. *Chocolate, Cocoa and Confectionery* 2nd edn., AVI Publishing Company Inc., Westport, Connecticut (1980).
2. Lees, R. *A Basic Course in Confectionery*. Specialised Publications Ltd, Surbiton (1980).
3. Lees, R. and Jackson, B. *Sugar Confectionery & Chocolate Manufacture*. Leonard Hill Books, [Blackie and Son Ltd., Glasgow] (1973).
4. Cook, L.R. (revised by E.H. Meursing) *Chocolate Production and Use*. Harcourt Brace Jovanovitch, New York (1984).
5. Kleinert, J. 'Some aspects of cocoa bean roasting'. 20th *PMCA Production Conf.*, Pennsylvania (1966).
6. Ley, D. 'Comparison and judgement of conching systems'. *37th PMCA Production Conf.*, (1983); *Manuf. Conf.* (1983) (May) 68–71.
7. Hoskins J.C. and Dimick P.S. 'The conching process' *36th PMCA Production Conf.* (1982), (Paper No 6420; Pennsylvania Agricultural Experiment Station).
8. Carlin, J.T., Chang, S.S., Hsieh, O.A, Lin Sun Kwang, L., Cri-Tang, Ho. 'Comparison of the

volatile flavour compounds of cocoa butter obtained from roasted and unwashed cocoa beans. *36th PMCA Production Conf.* (1982).
9. Clyne, E.J. *A Course in Confectionery.* Specialised Publications Ltd, Surbiton (1955).
10. Williams, C. Trevor. *Chocolate and Confectionery*, 1950, Leonard Hill Books [Blackie, Glasgow and London] (1950).
11. Jensen, H.R. *Chemistry, Flavouring and Manufacture of Chocolate Confectionery and Cocoa.* J & A Churchill Ltd, London (1931).
12. Maitland, C.E.C. *Silesia Confiserie Mannual* **2**(3) (1973) Silesia/Gerhard Harke KG, Neuss, FRG, 73a.

14 Pipes, pumps and mixers

R.B. NELSON AND S.T. BECKETT

14.1 Introduction

As chocolate manufacturing plant has increased in size, the transport of the product and some ingredients down pipes has become more and more important. Wheeled tanks and the transport of solid blocks of material have tended to be replaced by long lengths of jacketed pipe and temperature-controlled storage tanks. With pipes perhaps 200 m or more long it is imperative that the pump and pipe diameter are designed to give the required flow. Liquid chocolate is a non-Newtonian fluid, and this means that both the yield value and plastic viscosity (see Chapter 9) must be taken into account when designing piping and pumping systems. The first part of this chapter illustrates how relatively simple viscosity measurements can be used to calculate the required plant size.

Many different types of pumps exist, each with its own advantages and disadvantages. In addition, the various applications in the chocolate industry, ranging from metering cocoa mass to pumping tempered chocolate, mean that many types are employed in a single manufacturing plant. The second part of this chapter reviews some of these pumps and their applications. Finally, some of the principles involved in mixing chocolate, both in the conche and the storage tank, are indicated.

14.2 Viscosity and viscometry

In the confectionery industry a knowledge of viscosity is essential to appreciate the characteristics of the many diverse ingredients. The viscosity of a fluid becomes of particular importance when it is necessary to create movement in a handling situation. Whether a fluid needs to be shaken, stirred, spread or pumped into mixing applications, depends upon recipe and process requirements.

14.2.1 *What is viscosity?*

Viscosity may be easily perceived in an operating situation, but is difficult to quantify in meaningful way except by specialized instruments. What then is viscosity? Probably the simplest explanation is 'the resistance to motion when stirred or poured', and this may be referred to as *consistency.*

When specifying requirements, say to a pump supplier, it is always necessary to quote the expected range of viscosity and temperature variations against the flow rate required, and any pressure limitations on sensitive ingredients and equipment.

Fluids are often thought of as simple liquids but, through the whole spectrum of the biscuit, baking and confectionery industries, the fluids used may be far from simple. In fact they can be very complex, varying greatly in composition and viscosity characteristics. It is these characteristics that need careful consideration. As a reminder, some of the fluids are steam, brine, ammonia, water, sugar solutions, all types of milk products, glucose syrups, concentrates, various liquors, gums, chocolate, praline pastes, caramels, toffee, dough, flour batters, chewing gum, fats of all kinds, etc.

How is viscosity quoted to a pump manufacturer? Can the figures be found in reference books? In some cases they can, but for the more complex non-Newtonian liquids figures are usually scarce. Viscometry units can be confusing: absolute units in poises, apparent viscosity in poises, Casson plastic viscosity in poises, Bingham viscosity in poises, kinematic viscosity in stokes, etc. Low-viscosity solutions may be characterized in centipoises, centistokes, seconds, and many more, indicating the bounds of a controversial subject. To confuse further, poises are now converted to pascal/seconds (Pa s).

Many problems exist in pumping and mixing schemes, such as pump seals leaking due to excess pressures, giving premature failure! Excess pressures are

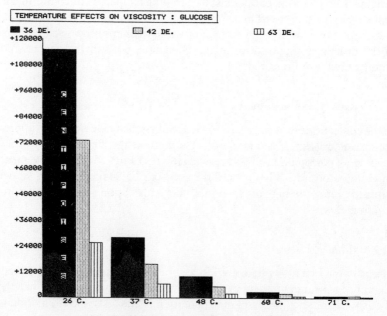

Figure 14.1 Temperature effects on viscosity.

almost certainly due to badly sized pipes or too great a flow rate. In some cases heating the liquid to a higher temperature will alleviate the problem. To illustrate this, Figure 14.1 shows how glucose syrups change in viscosity with temperature.

It is most apparent that some knowledge must be gained of product viscosity and flow rates and the use of suitable equations which enable accurate predictions to be made. It is extremely difficult to mentally assess pipeline pressures at various flow rates and viscosities. In calculating various flow rates at specific viscosities, accurate predictions can be made of pipe sizes and power ratings, thus avoiding later rework, delays and perhaps even more alarming consequences.

14.2.2 Newtonian and non-Newtonian liquids

The consistency or viscosity of a liquid, i.e. the internal friction resistance to movement, was mathematically described by Newton. He saw that the greater the surface area of a liquid, the greater the frictional force (F) in that liquid. He also stated that F was directly proportional to the 'velocity gradient' at the part of the liquid under consideration (refer to Figure 14.2). If v_1 and v_2 are the velocities of x and y, and h is the distance between them, the 'velocity gradient' between the liquids is expressed as $(v_1 - v_2)/h$. The 'velocity gradient' can be written as (cm per sec)/cm, or as 'sec-1' or s^{-1} (reciprocal seconds). If A is the surface area of the liquid, then the frictional force F on the surface is expressed as $F = \eta A(v_1 - v_2)/h$, where η (eta) is the liquid coefficient of viscosity. For confectionery products the units of viscosity (η) are usually centipoise, poise or pascals.

Newton stated that the coefficient of viscosity remained 'static' at varying flow rates, but this only applies to Newtonian fluids at a fixed temperature. The Newtonian fluid viscosity η is defined as the ratio of shear stress τ (tau) to the shear rate D, i.e. $\tau = \eta D$ or $\eta = \tau/D$. If η is a constant and does not vary with changing values of D, then the fluid is Newtonian.

A Newtonian liquid is therefore one for which the graph of shear stress plotted against the rate of shear is a straight line. To check if a liquid is Newtonian, double the rate of shear D and read the viscometer. This should

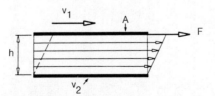

Figure 14.2 Simple shear diagram
v_1 = velocity (cm sec^{-1}); A = surface area of plane (cm^2); F = force in dynes;
v_2 = velocity in lower plane (cm sec^{-1}).

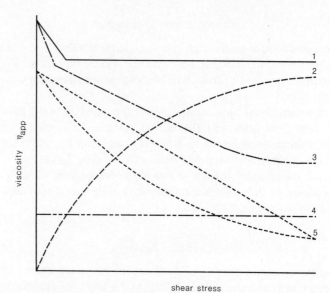

Figure 14.3 (*a*) Comparison of Newtonian and non-Newtonian plots. 1. Plastic: when the yield point is exceeded, fluidity results, e.g. chewing gum base. 2. Dilatant (time-independent): the apparent viscosity increases as rate of flow or shear rate increases, e.g. chocolate. The opposite of pseudoplasticity. Note the curve is not linear. 3. Pseudoplastic (time-independent): when the yield point is exceeded, fluidity results. The apparent viscosity decreases as the rate of flow increases. 4. Newtonian; shear is directly proportional to flow rate. 5. Thixotropy (time-dependent): the apparent viscosity decreases as the shear or flowrate increases. 6. Rheopexy (time-dependent): is the opposite of thixotropy, i.e. time-dependent thickening.

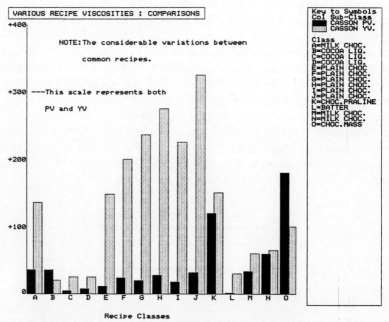

Figure 14.3 (*b*) Viscosities of common (non-Newtonian) chocolate recipe types.

indicate a doubling of the shear stress, τ. If this is not the case, then the liquid cannot be Newtonian. Figure 14.3(a) compares the relation between various more complex liquids which do not follow the rules for Newtonian liquids. Figure 14.3b compares viscosities of common chocolate recipes. (It is important that viscosity measurements should take place under laminar flow conditions for accurate predictions using recognized formulae—see 14.3.1.)

Other, more complex, fluids have varying coefficients (apparent viscosity) at differing flow rates, and thus the basic equations differ in power and wall friction calculations for pipelines and mixing applications. Chocolate is one of these complex, i.e. non-Newtonian fluids.

To illustrate how to calculate the force F between two layers of liquid, we can take two examples of Newtonian liquids: water and glucose solution (corn syrup). Assume that the layers are 0.15 cm (0.06 in), apart, water is at a temperature of 15 °C (59 °F) and the coefficient of viscosity is 0.011 poise. The glucose (36DE) is at 37 °C (99 °F), the coefficient of viscosity is 280 poise and the relative velocity between the two layers is 1.5 cm/s (0.6 in/sec). The surface area under consideration is 12 cm^2 (1.9 sq. in).

For

$$\text{water at } 15\,°\text{C (59 °F)}, \eta = 0.011 \text{ poise}$$
$$F = 0.011 \times 12 \times 1.5/0.15 = 1.32 \text{ dynes.}$$

for

$$\text{glucose solution at } 37\,°\text{C (99 °F)}, \eta = 280 \text{ poise}$$
$$F = 280 \times 12 \times 1.5/0.15 = 33600 \text{ dynes.}$$

From the results of the calculations it can be clearly seen that there is an enormous difference between the power needed to move these two liquids at the same velocities, even though the glucose solution is much warmer. Considering the glucose, the flow rate and viscosity are the factors most strongly affecting the required pump power. What can be done to reduce the power necessary? Referring to Figure 14.1, it is possible to raise the temperature of glucose, thus affecting a considerable reduction in viscosity: for 42DE at 26 °C (79 °F) $\eta = 730$ poise, but raising the temperature to 48 °C (118 °F), $\eta = 52$ poise. This alone can save costly updates in pipe or pump sizing, but, to sound a note of warning, some temperature-sensitive ingredients cannot be heated beyond certain limits without causing degradation or changes in efficacy (chocolate is one).

14.3 Calculations on non-Newtonian liquids

It is often asked why the Poiseuille equations cannot be applied to masses similar to chocolate or to chocolate itself. Although the poise is named after Poiseuille, the equations derived are not suitable for non-Newtonian liquids since the formula does not take into account the fact that non-Newtonian liquids exhibit the phenomenon of a 'yield value' (YV). Neither do they take into account the important factor of 'pipewall shear stress', or, in the case of Newtonian liquids, the friction factor f. Usually the d'Arcy equation is used in combination with a modified equation to calculate the friction factor. For the record, the Poiseuille equation is as follows:

$$V = \frac{\pi R^4 P}{8\eta L}$$

where V = flow rate
     R = radius of pipe
     L = length of pipe
     P = pressure
     η = viscosity

The International Office of Cocoa and Chocolate (OICC) recognized and adopted the Casson model equation, which is as follows:

$$(1 + a)\sqrt{D_N} = \frac{1}{\sqrt{\eta_{CA}}}[(1 + a)\sqrt{\tau_1} - 2\sqrt{\tau_{CA}}]$$

where η_{CA} = Casson plastic viscosity (poises)
     D_N = shear rate (for $D_N > 1$) at inner cylinder (s^{-1})
     τ_1 = shear stress at surface of inner cylinder (dynes cm^{-2})
     τ_{CA} = Casson yield value (dynes cm^{-2})
     a = ratio of radii of inner and outer cylinders.

Probably the yield value is the least understood unit. It is functionally referred to as the limiting value of stress below which the liquid cannot be sheared or moved. Visually the yield value is seen as the capability of the liquid to stand up without collapsing; the higher the yield the higher the liquid will stand. An extruded liquid ice-cream cone is a good example of a high-yield-value material. The range of yield values within chocolate is about 50–1000 dynes cm^{-2}. In chocolate manufacturing plants, the value produced is controlled by refining techniques in conjunction with specialized emulsifiers to give the desired results.

14.3.1 Moving liquid chocolate

The preferred method of moving chocolate and similar materials is by pump through heated pipelines, although earlier in the century pumping was thought to be detrimental to chocolate. Since chocolate is a non-Newtonian liquid, the pumping calculations necessitate the acquisition of viscometry figures for the types of chocolates which have to be pumped. Fincke and Heinz (1) produced the following formula to calculate the bore of pipes and the pump power required.

$$V = \frac{\pi R^3}{\eta_{CA}}\left[\frac{\tau_R 4}{4\,7}\sqrt{\tau_{CA}}\sqrt{\tau_R} + \frac{\tau_{CA}}{3} - \frac{\tau_{CA}^4}{84\tau^3 R}\right]$$

where V = output (ml sec.$^{-1}$)

η_{CA} = Casson plastic viscosity (poise)

τ_{CA} = Casson yield value (dynes cm^{-2})

R = radius of the pipe (cm)

τ_R = shear stress at pipewall (dynes cm^{-2})

$= pR/2L$ (p = pressure in dynes cm^{-2}, L = pipe length in cm).

It is presently accepted that the measurement range for a viscometer using the Casson equation is in the shear rate range of 5–60 reciprocal seconds (OICC). As shown earlier, the viscosity of liquids varies greatly with temperature fluctuations; even water varies 1–3.5% per degree Celsius. Therefore, temperature accuracy is very important for repetitive viscosity measurements and these should be carried out under laminar flow conditions (i.e. less than Reynolds number 2000), and at the temperature expected in the pipeline. Viscous liquids usually never achieve turbulent flow conditions, except during intensive high shear mixing, because the power requirement is so high.

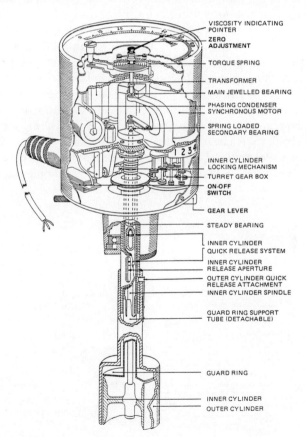

Figure 14.4 The Ferranti portable viscometer (Ferranti plc).

For calculations it is essential to use accurate viscometer figures in order not to over-or under-engineer to pumping and piping designs. The selection of a viscometer is also important, for there are many types and principles available. However, to obtain at least the two values at two different shear rates necessary for solving the Casson equation, the range of viscometers is narrowed down to a multispeed instrument, which usually means a form of the rotary cylinder type. Figure 14.4 shows a Ferranti portable hand-held instrument, capable of sensitive readings, with speed-changing facilities. The rotary paddle type of viscometer for measuring torque, which simply fits into a vessel without the usual cylindrical rings, will only satisfactorily measure Newtonian liquids. The cone and plate viscometer is used to measure non-Newtonian liquids and is available for multispeed tests. These viscometers are all batch instruments designed for laboratory-type skills due to their sensitivity and delicate mechanisms. Chapter 9 provides details on how to prepare samples and the methods of averaging readings to minimize errors. In-line viscometers are somewhat unreliable for non-Newtonian liquids unless they can be twinned up for microprocessor analysis, since the Casson calculation requires two readings at two flow rates; however, one such instrument is shown in Figure 14.5, with built-in processor analysis. In-line devices depend on various principles: vibrating elements, pressure devices, differential pressure orifices, rotary elements emulating the batch viscometer. In general, these units measure apparent viscosity at the current flow rate and cannot be used to measure 'real-time' on-stream Casson values.

A summary notes of the basic formulae after and terms for concentric cylinder viscometry is given in Table 14.1.

The Hagen–Poiseuille equation for differential pressure devices is given below, and it can be seen that there is now allowance for pipewall shear stress

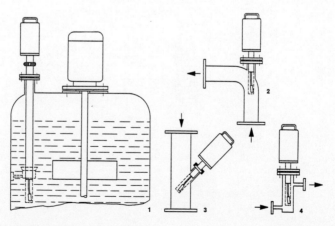

Figure 14.5 (*a*) Typical installation of a process viscometer-type convimeter.

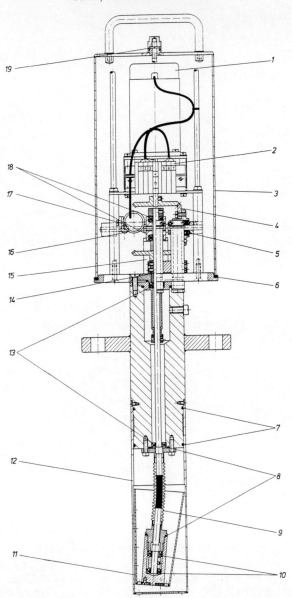

Figure 14.5 (*b*) The Convimeter. 1, Geared motor, with gear control; 2, terminal, explosionproof; 3, cover; 4, bevel driving pinion; 5, 2 ball bearings, 6/19 mm φ, for nylon bevel driving pinion; 6, rubber gasket; 7, 2 O-rings, Viton, for seat of protection sheath; 8, 2 O-rings for connecting pieces metal bellow; 9, metal bellow, with connecting pieces; 10, 2 ball bearings 7/19 mm φ, for drive shaft, lubricated with special oil; 11, conical mantle, with 3 screws M 4 × 6 12, protection sheath, with 2 screws M 4 × 6; 13, 2 ball bearings 7/19 mm φ, for drive shaft, lubricated with special oil; 14, drive shaft, with Teflon sleeve; 15, dash pot, with Teflon sleeve; 16, metering spring, Nr. I, II, or III, with barrels; 17, inductive transducer, explosion-proof, with drive eye; 18, 2 ball bearings 7/19 mm φ, for dynamometer-drive shaft; 19, cap nut for cover, with spring washer and O-ring.

Table 14.1 Viscometry notes

Units

1 poise = 100 mPa s	1 dyne cm^{-2} = 0.1 Pa
1 poise = 0.1 Pa s	10 dyne cm^{-2} = 1 Pa
1 mPa s = 0.01 poise	
1 centipoise = 1 mPa s	
100 centipoise = 1 poise	
10 poise = 1 Pa s	

Coefficient of viscosity

η = ratio of shear stress τ to rate of shear D

$\eta = \tau/D$ or $\tau = \eta D$ in poise or mPa

η_{pl}, poise

η_{app}, apparent poises

η_B, Bingham poise

η_{CA}, Casson poise

For a Newtonian liquid, η is directly proportional to D, i.e.

$\eta = \tau/D$ is a straight-line equation.

For a non-Newtonian liquid, η is not directly proportional to D and may fit various model equations. The Bingham and Casson equations are well-known equations for chocolate and similar masses.

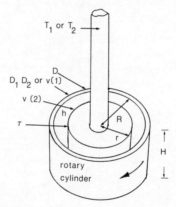

T_1, T_2 torque reading of shaft

D shear rate in units of reciprocal second (s^{-1}) (4–60 s^{-1}, OICC)

F fricitional force $= \dfrac{\text{area} \times \eta \times (V_1 - V_2)}{h}$

τ = shear stress in units of dynes cm^{-2} at the surface of the inner cylinder

Ideally, viscosity measurements should be made in laminar flow conditions, i.e. Reynolds number < 2000 (no turbulence)—temperature and stabilization stir times are important

Note: water varies in viscosity from 0.35 cp (0.35 mPa s) at 80 °C to 1.7 cp (1.7 mPa s) at 0 °C

Casson plastic viscosity η_{CA}:

$$\eta_{CA} = (\sqrt{T_2} - \sqrt{T_1})/(\sqrt{D_2} - \sqrt{d_1})^2 \text{ poise}$$

Casson yield value τ_{CA}:

$$\tau_{CA} = [\tfrac{1}{2}(1 + \alpha)(\sqrt{T_1} - \sqrt{D_1}\sqrt{\eta_{CA}})]^2 \text{ dynes cm}^{-2})$$

Bingham plastic viscosity η_B:

$$\eta_B = (T_2 - T_1)/(D_2 - D_1) \text{ (poise)}$$

Bingham yield value η_B:

$$\tau_B = \tau_1 - D_1\eta_B \text{ (dynes cm}^{-2})$$

or the yield value. This equation returns only one value and therefore is effective only for Newtonian liquids.

Hagen-Poiseuille equation:

$$\Delta P = \frac{0.128 \times L \times Q \times \eta_{\text{app}}}{\pi D^4}$$

ΔP = differential pressure (mbars)
Q = flow rate (m^3/s)
η_{app} = apparent viscosity (dynes/cm^2)
D = diameter of pipe
L = length of pipe.

14.3.2 *Non-Newtonian pipeline calculations*

Assuming that we now have two viscometer readings, it is possible to cut out the laborious task of manual calculation of the Casson values by using a computer program. Once the Casson PV and YV have been obtained, a prediction can quickly be made, making use of a modified version of the Heinz and Fincke formula mentioned earlier. It is then possible to indicate the pipewall shear stress, flow velocity, pressure drop, pipe size allied to the flow rate and pressure drop, and the power for a pump drive. The power requirement for a pump has to take into account all frictional inefficiences such as gearboxes, hydraulic couplings, or belt drives, and pump internal friction, which can vary greatly between pumps. Further allowances have to be make for start-up conditions, and for shut-down periods over holidays, which can cause sedimentation problems, thus raising pressure drops. (Such a program is available from the author.)

Once a pipeline has been calculated, the top limit on pipeline pressures is limited only by the pump capability. The authors' personal preference is to see a pipeline running under 6.9 bar (100 psi). This avoids undue pressure in the pumping elements and the possibility of caramelizing milk chocolate or other temperature-sensitive products. Also, plate-type temperers cannot take high pressures, and lines to a temperer should preferably be under 4.14 bar (60 psi). Using these principles ensures a reliable prediction, especially when long pipelines using viscous chocolate are proposed.

14.4 Pumps

When choosing a pump for transporting chocolate ingredients along pipes, at least some of the following criteria are likely to determine the most suitable type.

(i) The use to which the pump is to be put, e.g. metering cocoa mass or transporting chocolate over a long distance.

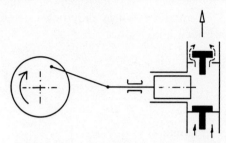

Figure 14.6 Positive displacement reciprocating plunger pump. $V = A \times L \times n$, where V = theoretical volume flow, A = plunger length, n = stroking speed, L = stroke length.

(ii) The pressure drop across the pump: this is very important in the case of tempered chocolate or pipes of inadequate diameter.

(iii) The wear rate: will the continuous operation of the pump result in excessive wear due to abrasion by the solid particles within the material?

(iv) The quality and position of seals in the pump. This is linked with (ii) and is very important during continuous operation, where leaks would be not only unsightly but also expensive in loss of product, and pose a hygiene problem.

(v) The temperature control and shear developed in the pump. Heating or shear may affect the taste (e.g. by causing caramelization) or alter the flow properties by de-tempering the chocolate, etc. Special chocolates, such as ones containing fructose or sorbitol, must be maintained below 40 °C during the processing.

(vi) Price and maintenance. This of course applies to any machinery. It is often possible to save money in installation, only to lose a greater sum in maintenance or loss of ingredients.

Many different makes of pumps exist in the confectionery industry, but most correspond to one of the following types:

(i) Positive-displacement reciprocating plunger pumps
(ii) Peristaltic pumps
(iii) Vane pumps
(iv) Gear pumps (positive displacement rotary)
(v) Screw pumps
(vi) Pawl pumps
(vii) Mono pumps
(viii) Centrifugal pumps.

(i) *Positive-displacement reciprocating plunger pumps.* The construction of this type of pump is illustrated in Figure 14.6. As the piston draws back, material is pulled through the lower valve to fill the space created. On the return stroke, the upper valve opens, releasing the product. By adjusting the stroke frequency and length, an accurately controlled volume is moved down

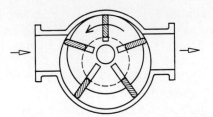

Figure 14.7 Mode of operation of vane pump.

the pipe. This allows this type of machine to be frequently used as a metering pump for cocoa mass and butter. In this case the quantity of material to be moved is relatively small, and the viscosity is low. At higher viscosities the flow drops below the theoretical volume flow as cavitation occurs, and the flow rate no longer varies directly with the stroke length and frequency. Calibration then becomes a major problem. Where large particulate matter is present there is a possibility of the valves being blocked open.

These pumps are, however, temperature-controllable and are a very useful alternative to 'loss in weight' tank/transducer systems, for the addition of liquid ingredients to chocolate.

(ii) *Peristaltic pumps.* These pumps push material along by pressure on compressible pipes. Although used as metering pumps in some industries, they are not normally employed in chocolate-making. This is, in part, due to the limited life of the flexible tube, which must also be of food-grade material. Normally these too are low-volume operating pumps.

(iii) *Vane pumps.* These can be used for 'intermediate'-viscosity material, for example, fully manufactured chocolate in a ring main or an enrober system. Their mode of operation is illustrated in Figure 14.7. As the off-centre shaft rotates, the vanes push the chocolate towards the outlet. The chocolate cannot keep going round, because the centre shaft comes in contact with the outside wall of the pump as the vanes retract. The vanes are kept against the outside wall either by centrifugal action or by a 'perfect geometry' central ring.

With the relatively large sample volume within the pump, high temperatures and pressures are unlikely to develop. This makes them suitable for tempered

Figure 14.8 Mode of operation of gear pump.

chocolate, as they will not greatly affect the crystalline state. They are less suitable for applications which involve very high throughputs or large pressure drops, e.g. pumping chocolate over very long distances.

(iv) *Gear pumps (positive displacement rotary pumps)*. This type of pump can be used for relatively viscous material. Many different shapes of gears, rotors or lobes exist. They normally operate via a driven gear pushing a second gear, both moving the chocolate or ingredients around the outside (Figure 14.8). Problems can arise, however, if relatively large quantities of material are to be transported, as this involves the gears moving at higher speeds, making temperature control more difficult. In addition, high shear will occur between the gears, with associated taste, liquidity and crystal phase problems, also causing increased wear on the pump itself. The maximum pressure occurs near the outlet, which in some designs means that this high-pressure zone is next to the drive shaft. In this case it is imperative to ensure that the seals are adequate to avoid the chocolate or ingredients leaking from the pump.

(v) *Screw pumps (spindle pumps)*. The problem of having the highest pressure next to the drive shaft has been overcome in the screw pump. Here the maximum pressure is at the outlet, but the drive mechanisms can be situated in a low-pressure area. Screw pumps normally have one, two or three rotors. Single rotors with a shaped stator tend to have a low efficiency and are unable to operate against high back pressure. A twin-screw system is illustrated in Figure 14.9. As the second screw system is driven by a gear wheel outside the pumping volume, there is no metallic contact between the screws or spindles,

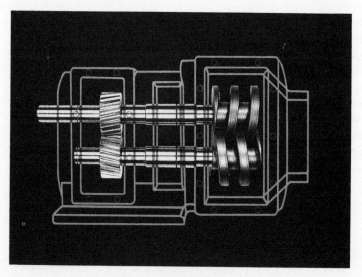

Figure 14.9 The Frisse spindle pump (Richard Frisse GmbH).

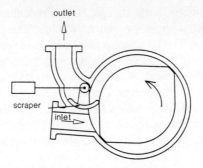

Figure 14.10 Mode of operation of pawl pump.

which means that the pump can run dry. In addition, lumps in the material of up to 5 mm (0.2 in) diameter can be coped with in some designs.

When the mass is pumped at a low velocity, good temperature control can be achieved. Very viscous material can be pumped, and, because of its ability to cope with large particles, this type of pump can be used between the mixer and pre-refiner in the two-stage refining process (see Chapter 6). Where high back pressures exist or exact temperature control at high throughputs is necessary, alternative types of pump may be required.

(vi) *Pawl pumps.* This type of pump is frequently used as a circulating pump in enrobers where the tempered chocolate must be maintained at an exact temperature. This is achieved by having a high conveying volume within the pump but operating it at a low number of revolutions per minute. Using jacket heating, pumps of up to 10 tons/hour have been constructed. The mode of operation is illustrated in Figure 14.10. The central rotor is shaped so that two volumes of material are present within the pump. As the rotor turns, the material is pushed round the pump until it comes against a scraper. This is pressurized against the rotor and prevents the material circulating back, by directing it through the outlet.

(vii) *Mono pumps.* These pumps can also deal with highly viscous material containing relatively large particulate matter. They operate at relatively high speeds (> 500 rpm), however, and do not normally incorporate temperature control. The main pumping elements are a stator (usually rubber) in the form

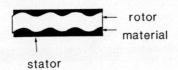

Figure 14.11 Schematic section through a mono pump.

of a double internal helix, with a single helical metallic rotor which rotates within the stator with a slightly eccentric motion. The pitch of the stator is twice that of the rotor. The latter maintains a positive seal along the length of the stator and this seal progresses continuously through the pump, giving uniform positive displacement. A schematic illustration of a section through the rotor and the stator is given in Figure 14.11.

During continuous operation in a large-scale chocolate plant, wear of the stator may become a major problem. Mono pumps can, however, be very useful and versatile in small-scale or pilot plants, or for sweet centres such as fondants.

(viii) *Centrifugal pumps.* In the confectionery industry this type of pump is normally restricted to the transport of thin ingredients such as milk or sugar solution. These pumps operate using a high-velocity spinning disc. The vanes on the surface of the disc direct the product towards the output at the circumference of the disc. The feed point is normally near the centre of the disc. This type of pump is unable to deal with the higher viscosities of most chocolate masses.

14.5 Mixing chocolate

Although cocoa butter is a Newtonian fluid and time-independent (see below), the chocolate itself exhibits non-Newtonian behaviour owing to the solid particles which contribute approximately 70% of the masse. Molten chocolate is, in many cases, time-dependent, i.e. the shear stress or the apparent viscosity of the chocolate decreases with time at a particular shear rate (2). This time effect can be expressed mathematically by an equation of the form

$$\tau = Kt^{-m}$$

where τ = shear stress
    t = time
    m = index of time dependency
    K = constant.

The value of m was found to increase considerably with shear rate and to depend strongly upon the lecithin content. This has two implications. Firstly, in the final mixing stage, normally the conche, it is important to process long enough to obtain the thinnest possible chocolate. The time at which this is reached will depend upon the point and amount of addition of lecithin and also the shear rate. Several conches, e.g. the Frisse and Tourell designs (Chapter 8) have been designed to have a final higher shear rate stage. Secondly, it should be remembered that most viscometers subject the chocolate to considerable shear breakdown. It is therefore important to have a standardized method of preparing and measuring samples when comparing different processes and recipes. It is often desirable to shear for at least 15 minutes at an appropriate shear rate before carrying out viscosity measurements.

When calculating the power required by the mixer or conche to process the chocolate, account must be taken of the non-Newtonian nature of the material. Frequently, higher current ratings are required than would be expected from Newtonian material, especially in the early stages of processing. This may in part be due to a turbulent conditions existing in the mixer, whereas viscometric measurements are made under laminar flow.

14.6 Storage of liquid chocolate

This is related to chocolate mixing in that any tank containing liquid chocolate requires to be agitated. Failure to do so frequently results in fat separation from the chocolate. In extreme cases the sediment can be extremely hard and may have to be broken up with pickaxes for removal from the tank. The speed of stirring depends largely upon the volume of material involved and the storage period. Recipes and previous processing, however, play an important part: some chocolates become thinner, whilst others thicken up. Such instabilities may indicate that the conching time was insufficient to account for the time dependency described in the previous section. Alternatively, the stirrer speed or design of the storage vessel may be incorrect and some sedimentation may be taking place.

As in the pipelines, temperature control is important. At low temperatures the chocolate becomes thicker, making it difficult to stir and pump. There is also the danger of the chocolate setting up solid altogether. Solidified chocolate may wreck machinery, and at best it takes an extremely long time to melt again!

Chocolate temperatures of about 40 °C (104 °F) are suitable for bulk storage, but care should be taken to ensure that the temperature is uniform throughout. Hot or cold spots can give rise to difficulties. Not only do higher temperatures above 60 °C (140 °F) cause flavour changes, but in some forms of chocolate thickening occurs which may not be reversible on subsequent cooling and shearing.

14.7 Conclusion

In pumping or storing chocolate it is necessary to ensure the equipment has adequate temperature control and power. Relatively simple viscometric measurements can enable the engineer to determine the correct pipe bore for the rate of flow required. It is also necessary to choose a pump according to its application. In general, it is better to have a relatively large pump and operate it slowly, rather than operate continuously with fast-moving machinery in contact with the chocolate.

References

1. Fincke, H. and Heinz, A. *Fette und Seifen* **62** (1960) 77.
2. Elson, C.R. *The Flow Behaviour of Molten Chocolates,*. BFMIRA Research Report No. 173, British Food Manufacturer Research Association, Leatherhead (1971).

15 Instrumentation

I. McFARLANE

15.1 Monitoring process conditions

15.1.1 *General-purpose sensors*

Any production process has some requirement, usually quantifiable, for maintenance of process conditions which permit satisfactory operation. Temperature of air surrounding the product, for example, usually needs to be maintained within close limits. The sensing elements, which are used for monitoring process variables in the hostile environment of a production plant, differ significantly from the sensors used for laboratory measurements. The differences are partly mechanical—the production plant environment being less frequently cleaned and less regulated than the laboratory—and partly functional. It is always difficult and sometimes impossible to standardize or recalibrate sensors which are built into production equipment. Process sensors thus need to be free of drift, or else capable of indirect standardization by methods which do not require the sensor to be removed from the process equipment. Process sensors, on the other hand, do not usually need as wide a dynamic range as laboratory sensors, because they are designated for a specific application.

In-line measurement of temperature, pressure and flow is required in a wide range of industrial applications, and process sensors for these variables have been developed to meet various cost and performance criteria. The reliability of some forms of process sensor is well documented, particularly in connection with safety requirements of the aerospace and nuclear power industries. Industrial process engineering has thus benefited from the development work on instrumentation in those other sectors of industry.

15.1.2 *Temperature*

Duncombe (1) of the UKAEA has reviewed instrumental techniques for temperature measurement in hostile environments. Stainless-steel sheathed and mineral-insulated thermocouple assemblies are recommended for high-temperature applications; when these are selected in sizes from 1 to 4 mm (0.04–0.16 in) outside diameter, they are flexible enough to allow simple installation in any part of a heat process.

Thermistors, thermocouples and platinum resistance thermometers are all readily available in the form of probe assemblies, suitable for long periods of unattended operation on process plant; ac resistance bridge circuits and amplifiers are necessary to convert the signals from thermistors and resistance thermometers into standard ranges for transmission, display and data logging. A different form of circuit is needed for thermocouples, which is based on the measurement of the electrical potential arising where two dissimilar metals are joined together. Thermocouples always need 'cold junction' compensation for the temperature at the termination point, where the dissimilar metals are connected to an amplifier. Bentley (2) has described the factors affecting calibration and performance of all three types of temperature sensor. Typical tolerances for thermocouples in the range 0–200 or 0–400 °C (32–400 or 32–750 °F) are about ± 1% of full scale, and these tolerances are greater than the equivalent tolerance for platinum resistance thermometers by about a factor of 10. Thermistors, which have the two disadvantages of needing individual calibration and of being self-heating when in use, also have calibration tolerances of the order of ± 1% of full scale, over a working range from − 80 to + 300 °C (− 110 to 600 °F). Platinum resistance thermometer probes are therefore recommended for immersion applications, such as in tempering equipment, where it is necessary to transmit measurements which are accurate to within tenths of a degree.

15.1.3 Pressure

Pressure measurement can usually be made using one of the many types of transducers in which the force, generated by the pressure acting on an elastic element, causes a mechanical deformation. Elastic deformation is seldom linear throughout a wide working range, and greater accuracy can be achieved using the force balance principle—a force is generated, which opposes the pressure causing the deformation, to restore the deformed element to its unstressed position. The magnitude of this restoring force is then sensed by some other means. The extra complexity adds to the cost, and introduces more ways in which the system can fail. A more simple way of extending the range of linear response is to use corrugations, bellows or stacked capsules.

A special case of a deforming shape is the Bourdon tube. These tubes are supplied in various shapes and styles, from a simple C-shape to styles which are spiral or helical. Bourdon tubes have the advantage of high burst pressures relative to their sensitivity.

Elastic diaphragms and other deforming shapes are usually made from metals, including brass and beryllium copper, stainless steel, monel or tantalum. Some modern gauges use a simple wafer of silicon, which has nearly perfect elastic behaviour.

A number of methods are used for converting elastic deformation to an electrical or pneumatic signal. The most common method involves the use of a strain gauge, which consists of a set of thin film resistors deposited together in a flat shape. When bonded to a deforming surface, the resistors act as a resistance bridge, giving an out-of-balance signal proportional to deformation. Strain gauges are suitable for flat diaphragms, but if the pressure element is a bellows or Bourdon tube, the displacement is measured by applying the motion to the wiper of a potentiometer or to a linear variable differential transformer (LVDT).

In all pressure sensors the sensitive element must be protected against overpressure, excessive temperature and corrosion. Protection sometimes requires a pipe containing dry air or a suitable fluid, to which the pressure to be measured is applied via a further diaphragm. Flush-mounting diaphragms, available for vessels or pipelines in food factories, are crevice-free and capable of withstanding steam cleaning and other forms of in-place cleaning methods.

Pressure gauges are calibrated either in units of absolute pressure, relative to vacuum, or in units of differential pressure (loosely described as 'gauge' pressure). In many applications it is the differential pressure which is important for process control. Differential pressure sensors are frequently used to monitor the hydrostatic pressure of fluid in a vessel (as a convenient way of transmitting a signal proportional to level or contents), and also to measure the pressure drop across a restriction in fluid flow, as a means of monitoring flow rate. This is only one of several forms of flow sensing, which is the subject of the next section.

15.1.4 *Flow*

Some form of mass or volumetric flow measurement is required in all processing, and many different types of flowmeter are available. Most of the lower-cost meters are based on measurement of the pressure difference before and after some restriction in flow. The restriction takes various forms, for example:

(i) *Circular orifice* (concentric with pipe): suitable for low-viscosity liquids and most gases, and with the flow rate determined from the square root of the pressure drop, in accordance with the fundamental physical law of the conversion of potential to kinetic energy. In practice, simple circular apertures have numerous disadvantages, being liable to accumulate deposits, and being affected by upstream fittings and obstructions which cause swirl and thus distort a uniformly turbulent velocity profile. Another disadvantage is the waste of energy associated with the irrecoverable pressure drop at the orifice.

(ii) *Venturi tube*: Pressure drop is not necessarily irrecoverable, and the pressure loss can be very significantly reduced, while retaining the same fundamental relation between flow and pressure drop, by tapering the inlet and outlet to the orifice (Figure 15.1). Note that the lower pressure is measured at the orifice of this 'Venturi' form of tube, and the pressure may

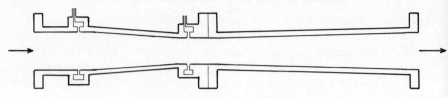

Figure 15.1 Venturi tube flowmeter.

be sampled at multiple points around the circumference, giving an average pressure reading and reducing the need for a long lead-in pipe run (3). The Venturi tube is suitable for dilute aqueous suspensions such as waste water.

There are other forms of restriction used for the purpose of measuring flow, such as a flow nozzle (mainly used for steam). The class of meters known as area meters, in which the variation of an opening is automatically adjusted by the motion of a weighted 'float', also operate on the same underlying principle as the orifice plate. The float is usually arranged to rotate. Some low-cost flowmeters simply consist of a rotating float in a vertical transparent tapering tube. Rotation which is driven by the fluid is the basis of turbine meters. These avoid the disadvantages associated with 'square root' characteristics, because when a turbine is driven by a fluid of sufficiently low viscosity it spins at a rate directly proportional to flow rate (since the kinetic energy of the rotor is proportional to the square of its angular velocity, which exactly cancels out the 'square root' law of pressure drop). Some turbines can be used for high-viscosity fluids, but the relation between rotor speed and flow rate becomes non-linear, and requires special calibration.

Vortex shedding at blunt obstacles in a pipe has become the basis of a class of flowmeters which are tending to replace orifice plates for fluids of viscosity of 20 cp or less, because the pressure difference is smaller. Accuracy of about 1% is obtainable over 10:1 dynamic range. Vortex meters are not suitable for viscous fluids, because the rate of vortex formation is proportional to flowrate only at Reynolds number values of 10 000 or greater. The Reynolds number is given by

$$N_{\mathrm{Re}} = D\frac{V}{\mu}\rho$$

where D = pipe diameter in feet
V = fluid linear velocity in $\mathrm{ft\,sec^{-1}}$
ρ = fluid density in $\mathrm{lb\,ft^{-3}}$
μ = fluid viscosity in $\mathrm{lb\,ft^{-1}\,sec^{-1}}$ (equivalent to cp/1488).

Magnetic flowmeters use the principle shown in Figure 15.2. They offer no obstruction to flow. Coils produce an alternating magnetic field in the fluid, and the motion of the fluid induces a voltage at the electrodes which is a function of fluid velocity only. The voltage is unaffected by the properties of the

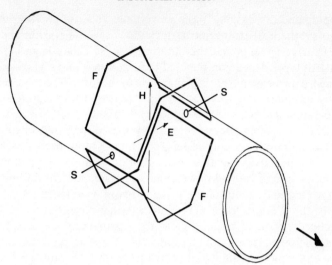

Figure 15.2 Magnetic volumetric flowmeter. *F*, field coils; *S*, sensing electrodes; *H*, induced magnetic field; *E*, induced electric field.

fluid, and the magnetic flowmeter is therefore a volumetric rather than a mass-flow sensor. The fluid must be non-magnetic, and have some electrical conductivity. The conductivity of chocolate is too low for any commercially available magnetic flowmeter to be used for chocolate flow measurement, but meters of this type are popular for dairy applications, partly because they are easy to keep clean.

Ultrasonic flowmeters are another class of meter which offer no obstruction to flow. For fluids in which the attenuation of sound is sufficiently low, changes in transit time along the direction of flow provide a simple indication of fluid velocity. The sound velocity is affected by temperature, and any entrained air bubbles may cause scattering of sound waves in fluids in which sound is normally readily transmitted. Problems arising from temperature changes and scattering effects are avoided in ultrasonic flowmeters based on the 'Doppler' principle. This uses frequency shift in echoes of sound from particles or bubbles carried in the fluid stream. Doppler meters are inexpensive, but the accuracy attainable is only in the range from 2–5% of full scale. Ultrasound is severely attenuated in molten chocolate, and these meters are not suitable for measurement of chocolate flow. They may, however, be used in the manufacture of some of its components.

There are only two forms of chocolate flowmeter in general use, and both involve obstruction to the flow—positive displacement meters, which are volumetric, and vibrating tube (Coriolis force) meters which measure mass flow. There is also the indirect 'loss-in-weight' method.

Positive displacement flowmeters continually divide the fluid into portions by trapping and discharging a known volume. This can be achieved using a variety of mechanical arrangements, including oscillating piston, sliding vane, oval gear arrangements and the nutating disc (so called from its 'nodding' motion). All types depend on the fluid to form a seal, and high viscosity is an advantage. The measurement is volumetric in the form described, but there is an equivalent mass flow technique is which each portion is weighed. This is not in common use, because if weighing systems are to be provided it is more convenient and useful to devise a 'loss-in-weight' dispensing system. Here the rate of loss fluid from a supply vessel is continually monitored. Loss-in-weight systems have the advantage that their calibration can be checked by topping up the supply vessel with a known quantity (or temporarily adding a known weight) while the flow out of the vessel is maintained constant.

Finally, the 'Coriolis force' mass flowmeter has recently become available, giving chocolate manufacturers their first opportunity to use a true continuous mass flowmeter for molten product (4, 5). The device measures the force produced by a moving fluid while undergoing an externally applied acceleration (Figure 15.3). The fluid is passed through a U-tube (with an inevitable pressure drop), and vibration is induced in a direction normal to the plane of the tube.

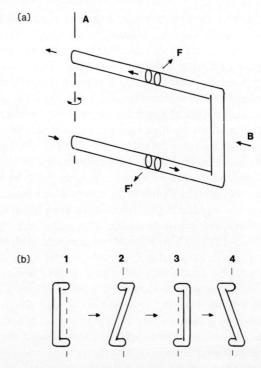

Figure 15.3 Coriolis force mass flowmeter. A, axis of induced oscillatory motion; B, direction of view for sequence below; F, F', direction of Coriolis force during phase 2 of the cycle.

With reference to Figure 15.3a, an object B at rest in a system rotating with angular velocity about axis A has angular momentum about the axis of rotation. Conservation of angular momentum requires that, if the object moves towards the axis of rotation, a force F exist perpendicular to the axis of rotation and to the direction of movement of the object. Note especially that the rotation in question is about an axis through the mounting points of the U-tube. In practice, this rotation is not continuous, but consists of induced angular oscillation of small amplitude. The force F is thus an oscillating force, and in the opposite direction to the force acting on an identical object moving away from the axis of rotation, in the other half of the tube.

With reference to Figure 15.3b, Tullis and Smith (4) showed that the phase of the twisting of the tube, superimposed on the induced side to side vibration, is such that the angle of twist is exactly out of phase with the side to side displacement. Tullis and Smith also show that the maximum angle of twist (at zero sideways displacement) is proportional to the total true mass flow in the tube. Thus if other parameters are constant, the angle of twist is a measure of mass flow in either direction. Various optical and magnetic arrangements are used to measure the twist in the tube. The flowmeters are readily calibrated with any convenient fluid using 'bucket and stopwatch', but it is necessary to take great care to provide antivibration mountings when installing this type of meter. Accuracy of the order of 0.25% is attainable, over a dynamic range of 25:1.

15.1.5 Reliability of process sensors

Manufacturers are becoming increasingly aware of the potential benefits of process automation. In some cases, early enthusiasm for the introduction of automation has been followed by scepticism on encountering problems with the reliability of automatic equipment.

Automation invariably requires input from plant sensors, and output to some form of regulator or actuator. Lees (6) has published an analysis of the reliability of sensors and control elements, using data collected from chemical plants and nuclear power installations. Failure rates are reported for the most basic types of process sensors as follows:

temperature sensor 0.35 faults/year
pressure   "   1.41
flow   "   1.13

Note that these figures are for installations where very high standards of maintenance apply. Equivalent figures for other sectors of industry, even allowing for improvements in sensor technology since the 1970s, would probably be significantly worse.

Lees (6) goes on to analyse the reported faults in terms of failure modes, and finds, for example, that a temperature trip amplifier for which the overall failure rate was 1.27 faults/year gave only 0.04 faults/year classified as 'fail-dangerous unrevealed'.

In the planning of any automation scheme, it is essential to make some assessment of the probability of various types of failure, and to make provision for situations with hazardous or economically serious consequences. The more complex the scheme, the more ways there are in which it can fail. In addition the difficulty of fault finding tends to rise exponentially with the number of sensors and actuators employed. As an aid to estimating the point at which an automation plan ceases to be viable, consider the following standardized man-hour ratings for instrument maintenance published by Upfold (7):

    temperature sensor   3 to 8 man-hours/year
    differential pressure cell 4 to 9
    flowmeter   4 to 12

Some advances in instrument technology have tended to reduce maintenance requirements per device, with self-calibration and self-diagnostic features being added to many types of sensor. These facilities reduce the burden on first-line maintenance services, but provision has to be made for checking the facilities themselves; a stage can be reached where the added complexity creates a burden of upkeep greater than that required for traditional devices.

15.2 In-line sensors for product quality

15.2.1 Demands of automation

In some sectors of manufacturing industry, for example petrochemicals, plants are operated with fully automatic control for normal production. A small number of supervisors are present to attend to alarm conditions, to make decisions concerning supplies of raw materials and allocation of finished product, to monitor the progress of production schedules and sometimes to optimize the process, for example by trimming the proportions of input and output streams to optimize economic performance and yield.

There are obvious advantages in leaving the routine work of plant operation to automatic devices equipped with suitable safeguards. It is likely that production plants in most sectors of processing will follow the trend towards full automation.

Raw-material variability will always be present in some form, and process conditions can only be adjusted to compensate for incoming variability if some measurement is made of product quality in-line. This measurement, partly analytical in purpose, requires a special class of sensor. The features required for in-line quality sensors include:

(i) *Performance*:
    Relevance of output signal to product quality

Adequate sensitivity and repeatability
Sufficient dynamic range and speed of response
(ii) *Durability*:
Few or no moving parts
Insensitivity to ambient temperature
Immunity to electrical interference
Protection against shock, dust and contamination
(iii) *Ease of use*:
Provision for testing and calibration
Hygienic design (crevice free or non-contact)
(iv) *Cost of acquisition and upkeep*:
Reasonable capital cost
Self-diagnostic failure indication
Infrequent routine maintenance.

This is a demanding set of requirements, and plant engineers have difficulty in obtaining sensors which satisfy them. The absence of suitable in-line sensors for product quality is frequently given as the explanation for food processing being less than fully automated, in spite of strong incentives for automation. The following paragraphs describe some of the few sensors which have gained acceptance for in-line quality monitoring.

15.2.2 Density

The principles of in-line fluid density measurement are described in various measurement and control textbooks, and some current types of sensor are reviewed by Langdon (8). One simple sensor consists of a container of known volume, filled with the fluid, being vibrated on a calibrated spring mounting to give a frequency which is a function of the enclosed mass. It is of course more convenient to allow the fluid to flow continuously in a pipe, and pipe resonators in the form of a U-tube are available from several manufacturers. The two parallel sections of pipe are driven to vibrate in opposition in flexural mode, by magnetic field, like a double-ended tuning fork. The vibrational forces, as in a tuning fork, cancel out at the point of support. The resonant frequency is

$$f = \frac{f_0}{\left(1 + \dfrac{\rho}{\rho_0}\right)^{1/2}}$$

where ρ is the fluid density, f_0 is the resonant frequency when empty, and ρ_0 is a constant.

The U-tube sensor can have accuracy of about 0.1%, and is suitable for slurries. It is important, in this as in all density measurement, to take precautions against entrained air.

The flexible couplings needed for a single vibrating U-tube are a disadvantage for hygiene and maintenance. The same principle of opposed resonance can be used with the fluid contained in two parallel U-tubes with fixed ends, and with the fluid passing through both tubes. No flexible couplings are needed, but there are more bends, causing a greater pressure drop.

For some high viscosity fluids, U-tubes present too much resistance to flow; for these applications, the Solartron division of Schlumberger have developed a single straight-through smooth-bore transducer. This tube is vibrated lengthwise by excitation coils mounted outside the tube assembly. The tube is isolated from the case by flexible bellows, which also serve to prevent thermal expansion and plant pipe-work stresses from affecting measurements. The whole unit is enclosed within an all-welded case, and can be supplied with bellows of sinusoidal format which facilitate cleaning and conform with food hygiene standards.

Solartron comment that direct in-line calibration of density transducers is not practical. They recommend a zero check after flushing, cleaning and filling with dry air; to check the range they recommend temporary installation of a second transducer in series, pre-calibrated as a transfer standard.

An alternative and more compact type of density sensor was developed by Agar (9). A thin-walled metal cylinder is mounted within the pipe, on the pipe axis. Once again resonance can be induced, provided the fluid is not too viscous, by magnetic field applied externally. The cylinder is flattened alternately on axes at right angles. If the cylinder wall is very thin (to minimize the vibrating mass) it is possible to use this sensor to measure the density of pressurized gases.

15.2.3 Viscosity

With some geometrical rearrangement, induced vibrations can also be used for in-line viscometry. Early forms of in-line viscometer used an oscillating sphere suspended in the fluid at the end of a vibrating support. The damping of the oscillation by the viscous fluid was measured either by measuring the exponential decay of oscillation after removing the driving force, or by measuring the power needed to sustain oscillation. Both these techniques have practical disadvantages, and it is preferable to monitor the phase lag between the applied force and the motion of the sphere. If this phase difference is zero, at the driving frequency ω, then

$$\tan \phi = \frac{2\omega_0 \alpha}{\omega_0^2 - \omega^2}$$

where ω_0 is the resonant frequency with no fluid, and α is the damping constant, proportional to the square root of viscosity.

Schimanski (10) describes another type of in-line viscometer similar to the familiar laboratory type, in which a cylinder is rotated inside a fixed cylinder, with the gap between them filled with the fluid. The original 'MacMichael' viscometer, introduced in 1915, was of the concentric cylinder type, and modern versions of it are described by Bourne (11), in which he distinguishes between constant-speed 'MacMichael' type instruments and constant-torque 'Stormer' types. The force to maintain rotation is directly proportional to the area of shear between the cylinders, and is measured from the torque on the drive to the inner cylinder. This torque is only proportional to viscosity for Newtonian fluids. Other fluids like chocolate have viscosity which is dependent on shear rate, and thixotropic fluids possess a structure which breaks down under shear, giving viscosity which is a function of time under shear. The structure gradually rebuilds after shear is removed. Schimanski explains that for some fluids it is possible to extend the range of a viscometer from tenfold change in viscosity to two-hundred-fold change in viscosity by increasing speed automatically as viscosity falls. Alternatively, the speed change can be utilized to find the change in viscosity of non-Newtonian fluids with shear rate. The change is proportional to a power of the shear rate.

15.2.4 *Composition*

Of all the in-line measurements which might be made, the automatic monitoring of product composition is the most fundamental. Although great care is taken in process design to ensure the accurate make-up of recipes and formulations, there are numerous ways in which deviation from specified formulation can occur. Some of the more important sources of error are:

(i)   Dispensing inaccuracy: dispensing equipment may develop systematic errors after initial calibration, and these are often difficult to detect
(ii) Mechanical faults: feeding systems and check-valves may fail totally or partially, permanently or intermittently
(iii) Physical separation: constituents may tend to separate, particularly two-phase mixtures, or powders of different density
(iv) Incomplete flushing: after a change to a different grade of product, when the processing equipment is not emptied and cleaned between grades.

These considerations provide a strong incentive for monitoring the composition of finished product and of product streams at intermediate stages of processing. Such monitoring can never be total, but some forms of partial monitoring are available at a cost which is not prohibitive. It is always necessary to ensure that a representative portion of the product stream is tested; this applies to all process monitoring, in food production as in all the process industries. Techniques for sampling of multicomponent streams have been reported (12).

The principles of several types of in-line composition sensor are given in a review by Kress-Rogers (13). Near infrared (NIR) analysis can be carried out in-line to monitor the proportions of a broad class of food constituents such as

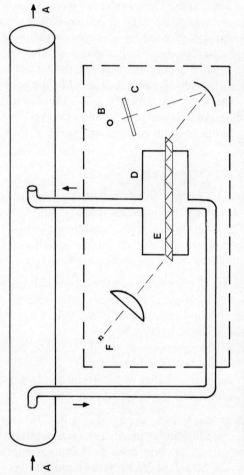

Figure 15.4 In-line NIR analyser for CO_2 (14). *A*, process stream; *B*, source; *C*, filter; *D*, sample cell; *E*, sapphire crystal; *F*, detector.

moisture, fat and protein, if is possible to extract a sample from the process stream for presenting to an analyser window. For solid samples, NIR analysis is carried out using surface reflectance; for fluids, cells are available which utilize the technique of multiple internal reflection. Frant and LaButti (14), describe a sensor for monitoring the sugar and carbon-dioxide in a carbonated beverage by this method (Figure 15.4).

Measurement of NIR surface reflectance gives an indication of surface moisture, but moisture gradients are common in all bulk materials during processing, and other techniques may be preferable for in-line measurement of total moisture content. Sensors based on measurement of changes in electrical conductivity or capacitance are widely available, but these are affected by the concentration of ions. The absorption of microwaves is determined by complex permittivity alone, and this is a strong function of moisture content. Microwave moisture meters are widely available, and are useful for monitoring various forms of material. They are inevitably affected by product density (15), and problems can arise from distortion of the microwave field if the product is in the form of shaped pieces rather than a continuous stream. A development of the microwave sensor (16), in which both absorption and phase shift are measured, provides a method for density-independent moisture measurement in food powders and grains.

Ultrasonic velocity is a function of concentration is some fluids, and has been used to test some juices, syrups and sauces (16), but the technique is restricted by the severe attenuation of ultrasonic propagation in viscous two-phase fluids. Wider application has been found for in-line refractive index measurement as a means of monitoring the concentration of sugar syrups.

Further progress with in-line composition sensors can be expected from current work on electrochemical sensors, also discussed by Kress-Rogers (13). pH electrodes are already well established for some specific applications in the food industry, such as the control of the setting of fruit jam. Newly developed ion-selective membranes will enable the introduction of further special purpose in-line sensors. Another form of pH sensor can be made by omitting the insulating layer from the gate electrode during the manufacture of field-effect transistors (FETs). The exposed surface can then interact with hydrogen ions in any solution in which the device is immersed. If ion-selective membranes are deposited over the gate, the device becomes an ion-selective FET (ISFET) (17). This concept has been extensively developed, and new forms of sensitivity have been introduced, based on enzymes immobilized in the deposited layer. A device sensitive to penicillin, for example, was reported by Caras and Janata (18).

Immobilized enzymes are also used in sensors which measure the heat evolved in an enzyme-catalysed reaction. This thermal bioanalyser type of probe is described in a review by Mosbach and Danielsson (19).

15.2.5 *Checkweighing*

The accuracy and speed of in-line checkweighers have been improved by two significant technical advances.

Firstly, weighing mechanisms have been developed with an output which increases linearly with applied force, enabling checkweighers to have 'full live range'. Formerly it was necessary to weigh with counterweights, and it was only possible to measure over a small range, without being able to monitor small changes in tare weight of the supporting platform. Accuracy can be improved if the tare weight (the weight registered when the platform is unladen) can be checked automatically each time the weighing platform becomes empty. Cells which have sufficient dynamic range to measure laden and unladen states without use of counterweights are known as full-live-range cells. The first full-live-range load cells were carefully constructed flexure systems, with appreciable displacement and mechanical inertia.

The second technical advance came with 'zero displacement' load cells. Strain gauges are now used to measure very small displacements in a specially-shaped block of metal when a load is applied. Load cells of this form retain the virtue of full-live-range, with the added advantage of greatly improved speed of response. They can be damaged by overloading, and it is hard to protect them against this because, in contrast to the flexure method, overloads cannot be prevented by a buffer or mechanical stop. Some low-range cells are deliberately designed to flex sufficiently to permit a mechanical stop to be incorporated. Settling times of 40 milliseconds are possible in some low-range cells from Hymatic Industrial Controls and other suppliers, and this is fast enough to permit a checkweigher to operate at 10 times per second.

Load cells are also becoming widely used to achieve tighter control of batch quantities held or dispensed by bulk handling systems. A single load cell, in one leg of the structure supporting a hopper, can be used to indicate the mass of material in the hopper to within about 3% of the full scale reading. A cell in each leg, together with precautions to prevent side loading and to isolate conveyor ducts and electrical conduit attached to the hopper, can be used to increase the resolution to better than 1%. This is within the dispensing tolerance of most bulk handling systems.

15.2.6 *Foreign-body detectors*

Metal detectors are the most common example of in-line foreign-body detectors used in food processing. Product is carried on a non-metallic conveyor through an aperture around which are three coils. High frequency is applied to the centre coil, inducing an identical voltage in each side coil. If there is no metal present, the induced voltages cancel one another when the coils are connected in opposition. Any metal passing through the aperture distorts the field and causes a transient out-of-balance voltage. Magnetic ferrous particles provide strong coupling between adjacent coils, and thus the system is most sensitive to magnetic particles. Non-ferrous metals such as aluminium, copper and brass are less easy to detect, and stainless steel gives the weakest response of all metals. Phase-sensitive detection is sometimes used to discriminate between materials which give out-of-balance signals.

X-ray inspection provides an alternative method for detecting gross items in bulk material, and for material in sheet form there are a number of optical inspection systems which offer good resolution at high speed. Optical inspection systems acquire data either from TV cameras or from 1- or 2-dimensional photodiode arrays. TV cameras have the advantage of being widely available, and readily interfaced. On the other hand, they are restricted to a set number of frames per second, and frame stores and analysers are complex and expensive. Diode array cameras can scan at rates up to 10 million points per second. If the product being inspected is passing through the field of view at a constant rate, a one-dimensional array of, say, 1024 elements is capable of providing 100% inspection with better resolution than is obtainable from a TV camera. The output from the array is easy to separate into data from successive scans for real-time analysis using dedicated low-cost systems. The dimensional stability of linear scanning is also superior to the stability of TV images.

Metal detectors are now used alongside checkweighers on most automated packaging lines. Optical inspection is not yet commonplace, but is likely to achieve widespread acceptance in the next few years.

15.3 Off-line sensors for maintaining quality during processing

15.3.1 QC for an automated plant

It is sometimes predicted that all necessary quality measurements will be made automatically in future, either directly in-line or by use of automatic sampling devices. It is clear that routine manual testing of samples in the factory laboratory is a declining activity, just as the number of manual operations in the production process itself is declining—in both cases, wage costs and employment overheads are increasing, while the costs of automatic testing and process automation are falling. One effect of the decline in routine laboratory testing is to allow staff more time for investigative activity. There will always be occasions when investigation is necessary, in connection with new ingredient supplies and with the introduction of new products; and of course technical services are needed for many aspects of support for production. Laboratory equipment is thus increasingly used for non-routine work, and instrumentation therefore needs to be adaptable and flexible in application.

Standard techniques such as soxhlet extraction for fat determination, or oven drying or Karl Fischer titration for moisture are widely found throughout the industry. The measurement of chocolate 'temper' by recording

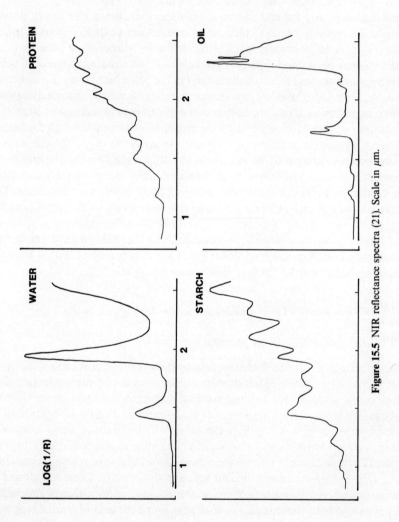

Figure 15.5 NIR reflectance spectra (21). Scale in μm.

the change in temperature of the chocolate as it sets is described in Chapter 11. The following sections, however, cover some of the more complex techniques used in the analysis of chocolate and its components.

15.3.2 NIR reflectance spectroscopy

Near infra-red (NIR) spectroscopy is widely used as a rapid and convenient method for assessing moisture content and other aspects of composition. 'NIR' is used to refer to the spectral region from 1000 to 2500 nm. At wavelengths longer than 2500 nm, spectrometers reveal a pattern of absorption bands with characteristics which identify functional groups in a sample. NIR spectra have a narrower spectral range and exhibit overlapping absorption bands, making them difficult to interpret. The relative weakness of absorption in this region does, however, allow measurements to be made without sample dilution.

Diffuse reflectance is usually measured with the sample placed against an opening in an integrating sphere. The main beam of illumination is directed onto the sample from an aperture diametrically opposite, while a reference beam for compensating for changes in source and detector characteristics can be introduced through the same aperture at an angle. In the popular types of scanning spectrophotometer, the illumination is provided from a grating monochromator. The integrating sphere needs to have reflectance as close as possible to 100% for all wavelengths. A commercial coating known as Halon is recommended by the US National Bureau of Standards (20) for use in the ultraviolet, visible and near infrared. Barium sulphate, suitable for ultraviolet and visible wavelengths, is not sufficiently reflective in the infrared. Powdered sulphur is almost 100% reflective in the infrared, and is slightly more reflective than Halon at 2100 nm.

Karl Norris of the USDA has published many papers on reflectance spectroscopy for composition analysis, and in a review of the technique (21) lists the following among the factors which affect reflectance data:

(i) Concentration and absorption coefficients of constituents
(ii) Particle size, shape and packing density
(iii) Refractive index of particles
(iv) Sample temperature.

Reflectance varies exponentially with concentration and absorption coefficient such that $\log(1/R)$ is approximately linear with concentration, where R is total diffuse reflectance relative to a perfect white diffusing surface. Reflectance increases as particle size decreases, while increasing the difference between the refractive index of particles and the fluid in which they are immersed also increases the reflectance. Sample temperature affects both the magnitude and wavelength of the water absorption bands.

The $\log(1/R)$ spectra of starch, protein, water and oil are shown in

Figure 15.5. The overlapping absorption bands are a typical feature of the near infra-red reflectance spectra of composite food materials. Quantitative composition data can be obtained more easily in such cases by taking the second derivative of the $\log(1/R)$ curve, and noting the amplitude of the sharp minimum in second derivative, which corresponds to the peak absorption wavelength of an absorbing component. The calibration procedure still requires multiple regression analysis of data from large numbers of samples of known composition. Manufacturers of NIR spectrophotometers provide software to assist selection of wavelengths to be included in the regression. Norris (21) recommends validation of the calibration by predicting the composition of another set of samples.

15.3.3 *Gas chromatography (GC)*

The basis of chromatography is the separation of sample constituents achieved by passing the sample through a medium in which the constituents are dissolved or absorbed at different rates, and recording the emergence of the constituents at different times after passage through the medium.

If the medium is packed in a column, there are many possible flow paths, and retention times are spread over a broad range. More selective results are obtained with the stationary medium within a capillary maintained at a preset temperature. Jennings (22) has pointed out the advantages of capillary systems compared with packed columns. Capillary systems are preferred partly because the vapour pressure of a solute is an exponential function of absolute temperature; in a column, the low thermal conductivity of the packing material tends to give an uneven temperature distribution in the column.

For food applications, Jennings distinguishes between 'total volatile analysis', which includes those volatiles dissolved in fatty components or adsorbed on particles, and 'headspace analysis' of the vapour in contact with the food. Preconcentration is recommended, by extraction, distillation and the use of adsorbants such as activated carbon. Gas chromatography has been used extensively to test the suitability of packaging materials for food applications, by revealing the presence, for example, of residual solvents in the material.

Volatile solutes have a tendency to move faster than the carrier gas, and distort the peaks in the chromatogram. This tendency is reduced by the use of lower coolant temperatures. To illustrate the sensitivity which can be achieved by using a liquid-nitrogen-cooled trap to delay the headspace volatiles on introduction to a capillary, Jennings (23) shows chromatograms, produced by injecting 1 ml of headspace overlying 2 g of cocoa at 65 C into a 30 m column of 0·25 mm fused silica capillary, and programming the oven to ramp from 50 to 200 °C at 10 deg/minute. The detailed pattern of the chromatogram is significantly clearer than the equivalent pattern obtained when using dry ice as the coolant. Corradi and Rizzello (24) used GC to analyse the lipid fractions of cocoa nibs and cocoa shells for acidic components, to reveal differences between nibs and shells.

Ziegleder (25) describes several applications for gas chromatography to identify volatile constituents liberated during cocoa processing. More than 360 volatile constituents have been identified. The amino acids 3-methylbutanal and 2-methylpropanal represent a substantial component of the flavour precursors formed during the fermentation of cocoa beans. Insufficient fermentation is indicated by a high proportion of pentanon-2. The ratio of 3-methylbutanal/pentanon-2 thus offers an index of degree of fermentation.

For indication of degree of roasting, Ziegleder finds methylfuran a good indicator but difficult to analyse because of its low concentration. 3-methylbutanal is preferred, with the unchanged peak from diacetyl providing a reference.

If broken nib roasting or roasting of a thin layer of cocoa mass is carried out, head space measurements of 3-methylbutanal provide a valuable aid to maintaining the critical temperatures required for this method of processing (see also Chapter 5).

Ziegleder's review concludes with details of the analytical procedures recommended for gas chromatography work in this application.

15.3.4 *Liquid chromatography*

High pressure liquid chromatography (HPLC) is carried out in closed systems, with pumps for rapid circulation of reusable solvent at pressures between 30 and 200 bar. Retention times of a few minutes are usually sufficient to separate the bands from different components. The retention time t_R is expressed as

$$t_R = t_0(1 + k')$$

where t_0 is the transit time of the solvent and k' is the 'capacity factor' of the column (26). The bandwidth t_W is expressed by the plate number N, such that

$$N = 5.16(t_R/t_W)^2$$

N is approximately constant in given conditions, and is thus a measure of column efficiency.

Resolution of two adjacent bands 1 and 2 is defined as

$$R = \frac{2.(t_{R1} - t_{R2})}{t_{W1} + t_{W2}}$$

and in terms of parameters which are set for a given system:

$$R = \frac{1}{4}(\alpha - 1).N^{1/2}.\frac{k'}{1 + k'}$$

where α is the separation factor. N increases with column length, and decreases with solvent velocity. k' is varied by changing solvent strength.

UV or visible photometers with logarithmic response are commonly used for detection in HPLC. If the solute is fluorescent, then fluorescence detectors offer greater sensitivity, but they may be non-linear. Differential refractometry is another detection method, suitable for all solutes.

Columns are packed with rigid solids, gels, or porous or particulate materials. If the latter, then pore or particle size should be as small as possible, to give a high value for N. Functional groups are bonded to the packing material in bonded-phase chromatography (BPC). In normal phase BPC decreasing the solvent polarity increases k', while in reverse phase BPC an increase in k' is observed as solvent polarity increases. Solvent mixtures of modified hydrocarbons are used in normal phase BPC, while water is the base solvent in reverse phase. Log k' varies as $1/T$ (Arrhenius characteristic) with halving of k' for approximately 30 °C increase in temperature. Normal phase BPC is carried out at ambient temperatures, while reverse phase BPC requires 50–80 °C (120–180 °F).

Kirk (27) comments that HPLC has advantages over other chromatographic methods for quantifying water- and fat-soluble vitamins in foods, and can be applied to the separation of vitamins using a wide range of physical and chemical parameters. Isomeric forms of nutrients can be separated.

Lehrian, Keeney and Lopez (28) proposed a method for using HPLC to identify phenols associated with the 'smoky' flavour of cocoa beans, and this method has been used successfully on cocoa mass from shelled unroasted beans. Ziegleder and Sandmeier (29) have described the testing of cocoa for degree of roasting by HPLC. They used a detector sensitive to ultraviolet (280nm), and found 5 peaks in the chromatogram, corresponding to:

1. Pyrazine
2. 2, 3-dimethylpyrazine
3. 2, 5-dimethylpyrazine
4. 2, 3, 5-trimethylpyrazine
5. 2, 3, 5, 6-tetramethylpyrazine.

Peak 5 is well known to arise from raw cocoa. Peak 4 increases continuously with roasting time at 120 °C (248 °F), showing that trimethylpyrazine is suitable as an indicator of the degree of roasting. Peaks 2 and 3, from dimethylpyrazine, arise after roasting at 140 °C (284 °F) or after extended roasting. Thus HPLC can provide separate indication of degree of roasting and of over-roasting. This technique has been used to show the importance of particle-size uniformity when nibs are roasted; small pieces rise more rapidly in temperature than the larger pieces, and become over-roasted on drying out, while the heat is still penetrating the larger pieces.

Ziegleder and Sandmeier correlated the relative intensity of peaks 3, 4 and 5 with sensory evaluation, and found the ratios peak 3/peak 5 and peak 4/peak 5 both correlate with degree of roasting, the ratio peak 3/peak 5 being particularly indicative of perceived over-roasting (burnt taste). Beans, kernels, cocoa mass, granules or powder may be sampled for testing in this way. Degassing from a thin layer of cocoa mass provides a sample which can give useful information about processing when tested by gas chromatograph.

Kimmey and Perkins (30) have used HPLC to distinguish the triglycerides of cocoa butter from those of replacement fats. They used a column packed with octadecyl-bonded 5-μm silica spheres and a mobile phase of acetone and acetonitrile. Peaks corresponding to eluted components were collected and their composition determined by GC of the corresponding methyl esters. The combined procedures can be completed within 30 minutes, and the method is useful for estimating the compatibility of a fat as an extender or substitute for cocoa butter; it also reveals adulteration of fats or oils.

15.3.5 *Atomic absorption and emission spectrometry*

Atomic absorption, emission and fluorescence provide the basis for another group of food analysis techniques. These are used primarily in nutritional assays for determining traces of minerals.

Ihnat (31) describes the application of the basic instrumental configuration shown in Figure 15.6 to food analysis. The source is a discharge lamp producing a sharp line spectrum of the element of interest. The light beam is interrupted by a chopper, to which the detector is synchronized to enhance sensitivity and eliminate interference by emission at the same wavelength from flame excited atoms. Attenuation of radiation passing through the flame provides a measure of concentration of the element in the sample, atomized by the burning of the fuel-oxidant mixture. There is an obvious requirement for identification of elements essential to the diet on the one hand, and of toxic elements on the other.

Absorption and emission wavelengths for major and trace elements which are nutritionally essential are shown in Table 15.1*A*, and wavelengths for five

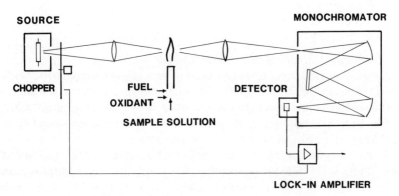

Figure 15.6 Atomic absorption spectrometry (31).

Table 15.1 Strongest absorption and emission wavelengths. Data from Ihnat (31)

Element	Absorption at	Emission at
A. *Essential elements*		
Sodium	589.00 nm	589.00 nm
Potassium	766.49	404.41
Magnesium	285.21	279.55
Calcium	422.67	393.37
Boron	249.68	249.77
Vanadium	318.34	309.31
Chromium	357.87	205.55
Manganese	279.48	257.61
Iron	248.33	238.20
Cobalt	240.73	238.89
Copper	324.75	324.75
Zinc	213.86	213.86
Selenium	196.03	196.03
Molybdenum	313.26	202.03
B. *Toxic elements*		
Nickel	232.00	221.65
Arsenic	193.70	193.70
Cadmium	228.80	214.44
Mercury	253.65	194.23
Lead	283.31	220.35

toxic elements are shown in Table 15.1*B*. In absorption spectroscopy all principal spectral lines originate from the atomic ground state, but for plasma emission the lines often correspond to transitions between energy levels within singly ionized atoms.

A sample solution is mixed with a fuel mixture to provide the flame used in absorption spectrometry. Fuel combinations and maximum temperatures observed for stoichiometric mixtures are as follows:

	°C	°F
Air/acetylene	2400	4350
Nitrous oxide/acetylene	2800	5070
Air/propane	1800	3270
Air/hydrogen	2000	3630

Electrothermal atomizers reach 3000 °C(5430 °F).

Plasmas obtain energy from collisions induced by electromagnetic fields. A quartz plasma torch surrounded by an induction coil is fed with seed electrons introduced into the argon stream by collision. A plasma at about 8000 °C (14500 °F) forms above the torch, which excites the sample introduced into the discharge. The emission spectra can then be monitored using a spectrometer.

Atomic spectra contain so many lines that analysis can be made difficult. The situation is worse for emission spectroscopy than for absorption. Ihnat (31) gives the example of the measurement of chromium at 267.716 nm, listing overlapping lines from 21 other elements between 267.595 and 267.829 nm.

Spectral bandpasses used in emission spectroscopy clearly have to be very narrow.

The preferred form of sample is a liquid. Solid samples require acid digestion or dry ashing. Clear digests are obtained with nitric/perchloric and nitric/perchloric/sulphuric acids in borosilicate glass. After dry ashing in a silica or platinum crucible, the ash is dissolved in dilute hydrochloric acid.

Liquid samples may require only dilution with water or solvent before analysis. Trace elements require concentration by solvent extraction. Elements As, Se and Hg may be liberated as vapours for introduction to the spectrometer. Standards are prepared in the same solvent as the sample, and are used for preliminary adjustment of light source, flame composition and spectrometer aperture.

As a combination of techniques, it is possible to use a spectrometer as a detector to identify components in the effluent stream from a gas or liquid chromatograph. One further analytical spectrometer is the mass spectrometer, used to confirm the identification of an organic compound. The sample is ionized by bombardment with high-energy electrons, and the ions are accelerated through electric and magnetic fields where they are separated according to mass and velocity.

15.3.6 Testing for bacteria

Bacteria require moisture to sustain growth, and the osmotic pressure in food materials (on a scale of water activity) determines whether conditions are favourable for bacteria. The osmotic pressure depends on the concentration of dissolved materials, such as salts, sugars, organic acids and their ions. Beuchat (32) reports that, with only a few exceptions, bacteria require water activities of 0.91 or higher. Moulds may grow at water activity as low as 0.8, and osmophilic microbial growth occurs at water activities lower than 0.5.

Food manufacturers need to assess bacterial activity by methods that are more rapid than the conventional plate-count techniques, which take several days. One rapid method is the Lumac method, based on bioluminescence. Another widely used method is the microbial monitoring technique employed by Bactomatic, in which a sample is mixed with a medium and placed in a temperature-controlled impedance sensing cell. The time taken for the number of organisms to reach 1 M/ml is measured. The impedance detection time (IDT) method has been used to determine the quality of pasteurized whole milk (33). 5 ml samples of milk were added to 5 ml of broth, and pre-incubated at 18 °C for 18 hours. After that time, each sample was inoculated and further incubated at 18 °C or 21 °C during impedance measurement. 18 °C was selected as the temperature which would prevent one group of organisms outgrowing others during preliminary stages, while 21 °C was used to obtain accelerated growth and hence shorten detection times. Results at both temperatures were found to give close correlation with shelf-life tests. The results were available in 25 to 38 hours after processing. Rapid techniques which are easy to operate and require no special skills have obvious advantages in industrial applications, and are being adopted for routine testing in most food processes where microbiological hazards are present.

15.3.7 Particle size determination

Traditionally in the chocolate industry quality control has been limited to a measurement of the largest particles. This has been to ensure that the product does not taste gritty to the consumer. These single-point measurements have been carried out by relatively simple instruments, such as sieves, micrometers or microscopes. All these techniques rely heavily upon the skill of the operator, and although the results may be meaningful in relative terms, they often have little meaning as absolute measurements. Of these, the micrometer is probably the most widely used instrument throughout the industry. Usually the chocolate is dispersed in a fatty or oily medium before being placed on the jaws of the micrometer. The sample preparation is critical, as is the pressure the sample is subjected to as the jaws close.

As the price of cocoa butter has increased, it has become more and more desirable to obtain the same flow rates in the chocolate while lowering the fat content. This can be achieved in part by adjusting the particle size distribution of the solid material (see Chapter 9). It has therefore become necessary to measure the full range of particle sizes. Initially this was often carried out by the Coulter counter. Here the particles are dispersed in a conducting medium, and passed one at a time between electrodes. The change in conductivity gives a measurement related to the volume of the particle. More recently, instruments based on the scattering of laser light by the particles have come into use within the industry (34).

Two types of machines are on the market. One, like the Coulter, passes the particles individually through the laser beam. The amount of light obscured by the particles is then monitored, and the individual signals combined to give a particle size distribution. Like the Coulter the size range covered is limited by the size of the orifice used to direct the particle past the detector. Several orifices may be required to cover the complete size range of chocolate. The more common method is to shine the laser through a suspension of the particles. The amount of light scattered into detectors, placed at the different distances from the axis, can then be used to determine the distribution. This usually covers a range from $2-180 \mu m$ ($8 \times 10^{-5}-7 \times 10^{-3}$ in). This can be extended, however, to $0.1 \mu m$ (4×10^{-6} in) by using information from light scattered at right angles to the axis.

Other devices working on sedimentation or centrifugal principles have also been used for chocolate. They tend in general, however, to be much slower to operate than the light scattering devices. In addition , the differences in density between the components may introduce added difficulties in interpreting the results.

15.4 Measurements of product attributes

15.4.1 *Alternatives to subjective testing*

All food products are eventually judged at the moment of consumption. Food manufacturer seek to ensure that the products give the consumer a sense of satisfaction; subjective sensory testing will always have a place in quality control, but increasingly sensory tests will be augmented by testing with instruments. The equipment and methods available for some forms of testing of product attributes are discussed in the remaining part of this chapter.

The aspects of testing described below are much more specific to chocolate and confectionery than the instrumentation described in section 15.3.

15.4.2 *Gloss*

Gloss is measured using a goniophotometer (Figure 15.7). Collimated light is projected on to the surface to be tested, and the strength of light reflected within a narrow angle is measured. By moving the detector in an arc about the point at which the incident light meets the surface, the directionality of the reflected light can be plotted as a goniophotometric curve. The detector moves in the vertical plane containing the incident beam and the normal to the surface at the point of measurement. The incident beam is usually projected at 30° or 60° angle of incidence, but for some purposes it is useful to measure at other angles—'sheen' for example is best revealed by using grazing incidence. Chocolate which has been formed with a sufficiently flat surface can be tested using this method, provided that the measurement can be completed before the incident radiation causes too much heating. A 60° angle of incidence is

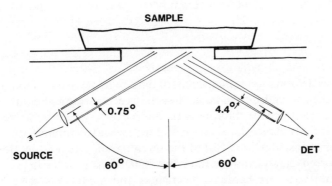

Figure 15.7 Specular glossmeter. Angles as for ASTM D523 (35).

Table 15.2 Gloss measurement procedures (34)

Test and specification	Application	Width of angle of incidence	Width of angle of view	Numerical scale
Sheen, 85 deg (ASTM D523)	Matt paint	0.75 × 3	4 × 6	Black glass = 100
Gloss, 75 deg (TAPPI T480)	Glassine paper	2.8 × 5.7	11.4 dia	Black glass = 100
Gloss, 60 deg (ASTM D523)	Paint, plastics, intermediate gloss	0.75 × 3	4.4 × 11.7	Perfect mirror = 1000
Gloss, 45 deg (ASTM D2457, PEI T2, 7, 18)	Porcelain	1.4 × 3	8 × 10	Perfect mirror = 1000
Gloss, 30 deg (ASTM E430)	Metal finishes	0.44 × 7	0.38 × 4	Perfect mirror = 100
Gloss, 20 deg (ASTM D523, TAPPI T653)	High gloss film and paint	0.75 × 3 or 1 to 2.5 dia	1.8 × 3.6 5 dia	Black glass = 100

ASTM: American Society for Testing and Materials, Philadelphia, PA
TAPPI: Technical Association of the Pulp and Paper Industry, Atlanta, GA
PEI: Porcelain Enamel Institute, Washington DC.

recommended for measurement of chocolate gloss (as for ASTM D523, see Table 15.2), with Illuminant C used as the source. Standard Illuminants are specified according to spectral distribution by the Commission Internationale d'Eclairage (CIE). Illuminant A is the light from a typical incandescent lamp, Illuminant B represents direct sunlight, Illuminant C represents average daylight from the total sky, and Illuminant D is daylight modified according to one of a set of specified colour temperatures. The range of light intensities in the reflectance measured at different angles of view is so large that it is customary to plot the intensity on a logarithmic scale. Standard procedures have been specified for gloss measurements used in other sectors of manufacturing industry. Table 15.2 summarizes some of the standard methods referred to by Hunter (35).

15.4.3 Texture

Numerous deformation testing methods have been devised for food products, many of them intended to simulate the actions of chewing and biting. Many of these are described by Bourne (11). Bourne gives a table of mechanical parameters with dimensional analysis of the measured variable, quoted from an earlier publication (36). Out of the seven parameters used by Bourne for texture profile analysis, three have the dimensions of force (mass × acceleration), and these are hardness, brittleness and gumminess. Two have the dimensions of work (force × distance) and these are adhesiveness and chewiness. The other parameters are springiness (or work/stress, with dimension of length) and cohesiveness (dimensionless ratio of maximum stretch/original length).

Methods for compression and shear testing and for measurement of melting and adhesion properties are described by Tscheuschner and Markov (37). For the compression test, cylindrical chocolate samples 11.28 mm (0.44 in) diameter and 100 mm (4 in) cross-section and 10 mm (0.4 in) high were cut at 30 °C (86 °F) (plain chocolate) or 28 °C (82 °F) (milk chocolate) and stored for at least 60 minutes at 20 °C (68 °F) before being compressed between plates of 100 mm (4 in) base area at a crosshead speed of 5 mm (0.2 in)/minute. Both types of sample show maximum stress at 19% compression. The maximum stress is reached at less compression if the crosshead speed is increased.

For shear testing, prism-shaped samples $20 \times 10 \times 5$ mm $(0.8 \times 0.4 \times 0.2$ in) were prepared as for the compression test. The test cell had a rectangular slot 20.2×4.2 mm $(0.8 \times 0.16$ in) into which a 20×40 mm $(0.8 \times 1.6$ in) plunger was driven at 5 mm (0.2 in)/minute. Tscheuschner and Markov comment that maximum shear stress values, which are lower than the maximum compression stress values, can be used as a measure of texture, but provide less information on structural characteristics than compression testing.

Cylindrical samples of the same size as those for compression testing were used for measurement of melting properties. Exactly 5 seconds after placing the cylinder, stored at 20 °C (68 °F) on a surface controlled at 37 °C (99 °F), a constant force of 1600 N (360 lbf) is applied to the top surface of the cylinder, and the rate of decrease in height of the cylinder is noted. Rates of 0.16 mm/s (0.006 in/s) and 0.23 mm/s (0.009 in/s) were observed for plain and milk chocolate respectively.

For the adhesion test, a sample of molten chocolate is placed on steel plate so as to cover a circle of 960 mm² (1.5 in²) total area to a depth of 0.5 mm (0.02 in). The outer retaining mould is removed, and a concentric circular plate of area 940 mm² (1.46 in²) is pressed on to the sample with a force of 1 N (0.22 lb force) for 10 seconds. The upper plate is then raised slowly at 11.3 mm/minute (0.45 in/minute) and the development of the rupture force recorded. The force increases linearly to a maximum value of the order of 5 N (1 lb force), when rupture occurs.

References

1. Duncombe, E. J. Phys. E. **17**(1) (1984) 7–17.
2. Bentley, J.P. J. Phys. E. **17**(6) (1984) 430–439.
3. Ginesi, D. and Grebe, G., Adv. in Instr. **40**(2) (1985) 1173–1194.
4. Tullis, J.P. and Smith, J. Paper no. 6.3, Fluid Mechanics Jubilee Conf., National Engineering Laboratory, East Kilbride, Glasgow (1979).
5. Plache, K.O. Transducer Technol. **2**(3) (1980) 19–23; (4) 5–6.
6. Lees, F.P. Chem. Ind. (1976) (March) 195–205.
7. Upfold, A.T. Inst. Technol. **18**(2) (1971) 46.
8. Langdon, R.M. J. Phys. E. **18**(2) (1985) 103–115.

9. Agar, J. Apparatus for measuring the density of a fluid. UK Patents 1175586 (1967) and 1542564 (1976).
10. Schimanski, H. *Control & Instr.* **17**(11) (1985) 115–117.
11. Bourne, M.C. *Food Texture and Viscosity: Concept and Measurement.* Academic Press, New York (1982).
12. Manka, D.P. (ed.) *Automated Stream Analysis for Process Control.* Vol. 1, Academic Press, New York (1982).
13. Kress-Rogers, E. *J. Phys. E.* **19**(1) (1986) 13–21; (2) 105–109.
14. Frant, M.S. and Labutti, G. *Anal. Chem.* **52** (1980) 1331A–1344A.
15. Nelson, S.O. *J. Microwave Power* **18** (1983) 143–152.
16. Zacharias, E.M. and Parnell, R.A. *Food Technol.* **26**(4) (1972) 160–166.
17. Bergreld, P., De Rooji, N.F. and Zemel, Y.N. *Nature* **273** (1979) 438–443.
18. Caras, S. and Janata, J. *Anal. Chem.* **52** (1980) 1935–1937.
19. Mosbach, K. and Danielsson, B. *Anal. Chem.* **53** (1981) 83A–94A.
20. *NBS Optical Properties of Pressed Halon Coatings.* Radiometric Physics Divison, National Bureau of Standards, Boulder, Co. (1979).
21. Norris, K.H. in *Modern Methods of Food Analysis*, eds. Stewart and Whitaker, AVI Pub. Co., Westport (1984) 167–186.
22. Jennings, W. *Advances in Chromatography* **20** (1982) 197–215 (eds. Giddings, Grushka, Cazes and Brown), Marcel Dekker, New York.
23. Jennings, W. in *Modern Methods of Food Analysis*, eds. Stewart and Whitaker, AVI Pub. Co., Westport, (1984) 319–338.
24. Corradi, C. and Rizzello, F. *Boll. Chim. Lab. Prov.* **33** (S4/S5) (1982) 559–571.
25. Ziegleder, G. *Choc Confect. Bakery Rev.* **7**(2) (1982) 17–22.
26. Snyder, L.R. and Kirkland, J.J. *Introduction to Modern Liquid Chromatography.* 2nd edn., Wiley Interscience, New York (1979).
27. Kirk, J.R. in *Modern Methods of Food Analysis*, eds. Stewart and Whitaker, AVI Pub. Co., Westport (1984) 381–406.
28. Lehrian, D.W., Keeney, P.G. and Lopez, A.S. *J. Food Sci.* **43** (1978).
29. Ziegleder, G. and Sandmeier, D. *Deutsche Lebensmittel Rundschau* **79** (1983) 343–347.
30. Kimmey, R.L. and Perkins, E.G. *JAOCS* **61**(7) (1984) 1209–1211.
31. Ihnat, M. in *Modern Methods of Food Analysis*, eds. Stewart and Whitaker, AVI Pub. Co., Westport, (1984) 129–166.
32. Beuchat, L.R. *Cereal Fds Wld* **26**(7) (1981) 345–349.
33. Bishop, J.R., White, C.H. and Firstenberg-Eden, R. *J. Food Protection* **47**(6) (1984) 471–475.
34. Robbins, J.W. *Manuf. Conf.* (1983) (June), 185–190.
35. Hunter, R.S. *The Measurement of Appearance*, John Wiley, New York (1975).
36. Bourne, M.C. *J. Food Sci.* **32** (1967) 601–605.
37. Tscheuschner, H.R. and Markov, E. *J. Texture Studies* **17** (1986) 37–50.

16 Packaging

A.V. MARTIN and N. FERGUSON

This chapter is divided into three main sections. The first deals with the various forms in which chocolate confectionery is presented and packaged, the second with the materials used in packaging chocolate confectionery and the third takes a brief look at the quality assurance methods applied to these materials.

16.1 Forms of chocolate confectionery

16.1.1 Moulded chocolate blocks

As might be expected with a product which has existed comparatively unchanged for so many years, the basic packaging requirements of chocolate are relatively simple. It needs protection against handling, dirt, taint and insect infestation as well as against moisture. For many years this has been afforded by aluminium foil (or its predecessor, tinfoil) with a paper band or overwrapper to provide display appeal and carry the ever-increasing volume of legally required product information (Figure 16.1).

There seems to be little likelihood that regular-shaped moulded chocolate blocks will depart from this traditional packaging. Reductions in the gauge of foil, area of foil used and similar economies in the paper employed mean that this style of wrap is highly economical. No extra space is taken up with individual wraps when they are packed together in boxes or cases, so distribution costs are minimized. Most importantly, extensive market research continues to show that the style of wrap is associated by the consumer with this type of product, including the traditional flat segmented bars filled with fondant creams or praline and chocolate neapolitans.

A number of bars, generally 2–6 dozen, are packed together into a box known as an outer, usually made of solid cardboard, but 'E' flute corrugate is frequently used where extra strength is required. When this outer is to be used for counter display the bars are generally packed flat, but if it is for solely transit purposes they will usually travel better standing on end. Where extended shelf-life is required, heat-sealed foil provides a wrap. This is impervious to practically all the hazards of distribution but is expensive. Lesser protection which is frequently perfectly adequate in moderate climates can be provided by polyethylene-lining the outer or overwrapping it in barrier film.

For most markets a number of outers are put into a corrugated shipping

Figure 16.1 Traditional foil and paper-wrapped products.

case which may be glued or taped shut. Alternatively, a collation of several outers can be wrapped together in transparent stretch film for ease of handling, although the finished unit is less strong than the shipping case.

Many large customers increasingly insist on having the minimum of handling, the minimum of packaging materials to dispose of and the maximum product display. These requirements can be met by the use of large trays which can go straight on to the shelves, a number of such trays being cased or wrapped together.

An even larger transit and display unit is the pallet box, in which a large container, generally corrugated cardboard with its own integral pallet base, is bulk-filled with product. All the shopkeeper has to do is remove the lid and drop the front to achieve instant display.

16.1.2 Chocolate countlines

The other, and increasingly important, mainstay of the chocolate confectionery market, the so-called 'countline', has for many years been packaged quite differently from the moulded bar. Generally irregular in shape and enrobed rather than moulded, countlines do not lend themselves to packaging in unbacked foil.

Until relatively recently this type of product generally had a 'die-fold' or 'DF' wrap in which the wrapping material, usually a heavily-backed foil, waxed paper or glassine, was formed to a fixed size by a folding box. The size of package so formed had to accommodate the bars largest in all dimensions, so most tended to rattle around inside and the pack was prone to crushing.

Furthermore, die-fold equipment is relatively inefficient and slow and is unable to take full advantage of modern flexible materials such as polypropylene film.

The countline wrapping scene, however, has been revolutionized by the advent of the 'flow wrap', 'pillow pack' or 'fin seal', as the package produced by the continuous horizontal form-fill-seal machine is variously described. This takes a continuous web of flexible material and forms it into a package around the product by making a tube and sealing its edges together by heat and/or pressure as it progresses through the machine. The tube containing the product is then cut to a predetermined length and the open ends sealed, again by heat and/or pressure.

The continuous flow wrap, by varying the materials used, can give a very wide range of properties to the package. It can be used simply as a collation pack to hold a given number of pre-wrapped pieces where there is no requirement for barrier properties; alternatively, it can protect its contents against moisture vapour and oxygen, the air inside having been replaced by an inert gas such as nitrogen or carbon dioxide.

The requirements of the great majority of countlines do not, however, usually call upon all the possible protective resources of the flow wrap. It generally suffices for it to exclude dirt, moisture vapour and potential taint, and a wide range of materials can do this, ranging from thin coextruded polypropylene to film/foil/ionomer* laminates.

Currently, however, by far the largest proportion of countline production is wrapped in 'pearlized' film, i.e. white, cavitated polypropylene with cold seal (Figure 16.2). This material is bright and glossy and lends itself to colourful

Figure 16.2 Flow-wrapped countlines and multipack.

*A family of polymers containing both covalent and ionic bonds, especially Du Pont's 'Surlyn'.

and attractive presentation, with the major advantage of eliminating the 'optical staining' or 'show through' which can occur when chocolate-coated bars are wrapped in conventional film or paper. It also provides protection against light-induced damage, which can be a problem when treated nuts are involved.

Metallized film has never supplanted aluminium foil for conventional bar wrapping as it has never matched its essential dead fold* qualities, but it is in widespread use on flow-wrap machines where its striking appearance and protective qualities can be utilized, again primarily in conjuction with cold seal.

Some countlines are still twist-wrapped, unsealed, with longitudinal overlap and the film ends twisted. This does not really protect the product, except from dirt and handling, but it is an essential part of the product's character and thus not lightly dispensed with.

The outer packaging and casing of countlines is basically the same as that for moulded blocks except that a higher proportion are put into display outers and very few are transported on end (Figure 16.3). Because the ends of the wrappers are extended compared with closely wrapped foiled products, this form of wrap requires larger outers and cases; thus distribution costs are increased.

*When foil is folded it stays folded—it does not spring back, like paper or plastic films.

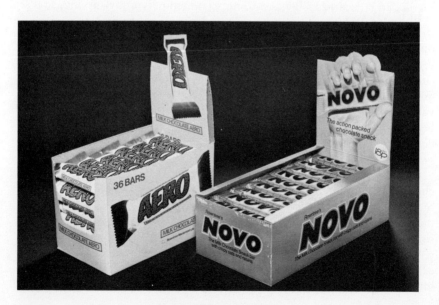

Figure 16.3 Display outers.

16.1.3 *Bulk chocolate*

Although most bulk chocolate is now transported in liquid form in tankers, an appreciable market still exists for block or pelleted chocolate. In both instances the preferred package is the polythene-lined multi-ply paper sack containing say 5 × 5 kg (11 lb) blocks or an equivalent weight of pellets or flakes, etc. To maintain stability it is advisable to ensure that the outer paper ply is treated against 'slip', and a polyethylene sheet between the transportation pallet and the load will reduce any possibility of moisture or taint from the wood affecting the product. Liquid chocolate can be poured into a case, generally corrugated, lined with a polyethylene bag. When the chocolate has set, the finished pack is very stable for transportation and the bag and chocolate can easily be removed for use.

16.1.4 *Boxed chocolates*

The variety of containers for boxed chocolates appears to be almost infinite, ranging from rigid handmade boxes with plinth bases, metal feet and tasselled lids to machine-erected single-walled cartons. The materials used run from flock paper, covered straw or chipboard, to metallized or film laminated carton board. Rigid plastic can be used to give a clear container showing off brightly-foiled sweets to best advantage. Alternatively, thin film in the form of a window can be combined with cartonboard to give the same effect, but in conventional carton style.

In contrast with the multitude of container forms, the inner fitments of chocolate boxes vary relatively little from box to box. Although they might one day make a comeback, the heat-formed glassine cups once universally used to hold individual sweets are currently to all intents and purposes completely replaced by thermoformed* plastic trays with an individual cavity for each sweet. These trays can be made from several different plastics in a variety of colours as well as crystal clear and gold or silver metallized. They can be made to reproduce the appearance of a layer of paper cups, or designed to give extra protection to particularly vulnerable centres, or to build up flatter sweets so that the whole layer appears of uniform height.

Other items which may be incorporated in the internal packaging of a box of chocolates are wave-embossed glassine or greaseproof layer sheets, and single- or double-faced fluted glassine or greaseproof to provide cushioning and prevent crushing of the sweets. Glassine-lined cellulose wadding can perform a similar function, perhaps more attractively but certainly more expensively.

Some nutty sweets such as pralines or coconut clusters exude fat to such an extent that they can permeate glassine or cellulose layers and require a sheet of impermeable film to prevent staining of the box itself; an alternative, more

---

*Produced by heating and softening plastic sheet which is then drawn by vacuum into a metal mould of the desired shape.

unusual, method of restraining fat exudation is the simple expedient of foiling the offending sweet. In addition to protecting the packaging from the sweet and, in some overseas markets with particularly extreme climates, providing a last protective barrier to the sweet, foil is frequently used on individual sweets for decorative purposes.

The principal means of protecting boxed chocolates from the atmosphere, and incidentally the box itself from scuffing and marking, is a film overwrap. This can take the form of a cellulose or plastic (generally polypropylene) overwrap with sealed lapover along the base and sealed envelope folds at the end. Modern materials and overwrapping machines make these overwraps very effective, but they are difficult to get into, a problem which can be avoided by the use of tear tape. Until recently most tear tape was applied by solvent or hot-melt adhesives, neither of which are particularly 'operator friendly', but a recently developed self-adhesive tape would appear to be much more acceptable and effective.*

Standard cartons of chocolate are frequently converted to seasonal use by the addition of a printed paper or film band with a seasonal theme which can be removed from unsold stock after the festival. The same objective can be achieved by the use of self-adhesive stickers, e.g. heart-shaped for St Valentine's Day. Such stickers can be made of a wide variety of materials, ranging from high-gloss paper to crystal-clear polyester.

When boxes of chocolates are combined together in an outer, a conscious decision needs to be taken as to whether they should travel flat or on edge. If there is any likelihood of a large soft-centred sweet coming under pressure it is often better to pack the carton on edge so that the thermoformed tray can relieve the pressure, provided it is made sufficiently strong to do so. In addition, in a double-layer box the tray should be so designed that the upper layer can be turned round to place a small sweet over a large one in the bottom layer.

In other respects the construction of outers and cases is much the same as for bars or countlines.

16.1.5 Twist wrapping

Gaily wrapped twist-wrapped chocolates have been a major part of the confectionery scene for many years. The basic materials used have become rather more sophisticated and the machinery faster, but fundamentally there has been little change. The materials used are still aluminium foil, backed or unbacked, plain silver or coloured, and film, tinted, clear or printed. The foil can be applied separately as an understrip or it can be laminated to the film (this can cause problems because the reel of material builds up in the centre where there is a double thickness of material). Twist-grade metallized paper is

*Supplied by Payne Packaging Ltd, Giltbrook, Nottingham, UK.

being used increasingly to add an extra dimension to the variety of wraps in an assortment, at a somewhat lower cost.

The most common form of twist wrap is the 'double end fantail' but a popular alternative is a combination of twist wrap at one end and folds at the other, giving a sachet-type wrap. This is particularly effective with fruit designs, for instance for strawberry creams where the body represents the sweet and the fan tail the leaves.

Because the attraction of a twist-wrapped assortment lies in its bright, glossy and colourful packaging, the container in, or from, which it is sold usually tends to be transparent or at least have a clear window. For many years twist-wrapped chocolates were sold from screw-topped returnable glass jars, but, in line with the modern trend to non-returnability, these have generally been replaced by clear PVC containers. The glass jar is still a popular container for this type of assortment, but as a retail unit because of its perceived re-use value as a storage container, and in rather more elaborate shapes.

Brightly printed round tins have also been used traditionally as both consumer and weight-out packages, the finish and quality of print and design making up for the loss of visibility. Tinplate, however, is expensive, and there are moves to replace it with metallized materials such as cartonboard laminated with metallized film or metallized plastic drums.

Twist-wrapped sweets also lend themselves to packaging in printed film bags, generally with display windows, produced on vertical form-fill-seal machines. The variety of films which can be used to make the bags is very wide, but at present polypropylene or polyethylene in one form or another is usually used. Such bags are frequently intended to be hung from pegs for display purposes, so the material, or configuration of the top seal, should be such as to enable a hole to be punched in it and the pack suspended without tearing.

16.1.6 *Easter eggs and other seasonal chocolate novelties*

Traditionally, Easter eggs and other seasonal novelties such as Easter rabbits, Father Christmas, chocolate bells, etc., have been foiled; taking advantage of the ability of foil to follow irregular contours and to be smoothed into shape as well as its decorative appeal when printed.

Foiling Easter eggs is relatively simple, either by hand or machine, but the designing of a piece of foil to go around Father Christmas requires the skill of a specialist to ensure that all the features on the foil match up with those of the chocolate figure. This can be achieved by using the printed foil, in heavy gauge, as the mould itself, or by lining the mould with the foil. The usual method is, however, to produce the figure first and then foil it, relying on the skill of the foiler to apply the properly designed foil accurately. Machinery is available to foil simple shapes and, where appropriate, to attach cords for hanging from Christmas trees (Figures 16.4, 16.5).

Figure 16.4 Rasch Type RG universal foiling machine.

Where long shelf-life and possibly adverse storage conditions are not involved, unfoiled novelties may be wrapped in clear or part-printed film to show off the quality of the product. Care must be taken to ensure that the film used is impervious to moisture and that conditions conducive to chocolate bloom are avoided. Strip metallized films are frequently used for this purpose as they are good barriers and highly decorative. They make good bags, but of course cannot be used to 'foil' as they have no dead-fold characteristics.

A wide variety of containers is available for presentation of foiled eggs and figures. The most usual is some form of open or windowed carton, and the range of shapes on the market is a striking tribute to the skill of the carton designer. Eggs are frequently sold in egg cups, printed beakers, wicker or plastic nests and baskets.

Although it is occasionally alleged that the retail trade is not above taking a hammer to eggs left after Easter and claiming 'breakages', there is no doubt

Figure 16.5 Foil-wrapped seasonal shapes.

that the most likely cause of damage is poorly-designed packaging. The most attractive cartons are useless if they cannot protect their contents from the rigours of the distribution chain, and it is often necessary to effect a compromise between design and strength. New high-quality boards and board laminates can enable designs to be used which were formerly impracticable, but at a cost.

It is always advisable to carry out carriage tests on packaging, designed as far as possible to reproduce the normal hazards of distribution, but with Easter eggs and novelties this becomes absolutely essential. It is also most important that such tests are carried out using exactly the materials which will be used in bulk manufacture and that these materials are clearly specified and checked. There have been many instances when cartons or platforms are made of different boards when printed from those supplied in plain form for testing.

16.2 Materials

16.2.1 Aluminium foil

This is the best barrier to water vapour and gas transmission available in flexible form. It is generally defined as fully annealed, soft temper metal of 99–99.5% purity, the remaining percentage being made up of silicon and iron with traces of other elements. Recent developments may have reduced the percentage of aluminium by incorporating manganese to give added strength (the new so-called 'special alloys'), but this is a secret so far well kept by the foil manufacturers.

Foil thickness can range from 5 μm to 20 μm (0.2×10^{-3} to 0.8×10^{-3} in) but for practical confectionery purposes the range is from 7–12 μm (0.3×10^{-3} to 0.5×10^{-3} in). Cost reduction has taken the form of the 'special alloys' already mentioned, rolled thinner on more efficient modern plant so that, for example, 8 μm (0.3×10^{-3} in) material can now be obtained that is stronger than the old 9/10 μm (0.4×10^{-3} in) foil and in some cases than traditional 12 μm (0.5×10^{-3} in).

Thinner gauges of foil are rolled 'double', giving the characteristic shiny and matt sides which appeal to artists and designers in different ways–some prefer the bright side for display value whereas others regard the matt face as conveying discreet quality.

For bar wrapping, foil can be specified in three main forms–plain unbacked, backed and coated. In most cases unbacked foil is used, but at times when added strength and crease resistance is required, e.g. when nuts protrude from the back of a bar, it may be necessary to laminate it to other materials.

The most popular backing material is paper in some form, as it is strong, easily printed and relatively inexpensive. Depending on the end use, and whether it is laminated to the outside or the inside of the foil, there is a wide choice of paper forms available, such as sulphite tissue, kraft (usually bleached) and glassine, with an equally wide selection of adhesives such as dextrin, polyethylene wax and hot melt. A widely used complex is, for example, 7 μm foil/3 gsm wax/20 gsm* tissue.

Cellulose and plastic films are used where extra barrier properties and puncture resistance are required and also on occasion to provide a high gloss finish. In this instance it is important to balance the relative gauges of material so that the "spring" of the film does not overcome the desirable dead fold of the foil. An example of such a complex is 9 μm (0.4×10^{-3} in) foil and 12 μm (0.5×10^{-3} in) OPP.**

When foil is coated, usually with vinyl or polyethylene in order to make it heat sealable, this has the added effect of filling in the minute pinholes unavoidable in thinner foil gauges. Although research has shown that most pinholes have little adverse effect on the barrier properties of foil, such filling in can only be beneficial.

Cold seal coatings are generally applied to paper foil laminates rather than to unsupported foil.

Provided all trace of residual rolling oil, which could go rancid under certain conditions, is removed, unconverted foil is perhaps the material least likely to taint a sensitive product such as chocolate. Converted foil (i.e. foil converted by

---

*gsm = grams per square metre
**OPP = oriented polypropylene in which clarity, strength and barrier properties have been improved by stretching the film under heat in a specific and controlled manner.

printing, coating or laminating) can create odour problems from coatings and adhesives as well as from printing inks, so it is important that full taint and residual solvent tests are carried out before such material is used.

Foil is most attractive when printed; this can be done by all the major printing processes, particularly gravure and flexography, and it is frequently overlacquered with transparent colours to make full use of its metallic visual properties. Embossing is also a particularly effective treatment for foil, although it does tend to reduce its tensile strength.

In its unconverted form, aluminium foil has a wide covering area, ranging from $46\,\text{m}^2/\text{kg}$ ($25\,\text{yd}^2/\text{lb}$) for $8\,\mu\text{m}$ ($0.3 \times 10^{-3}\,\text{in}$) to $31/\text{m}^2/\text{kg}$ ($17\,\text{yd}^2/\text{lb}$) for $12\,\mu\text{m}$ ($0.5 \times 10^{-3}\,\text{in}$) but coating and lamination in particular will drastically reduce this.

16.2.2 Paper and board

These are materials produced from the natural cellulose fibres found in trees. The production process broadly consists of taking the wood apart and subsequently rejoining the fibres, without their own binding materials, as an engineered matrix designed to perform a specific function. This pulping can be done either chemically (the organic binding material holding the fibres together is dissolved away) or physically (the wood is broken up and the binding material washed away by water). The former method produces strong, expensive paper or board, the latter relatively weak, cheap material, which is, however, perfectly adequate for many purposes, such as newsprint.

The finished slurry is dried out in sheet form and passed through rollers to make paper, or built up in layers to make board. The layers can be of different qualities, making a multilayer sandwich by including different materials such as reclaimed waste paper and pure fibres. If the board or paper is to be printed, it normally has a top layer of good-quality bleached fibres plus a coating of china-clay-containing pigment such as titanium dioxide and possibly an optical brightening agent. This is known as coated paper or board, and many different types and qualities are available.

Differing chemical techniques can produce glassine, where the fibres are completely destroyed, giving a very high degree of resistance to oil and grease in the final product, but making the paper relatively weak. It can be formed into rudimentary shapes by heat and pressure, producing, for example, paper cups into which chocolates can be placed, or very fatty product such as noisette deposited directly. Glassine is frequently used laminated to board, in order to protect it from fat penetration and staining. It can also be incorporated in a non-heat-sealed wrap to contain fat from chocolate bars in hot climates. Glassine in corrugated form is also in widespread use, ranging from 'sincor' (wave-embossed) to single- and double-faced fluted material, to provide cushioning inside cartons of chocolates.

Similar cushioning, but of rather coarser quality, can be achieved by using corrugate made from greaseproof or vegetable parchment. The latter is produced by the action of concentrated sulphuric acid on wood pulp, giving greater strength, along with grease resistance.

In many instances boxboard outers are packed into shipping cases for distribution, particularly where rail or sea transportation are involved, and here corrugated board is used. No other material offers corrugated board's unique combination of product protection, stacking strength, printability, light weight and relatively low cost.

In general, 'corrugate' is made of two plies of liner material separated by a layer of fluting or corrugating medium. Both liners and fluting can be treated to give wet strength, flame resistance etc. The fluting is formed from the flat material by huge meshed rollers—glue is applied to the tips of the fluting and the liners are then stuck on. Usually this is done on one long machine, which also prints the finished materials and cuts it into the appropriate shape for the finished case. The type of corrugate is described by the number of flutes per cm or foot. The most widely used version is 'B' flute, which has some 1.7 flutes per cm (51 flutes per foot). For smaller boxes where a finer material and more decorative surface are required, 'E' flute, with 3.2 flutes per cm (96 flutes per foot), is often used.

Depending on the use to which the finished box is put, many different qualities of material can be used for the liners, ranging from heavy-duty kraft down to glassine or greaseproof.

The traditional method of printing as an integral part of the process of corrugate manufacture, followed by creasing and slotting, results in a poor quality of reproduction by modern standards. There is therefore an increasing tendency to pre-print the outer liner before it is formed into corrugate. This enables more sophisticated printing methods* to be used, giving much better decoration, and illustrations can be achieved by the use of halftones. Variable information can be applied at the last moment using the actual corrugating equipment.

16.2.3 Regenerated cellulose film (RCF)

Until relatively recently RCF dominated the flexible packaging field. Like paper and board it is made from wood pulp, although generally of a higher quality: a high proportion comes from eucalyptus grown specially for the purpose.

A chemical process is used in which, as in paper making, individual fibres are put into solution. Instead of being spread out in layers, they are then chemically regenerated and passed through a slot to form a transparent film. Glycerol and various glycols are used as plasticizers and to add flexibility to the film.

---

*High-quality print methods such as gravure or litho can be applied to single-web paper or board but not to formed corrugate, as they would break down its structure.

In its natural state RCF has few characteristic properties other than transparency, flexibility and the ability to form a barrier to oxygen, provided it is dry. It cannot be sealed and it is very susceptible to moisture, expanding or contracting with changes in atmospheric humidity. This property was taken advantage of in wrapping fancy boxes before the advent of custom-designed shrink films. Plain RCF would be dampened and placed over the box so that when it dried out it gave a very tight, sparkling wrap. Unfortunately its shrinkage was uncontrolled, so either the film would split or the box crush, depending on which was the stronger.

In order, therefore, to make RCF a practicable proposition it has to be coated. Initially it was made heat-sealable and given some barrier properties by the application of a nitrocellulose coating. This material (MS) is perfectly adequate for most confectionery purposes. Where extra protection is required, a PVdC (polyvinylidene chloride) coating is used and the film is then characterised as MXXT, with the suffix 'S' for solvent applied, or 'A' for aqueous dispersion coated. Although it is more effective and avoids the problem of disposal of residual solvents, the reactivity of RCF to moisture makes the aqueous method of coating much more difficult. MXXT films generally have more sparkle than MS and more resistance to abrasion.

Other coatings or treatments can be used to make RCF permeable to some gases and not to others. This enables the atmosphere inside to be controlled to a certain extent for the benefit of some products. It can also be made extra flexible (PF) for twist wrapping, supplied uncoated where moisture pick-up is not a problem, e.g. most chocolate-covered sweets, and it can be coloured either intrinsically or in the coating.

Coated RCF is an ideal film for automatic packaging, as it is not subject to structural alteration because of temperature; however, it cannot be thermoformed. It offers a combination of rigidity with elasticity—it will stretch before it breaks—and it avoids problems associated with static electricity. Its tensile and burst strengths are good; a tear, once made, propagates easily, but this is actually an advantage when a tear strip is required.

Heat-seal temperature is not critical, as coated film has a relatively wide sealing range. It can be obtained in various thicknesses, and in general the thicker the film, the stronger it is; barrier properties, however, do not vary with overall thickness, as the coating weight (expressed in gsm equivalent units) remains the same. RCF, unlike most plastic films, is biodegradable so it appeals to environmentalists.

On the debit side, RCF, even when coated, does not have the shelf-life of plastic materials. It requires carefully controlled storage conditions, but even so will deteriorate with time.

Although banned in the USA, ethylene glycol (MEG) and diethylene glycol (DEG) have until recently been in widespread use in Europe as softeners in RCF. Severe restrictions on the amount of these additives migrating into foods have now been introduced throughout the EEC.

16.2.4 *Plastics films*

Despite all its properties, regenerated cellulose film has largely been supplanted by plastics films for flexible packaging. PVdC, used to coat RCF, is itself a plastic and can be obtained in film form as well as coating. The range of plastics materials is very wide and is increasing, but this chapter will confine itself to those which play a significant part in the packaging of chocolate products.

Polyethylene was the first major synthetic material to find a place in packaging and it is still the most widely used. The most outstanding characteristics of its commonest low-density form are low cost, flexibility, moisture protection, heat sealability and versatility. It can be made with a 'memory' so that it can be induced to shrink or stretch and remain shrunk or stretched or revert to its original form. It can be used as a coating, as a film, as an adhesive or in solid form.

On the other hand, polyethylene is not a good barrier to gas or taint and it is not easy to seal on its own: its surface is non-polar, so treatment with flame or corona discharge is necessary before it can be printed or induced to accept adhesive. Care must also be taken to ensure that polyethylene coatings or laminates are odour-free.

As a film, its principal uses are where toughness rather than clarity is required, e.g. as a stretch film for pallets or trays of heavy items. It can also be used as a barrier film on pallets. As a coating, it adds toughness, moisture barrier and heat-sealability to other materials such as paper and aluminium foil. In its opaque white form it can be used to coat very poor-quality grey board (partially derived from reconstituted waste paper and board) and give it a high-quality finish. As an adhesive, it can be used to laminate disparate materials and additionally provide a moisture and fat barrier where these materials are deficient in such properties.

Polyvinyl chloride (PVC) is most commonly used where clarity and sparkle are required in a film, rather than strength, e.g. for shrink wrapping fancy boxes. It has good resistance to oils and fats, but its water-vapour permeability is relatively high. Where protection against moisture is important, for example if a chocolate box to be overwrapped contains wafer-based sweets, a form of polypropylene shrink film would be preferable.

Much thicker PVC is used in thermoformed trays for packing chocolate assortments. Although in some parts of the world the old glassine cups are again being used, there is little doubt that the great saving in labour resulting from the use of thermoformed trays to locate and protect sweets in predetermined formation will encourage most manufacturers to continue using trays. The situation could be reversed, however, by the development of robotic packing (see Figure 16.6). PVC for use with food should be unplasticized to minimize the possibility of taint or migration from the packaging into the product.

The plastic which has had perhaps the greatest impact on the packaging of confectionery is *polypropylene* (PP). This is a close cousin of polyethylene; both are polyolefins derived from ethylene, but PP has an even wider range of applications. As a film it can be made very thin, down to $12\ \mu m$ $(0.5 \times 10^{-3}\ in)$; it is the least dense commonly-used plastic material, so it is very light, and the combination of both characteristics gives very high yields.* It can be coated with other materials or coextruded with them to give any combination of

*yield = area of material contained in a given weight, generally expressed as square metres per kilogram or square inches per pound.

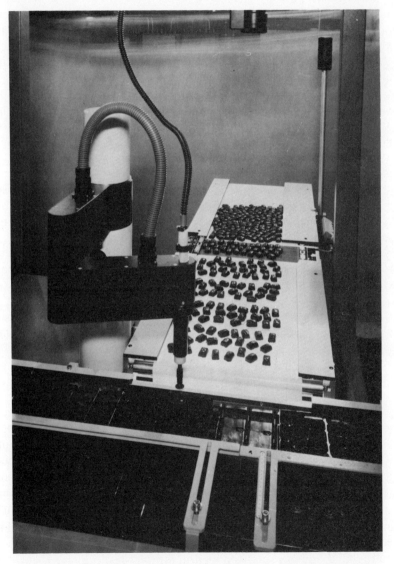

Figure 16.6 Otto Hänsel vision robot: pick and place unit (Otto Hänsel GmbH).

properties required in a package. It runs well on automatic packaging lines, having a low coefficient of friction with metal, and it has a wider sealing range than its competitors, provided the machinery has adequate heater controls.

In its clear form, polypropylene has largely captured the box overwrapping market, and most confectionery bags are now made from it. It can be used to make thermoformed trays, although these are more rigid and less 'user friendly' than the unplasticized PVC version. On the other hand, polypropylene shrink film, if rather more difficult to handle, gives more protection than PVC.

The material whose use has shown the most rapid growth worldwide is white opaque or 'pearlized' polypropylene, which has replaced glassine and other forms of paper as the principal countline wrapping material. As well as being light, thus giving a high yield, it is relatively rigid and crisp and it can be made either with high sheen or low, to look like paper. Perhaps its biggest advantage for use with chocolate-coated products is that it eliminates 'optical staining' (when wet or greasy marks show through), a constant problem with paper, glassine and transparent film.

Development work is continuing, several manufacturers striving to reduce density even further while increasing opacity and stiffness, without impairing the barrier properties of pearlized material. These can, however, easily be provided, albeit at a cost, by coating it with PVdC.

The other plastic film which has achieved relatively widespread acceptance in confectionery packaging is *polyester* (polyethylene terephthalate) which is very strong and clear. Its clarity is, unfortunately, impaired when the slip additives necessary to enable it to run on some machinery are added. It does not seal in its natural form, but this problem can be overcome by coextrusion or lamination to polythene for example. It does not tear easily, and is resistant to abrasion as well as to oils and fats. It is a fairly good barrier to moisture vapour and gas, and thus to taints.

Although it can be obtained commercially down to $12 \mu m$ (0.5×10^{-3} in) in thickness, polyester is still an expensive material, largely because of its other uses: recording tapes and in the electronics industry. As a packaging material it is probably most widely used in its metallized form, which greatly enhances not only its appearance, but also its barrier properties against moisture, gas and ultraviolet light.

While polyester is a particularly good carrier for vacuum-applied metallization, this technique and others are now used to add the above properties to oriented polypropylene and to coated cellulose film. As with pearlized PP, metallized polyester, used as a light barrier can retard UV-induced rancidity, thus extending shelf-life, and it also minimizes optical staining or 'show through'.

In many ways metallized films are superior to aluminium foil, which they closely resemble, but none has the same dead-fold properties, so they are primarily used on horizontal and vertical form-fill-seal equipment.

16.2.5 Cold seal

Although all these 'new' materials are capable of being heat-sealed by means of coextrusion, other coating methods, or lamination, there is no doubt that a revolution in packaging has recently come about through combining films with special properties and cold seal coating, based on natural rubber latex combined with resins.

The great advantage of cold seal is that all three parameters governing the sealing process (pressure, dwell time and temperature) are reduced. Without the necessity to control temperature, which can be particularly critical with plastic films, there are no restraints other than mechanical ones on the wrapping speeds which can be achieved. In addition, as heat is not required, the possibility of damage to chocolate products during machine stoppages is eliminated. Other benefits are reduced energy usage and wear on the sealing mechanism.

Not all films accept cold seal equally well, and it is important to ensure that both material and coating are fully compatible with each other and with the machinery on which they are intended to run. Few difficulties will arise if the cold seal is applied to a PVdC coating, but where the extra barrier properties of PVdC are not required, this could be a rather expensive way of achieving compatibility.

In order to ensure that the cold seal stays on the reverse of the film and does not cause 'blocking' in the reel, a release lacquer should be applied over the whole printed and unprinted surface of the material. Such lacquers are based on polyamide or nitrocellulose-modified polyamide, and may under certain circumstances be incorporated in the ink, as long as the ink covers the whole surface of the film. It is important to ensure that any inks used for overprinting or in line marking (for example of 'best before' information) can be applied over the release lacquer.

Some older horizontal form-fill-seal (HFFS) machines or flowwraps may require minor modifications to enable them to handle cold seal materials—for instance alterations to the infeed system may be required, and frequently cutting knives have to be adjusted to allow for the loss of expansion when heat is removed. More extensive alterations are required to enable vertical form-fill-seal (VFFS) machines to cope with cold seal; no doubt most new machines will include this facility. When using pattern-applied cold seal on opaque material, difficulties may arise in setting up the wrapping machine, or in checking presence and correct positioning of the cold seal on the area to be sealed. This can be overcome, however, by incorporating a light colour (generally pink or blue) in the sealant.

Tough plastic materials and cold seal have developed in parallel with the growing sophistication and availability of microprocessor controls. Together they have, in effect, removed the bottleneck frequently caused by the packaging element in an integrated production line, and at the same time have reduced operator supervision.

The use of cold seal means that wrapping machines can speed up or slow down in instant response to availability of product, whether they are wrapping naked bars straight from an enrober or moulding plant, or collations of wrapped bars from a number of primary wrapping machines. At speeds of some 600 or so per minute, automatic feed systems and automatic boxing are essential, as is the automatic splicing to register of the wrapping material. This avoids the need to stop the wrapping machine to renew the wrapping material.

16.3 Quality control

The integration of production lines such as those described above would hardly be possible without full electronic control of what is basically a series of mechanical operations. It also depends heavily on the quality of the packaging materials, and here such factors as print and sealant registration, particularly where the photoelectric cell (PEC) register mark is concerned, stretch, slip and accuracy of slitting, must be controlled to a much higher degree than before, and monitored by the end user.

This control of quality can involve a multiplicity of tests to be carried out before materials reach the production line. If agreement on standards can be reached with suppliers, it should be possible for them to carry out the bulk of the testing, with only spot checks being carried out by the user.

Nevertheless, a well-equipped packaging laboratory should be capable of carrying out at least the following tests:

Seal strength, tensile strength, burst strength
Coefficient of friction
Moisture vapour resistance (MVR)
Detection of residual solvents, e.g. by gas liquid chromatography (GLC)
Taint, using a panel of tasters
Print stability and resistance to abrasion
Scannability of bar codes.

Chocolate is particularly sensitive to external taint, and it is essential that every precaution be taken to exclude it. Most modern flexible materials are at least fairly good barriers to taint, and can be made very good by coating with PVdC, laminating or metallizing. The quality of seal is important: where the package is not sealed, external taint sources must be eliminated.

While it is appreciated that after the product has left the manufacturer's hands he has little control over the environment in which it may be kept and sold, there are many potential sources of taint in his own area of operations. These include residual solvents, or other likely contaminants in wrapping materials resulting from incomplete drying or oxidation of printing inks and varnishes, including sealants. Print can 'offset' on to the inside of reeled, or indeed any printed, materials and come into direct contact with chocolate. Incorrectly applied coatings or laminants such as polythene can cause taint.

In addition to its role in protecting its contents from hazards of climate and distribution, packaging is also instrumental in persuading the consumer to buy. It is therefore well worth taking pains to get it right.

References

1. Briston, J.H. and Katan, L.L. *Plastic Films.* (George Godwin, in association with The Plastics and Rubber Institute).
2. Hanlon, Joseph F. *Handbook of Packaging Engineering.* McGraw-Hill, New York (1983).
3. *Handbook of Cartonboard and Carton Test Procedures.* British Carton Association.
3. Paine, F.A. *The Packaging Media.* Blackie, Glasgow and London (1977).
4. Paine, F.A. and Paine, H.V. *A Handbook of Food Packaging.* Blackie, Glasgow and London (1983).
5. Paine, F.A. *Fundamentals of Packaging.* The Institute of Packaging, London (1981).
6. *The Practical Packaging Series.* The Institute of Packaging, London.

17 Non-conventional machines and processes

S.T. BECKETT

17.1 Introduction

Each chocolate manufacturer employs his own individual production techniques; larger companies may have several processes. Most, however, use conventional equipment while perhaps altering the order or the length of time of a process. Sometimes machinery, such as mixers and evaporators, is introduced from other parts of the food industry, and indeed one of the most widely used pieces of equipment, the roll refiner, is found in such diverse places as the paper and paint pigment industries. Apart possibly from the chocolate conche, most machinery is also used to manufacture non-confectionery products. The aim of this chapter, however, is not to catalogue individual items of equipment that have been introduced from time to time, but to review some of the published unconventional chocolate-making methods. This includes showing how they deviate from the standard ones, and giving some of their claimed advantages and possible disadvantages.

The development of novel chocolate-making machinery normally falls into one of three categories:

(i) Milling equipment to replace the five-roll refiner
(ii) Methods to shorten the processing time
(iii) Complete chocolate-making systems (including those where raw materials are processed to finished chocolate without supervision).

In addition, tempering and chocolate coating-machines are the subject of a large number of patents.

17.2 Milling processes

17.2.1 *Limitations of roll refiners*

The vast majority of chocolate manufacture over the past 30 years has involved a roll refiner. This breaks the solid particles to a size smaller than can be detected individually in the mouth. Frequently, these refiners consist of five cambered, water-cooled rolls up to 2.5 m (8 ft) wide. The chocolate passes between the four gaps, and the particles are crushed and sheared due to the pressure between the rolls and their differential speed (Chapter 6). The disadvantages of a five-roll refiner are:

(i) All the material passes through the gaps and differential crushing of the individual components may occur; for example, the sugar may be overcrushed in order for the cocoa nib to reach the required size.

(ii) In order for the chocolate paste to pass from one roll to another up the refiner it must have the correct range of 'textures'. If it does not, either the material passes between the rolls in thick layers and is not crushed correctly, or alternatively, it is thrown off the rollers by the centrifugal force. The latter creates a very dusty environment, and deposits very large amounts of material on to the trays under the refiners. This limits the range of particle size and fat content which can be processed on these machines.

(iii) The refiner is an expensive piece of machinery which requires relatively skilled operators to produce a product with a good particle-size distribution.

(iv) A refiner operates most efficiently in the range $> 30\,\mu m$ $(1.1 \times 10^{-3}\,inch)$ (1) and when the desired reduction in the maximum particle size is approximately 4–7 times (2). Therefore a combination with another type of mill for the bigger type of particles may be desirable. In addition, in certain materials such as cocoa mass and some dietetic sweeteners, platelets can form which may be only $25\,\mu m$ $(1 \times 10^{-3}\,in)$ in one dimension, but may be up to $100\,\mu m$ $(4 \times 10^{-3}\,in)$ in another (1).

Various alternative modes of operation have therefore been developed, for instance using more than one refiner, or the use of a two-roller refiner to pre-crush the chocolate masse before it is fed into the standard five-roll machine (3) (see Chapter 6). The two-roll machine is less sensitive than the five-roll to the texture of the ingredients as the feed material passes between only the two horizontal rolls.

17.2.2 *Alternative forms of mill*

In order to overcome the problem of one ingredient being overcrushed with respect to others, chocolate processes were developed whereby the individual ingredients were milled separately. This process was patented by Rowntree and Co. (4) in 1921, and adapted by many other workers (5, 6). Although the taste of the chocolate was different from that produced by the conventional process, the chocolate was claimed to be thinner at a given fat content owing to the better particle-size distribution. The process also enabled a wide variety of mills to be used, the choice often depending upon the fat content of the material and the size of the particles involved. Figure 17.1 shows a variety of mill together with the size range where they are most applicable.

Hammer or pin mills are able to crush sugar and milk powder to the final

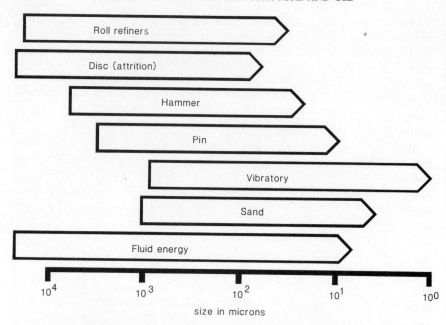

Figure 17.1 Range of application of different types of mill (1, 7).

size required for chocolate making, without requiring the use of a roll refiner. Although they produce some fine particles, these can be minimized by mill design and operation. In addition, some mills can be set independently for the sugar and for the milk, to give the best particle-size distribution for each ingredient. The high temperatures developed within this type of mill during operation will, however, melt most of the chocolate fat in the material being processed, and if the fat level is sufficiently high (this depends upon the material but is often about 10%) the particles become sticky and block the mill. In addition, high-fat materials do not easily break under impact. This problem can be overcome by operating at low temperatures. Frequently, liquid nitrogen is used to cool the mill or feed the materials. This retains the fat in a solid state, so that the particles break and yet remain in a form where they can be conveyed from the mill. The nitrogen also excludes oxygen, which is said by some workers (8) to impart a cheesy flavour with this type of grinding. Although this may operate satisfactorily in terms of engineering and the products, it is seldom economic.

For higher-fat material, such as the full chocolate recipe or cocoa mass, other types of mill are normally employed. Whilst 'disc' or other forms of attrition mill may be used to produce cocoa mass, their high operating temperatures would develop totally unacceptable flavours should sugar also be present. Ball mills, however, are used in cocoa-mass grinding systems, and also some chocolate-processing plants. For the latter, relatively long times are required to reduce the solid material to the final level of chocolate fineness. This means that efficient cooling systems are required, so that the chocolate is not burned. Also, very large machines or low throughputs are required to give the necessary residence time within the mill. This greatly increases the captial cost and running time of such a system. An alternative approach is to recycle material at relatively high flow rates. This is in fact claimed (9) to reduce the overall grinding time relative to a single-pass large mill, and is used in commercially-sold complete chocolate-making systems (see section 17.4).

Many variants of this mill are available commercially; some are known as ball, sand, pebble, perl, agitated media or rod mills, and each may have its specific application. Most of them, however, can be divided into one of two categories by their horizontal or vertical mode of operation. In the former a horizontal drum, usually cylindrical (or cylindrical/conical) and containing the grinding elements, often steel balls, is rotated on trunnions. In some mills the speed is less than 80% of that at which the charge would begin to centrifuge, but is high enough for the charge elements to tumble over each other without being lifted out of the bed. In another type (Figure 17.2) the charge may be lifted by using stepped liners, or by increasing the rotational speed beyond 80% of its critical value, or both. The lifted charge falls on to the rest of the charge and its contents, and grinds it by impact. The use of rods rather than balls is said to give a narrower particle size distribution (7) which should give improved flow properties.

The vertical type of mill is more commonly found within the confectionery industry. Here the material to be ground is often pumped as a slurry into the base of the mill, and it is discharged from the top via a screening device which retains the grinding medium (Figure 17.3). A rotating agitator device may have one of a variety of designs. Adjusting the rates of agitation and the throughput of the material enables the degree of crushing and the final particle size distribution to be controlled within a limited extent. A major disadvantage

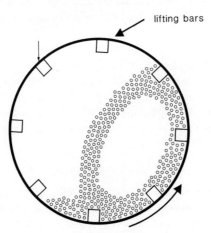

Figure 17.2 Horizontal mode ball mill.

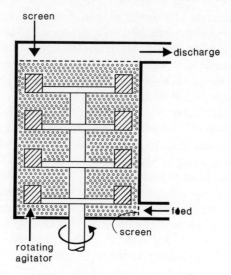

Figure 17.3 Vertical mode ball mill.

with this type of mill is that the material must be in slurry form before it can be pumped in. In general this means that a chocolate masse must have a relatively high fat content before it can be processed, and the use of these mills is consequently often limited to cocoa mass. Where the mill is being used for chocolate, however, it appears to be preferable to precrush the particles to a maximum size of less than 350 μm (0.01 in) using a two-roll refiner or similar device (10), before the material enters the ball mill. Another type of mill reported in the literature is the jet (or fluid energy) mill. Here the material is sent at a high velocity against a stationary block, or alternatively several high-velocity gas jets are used to make the particles collide with one another. For cocoa mass the beans must first be ground to almost final fineness, but these machines can be useful as a final processing stage, when the cocoa mass is roasted rather than the beans or nibs (see Chapter 5). The type of mill where the material is ground in two or more opposing air jets can also be used to grind sugar and milk powder, with or without cocoa powder (8). Normally some recirculation/classification system is required as not all of the particles are pulverized to the desired size during the first pass through the mill. In order to increase the efficiency of the mill, it is also desirable to precrush the sugar to smaller than 100 μm (0.004 in) before feeding it into the jet mill.

17.2.3 *Limitation of the size of the initial ingredients*

An alternative method of avoiding particle refining is to produce the sugar and milk solids in a form where they are already small enough to require only a limited amount of comminution in a hammer or similar mill, or are indeed fine enough to be incorporated into the finished product. These techniques usually incorporate a spray drier to dehydrate a sugar/milk solution, with or without cocoa material (11). A method patented by Tuross (12) describes how a narrow particle size range is achieved using a spray drier. Here the 'small' particles are redissolved and the oversized ones are crushed, in order to reduce this range.

A far older technique, used to a limited extent, is known as the 'boiled sugar process' (13). In this cocoa mass is added to a rapidly boiling, highly concentrated sugar solution. The sugar crystallizes spontaneously, producing crystals which are smaller than $30\,\mu m$ (1.1×10^{-3} in). The latent heat which is given out evaporates most of the residual moisture. For milk chocolate, concentrated milk is boiled with the sugar. The heat evolved during the processing caramelizes some of the milk, giving rise to a crumb type flavour in the final product. For both milk and plain chocolate the crystallized material is mixed in a heated pan or trough, and then passed through a refining roll. This is set to disperse aggregates and not to grind the sugar crystals, which are already small enough to be incorporated into the finished chocolate.

17.3 Methods to shorten the processing time

17.3.1 *Continuous conching* (see also Chapter 8)

In the majority of chocolate-manufacturing processes, once the cocoa beans have reached the factory the most time-consuming process is conching. This is normally a batch process, in which large quantities of chocolate are processed for long periods (up to several days). One method of overcoming the disadvantage of intermittent chocolate availability was to develop a continuous conching scheme (14, 15). Chocolate was then always available at a constant rate, reducing the necessity of large storage tanks.

One such machine, manufactured by Tourell, consisted of a series of vessels/tanks containing agitators capable of applying a shear rate appropriate to the viscosity of the chocolate. In each of the initial tanks the chocolate is mixed, and progresses along the chamber assisted by helical paddles. The period in each section is governed to a limited extent by an adjustable weir at the outlet. Following a minimum of two mixing stages, during which additional fat and/or lecithin can be added, the material passes through a relatively high speed mixing/shearing system. It is recommended that the lecithin should be added at this stage, in order to achieve the optimum effect from this surface-active agent. In one description of the system (14), part of the finished chocolate is recirculated to the first mixing tank where it is added to the newly ground ingredients. This is said to control the viscosity of the mixture in the earlier stages of processing without the requirement to add excessive amounts of fat or lecithin. The chocolate in fact remains within the conche for less than half the time of the traditional process. In order to reduce this time still further, other conche designers (15) have tried to speed up the flavour development process by subjecting the relatively low-fat refiner flake to a high shear while blowing conditioned air through the system. This is claimed to disperse aggregates quickly, while the rapidly changing surface enables the volatiles and moisture to escape. In the Konti-conche 420, two high-shear vessels in parallel treat the material from the refiner, followed by a mixing vessel for fat and surface-active agent addition. Finally, another high-shear device, this time without air, liquefies the chocolate. The use of the continuous conche has tended to be limited to a relatively few larger chocolate manufacturers where the continuous supply of a large quantity of chocolate is critical.

17.3.2 Flavour development machines

Conching is required to coat the solid particles with fat, develop flavour and remove moisture (Chapter 7, 8) the latter being important with respect to the flow properties of the chocolate. Whereas the liquefying and moisture removal can be relatively short, it may take many hours or even days before the best flavour is developed, particularly in plain chocolate. A widely used approach to reducing conching time is to try to separate the flavour development and liquefying processes. By pretreating the most highly flavoured component i.e. cocoa mass, conching times can be greatly reduced. Machines to alter the flavour of cocoa mass have been developed by Petzomat, Convap and Carle-Montanari (Chapter 5). In order to develop the flavour, some sugar may be processed along with the cocoa mass. In one process developed by Olezinkova and Bukin, cocoa mass and sugar syrup are mixed in proportions of between 1:1 and 1:2 with a residual moisture of 13–18%. This is then spray-dried after heating to 85–90 °C (185–195 °F), and is said not to require conching. The inclusion of the 5–10% sugar is claimed to accelerate the physiochemical processes, although its effect on the final chocolate viscosity is not stated. All these processes reduce the operation of conching primarily to one of liquefying the chocolate. With this is mind, devices such as the PIV have been developed (Chapter 8). Whilst cocoa mass treatment does reduce processing time and energy, it frequently produces a chocolate whose taste and texture differ from one where a non-treated cocoa mass has been conched for an extended period. This is particularly important in the case of plain chocolate.

The methods of cocoa mass treatment usually operate either on a continuous short-time, thin-film basis, or on a longer-period batch process. Each has its own advantages and disadvantages in terms of processing and quality of the final product.

17.3.3 Extruders

An alternative to pretreating the cocoa mass before adding the other ingredients is to subject an almost complete recipe to an additional physical process which will alter its flavour profile. The extruder is widely used in the food industry for applying high shear, fast heating or cooling, and for 'flashing off' certain volatiles. In the confectionery industry, the twin-screw extruder in particular is being employed in making licorice, toffee and jelly bean-type articles, and it is claimed that it can be used to provide a continuous process which will eliminate part or all of the conching procedure.

Two distinct types of twin screw extruder exist, namely co-rotating and counter-rotating systems. Their major characteristics are outlined in Table 17.1. Typical designs of the elements attached to the two rotating shafts are illustrated in Figure 17.4 for the two types of machine.

An extruder can be adapted by rearranging the elements within it, so as to give a series of shearing and degassing stages as required. The chocolate can be extruded either when it has been ground to its final fineness (16) or before the refining stage (17). Additions of liquid fat or surface-active agent can be made during the process, thus giving continuous conching. Where the ingredients have been ground, the chocolate is stated to be nearly ready for use. Where further refining is required, it is possible to liquefy the chocolate by employing another high shear (normally co-rotating) extruder, where final additions are also made.

The ability of the extruder to heat and degas liquid material means that it can also be used to pasteurize cocoa mass or nibs. Traditionally this has been carried out by thin-film or spray devices (Chapter 5). One system using an extruder, described by Werner Pfleiderer, adds water or steam into the extruder, where cocoa nibs are thoroughly wetted and slightly crushed. Pasteurization is carried out by applying a high temperature over a short time. The steam degasses at the machine outlet, taking with it many other

Table 17.1 Relative characteristics of co- and counter-rotating extruders

Co-rotating screws	Counter-rotating screws
Mass transport by positive conveyance and drag	Positive conveyance acts like a 'screw pump'
Longitudinally open flights leaves open path from feed to discharge	Radially and longitudinally closed flights. The screw flights form sealed chambers. Limits mixing but quickly generates pressure. Good transport of low viscosity material
Pressure generation depends upon the number and pitch of the reverse pitch elements, and very much upon the die exit. The through 'path' limits the upper pressure obtained	Pressure generation by the reduction of pitch length
Good mixer but less control over temperature and shear	Poorer mixer, but each small 'pocket' of material can have a more precisely controlled temperature regime. Within limits temperature and shear can be controlled independently

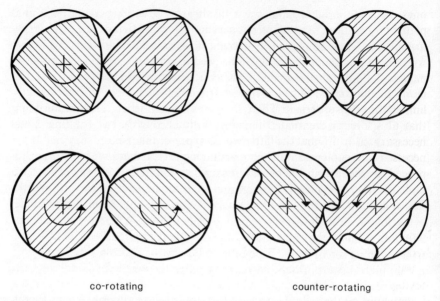

co-rotating   counter-rotating

Figure 17.4 Examples of cross-sections of co-rotating and counter-rotating extruder elements.

compounds, including some of the acidic ones which are detrimental to chocolate flavour. The high humidity during the process also raises the water activity (equilibrium relative humidity) of the material, thereby increasing the probability of destroying Salmonellae and other bacteria.

17.3.4 *Ultrasound*

A more controversial alternative to conching is to subject the chocolate or cocoa mass to ultrasound (8). This technique was proposed as early as the 1940s by Nordenskjold and Holmquist. The high-frequency accelerating and retarding forces were used to impart energy to chocolate and cocoa mass, and this was said to have the following effects:

(i) Acceleration of some chemical reactions without changing the form of the reaction, e.g. in ageing of wine and extraction of tannins from wine and cocoa mass
(ii) Freeing of gases from the mass
(iii) Formation of electrostatically charged oxygen and ozone, thus causing some oxidation
(iv) Homogenization of the ingredients.

These reactions may be beneficial to some types of chocolate but detrimental to others. For example, it is said that whereas oxidation may assist flavour development in plain chocolates it may give rise to an unpleasant taste in milk ones. Initially, in fact, treatment with ultrasound was more successful with plain or bitter chocolate, where it was generally found to reduce conching time from about 72 to 36 hours. Mosimann (8) in particular developed it for use with all types of chocolate and so that it could be used as part of a manufacturing process which required no conching at all; concluding that many people were using too long a treatment time (up to 120 seconds), he limited his exposure to ultrasonics to a fraction of a second. Mosimann found that ultrasonics accentuated flavours, including some bad ones, so it was necessary to ensure that the latter were not present in the feed material. It was necessary to de-aerate milk chocolate as part of the conche replacement procedure (see section 17.4). In addition, it was found that a longer exposure of milk chocolate to ultrasonics produced deterioration of flavour and an increase in astringency. The frequency of the ultrasound was also critical– 800 kHz was found to heat the chocolate and give inferior results, and although the most suitable frequency was found to vary with the chocolate recipe, it was normally in the region of 20 kHz.

Additional advantages of this type of treatment were said to be an improved texture in the final product, in that it was less sticky in the mouth, and also that it tempered more easily. In general, this approach to chocolate manufacture has not been widely accepted. It is interesting to note, however, that in the 1980s interest in the use of ultrasonic treatment in chocolate-making revived, only this time for a wide variety of uses from particle size measurement to its effect on tempered chocolate.

17.4 Complete chocolate-making systems

Many machinery manufacturers supply complete chocolate-making systems, each differing more or less from traditional systems. In addition, many research institutes connected with or sponsored by the confectionery industry have developed their own chocolate-manufacturing processes. This section reviews some of the more widely reported or novel systems.

17.4.1 McIntyre system

As has been described in other chapters in this book, chocolate manufacturing equipment is frequently complex, expensive and normally requires supervision. For the small sweet manufacturer the capital outlay may be prohibitive, and he may be obliged to buy his chocolate already manufactured. An alternative approach is to use a single machine which will grind milk powder, sugar and cocoa mass and conche it at the same time. A machine of this type is the L & D McIntyre (see Chapter 8). This consists of a drum with a serrated internal surface. Spring-loaded scrapers rotating inside the drum break the particles as they are forced between it and the outer wall. A water jacket allows for heating or cooling of the mixture, while circulation air removes moisture and other volatiles. Over a period it is possible to produce a chocolate of finished fineness. Batches weighing between 45 kg (99 lb) and 5 tonnes can be processed at a time. The major disadvantage of this type of machine is that it is difficult to optimize the flow properties and flavour of the final product. In addition, its overall throughput is frequently too small for larger chocolate manufacturers.

17.4.2 Ball mill systems

In section 17.2 it was noted that the ball mill offered a possible alternative to the roller refiner for breaking down the large solid particles within chocolate. This mill is in fact incorporated in several complete chocolate-making systems (10, 18) as well as many patents (9, 19). A typical system is illustrated in Figure 17.5. In order to obtain a final fineness, the chocolate mass must be passed several times through the ball mill. This requires one or more

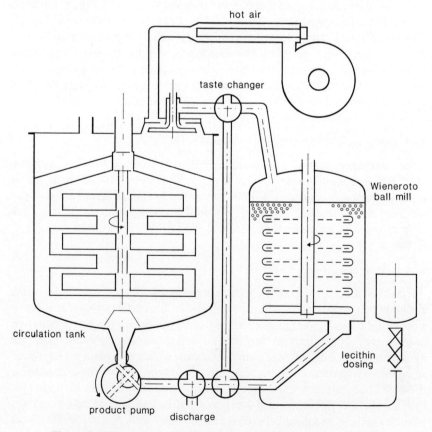

Figure 17.5 Wiener chocolate plant, using recirculation through ball mill.

holding tanks, which are often stirred and force-ventilated to enhance the conching reactions. These tanks are often of 3–5 tonne capacity and sometimes have a spray or centrifugal device producing a thin layer of material; this, together with the ventilation, aids moisture removal. This type of operation requires little supervision and once again has a throughput suitable for a smaller chocolate manufacture. Another advantage over many systems is that the product is totally enclosed, which aids hygiene and reduces the possibility of contamination. The ability to control flavour and chocolate flow is, however, somewhat limited.

17.4.3 Extruders

The extruder has already been described in section 17.3 as a machine with possible chocolate-making applications. A full chocolate-making system

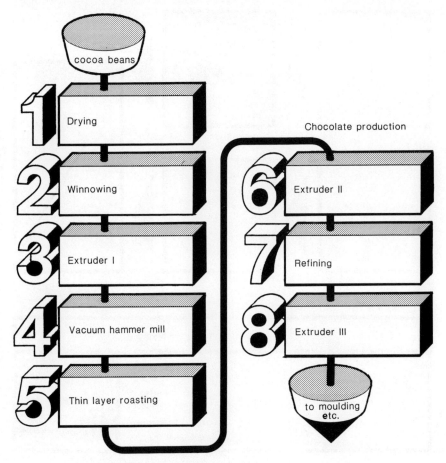

Figure 17.6 Werner & Pfleiderer systems for chocolate manufacture.

involving three extruders has been developed by Werner & Pfleiderer (17) (Figure 17.6). In this the beans are dried and winnowed before being slightly crushed under high moisture conditions in the first extruder, which pasteurizes them and gives some flavour development. The cocoa mass is then produced by grinding followed by thin-layer roasting. The second extruder is used to mix in the other ingredients, apart from the surface-active agent and some of the cocoa butter. It also completes the flavour development process. A refiner then reduces the solid particles to their required size, before the material is fed into the third extruder for the final additions and liquefying. The process is reported to take some 30 minutes to be able to operate at between 300 to 3 000 kg/h (660–6 600 lb/h) and to produce a wide range of flavours.

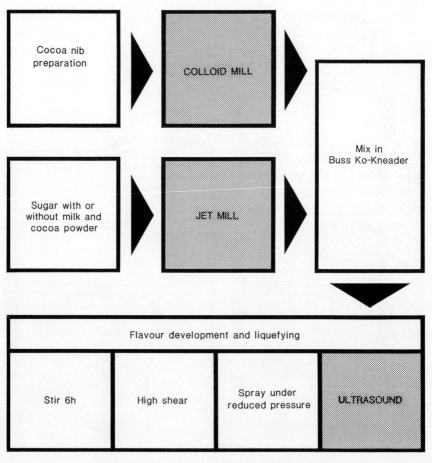

Figure 17.7 Mosimann process for chocolate-making.

17.4.4 *Mosimann process* (Figure 17.7)

An attempt by G. Mosimann (8) in the 1960s to develop a very original method of chocolate manufacture failed to live up to expectations. It is, however, interesting to look at his approach and the problems which arose.

In his process Mosimann started by investigating the use of ultrasound to develop chocolate flavour. It is, however, not desirable to alter part of the manufacture in isolation from the other stages, and eventually an entirely new process was evolved. The ultrasound itself required a lot of development work, from the point of view of the equipment, frequency and period of exposure to the energy, the latter being found to be dependent upon the type of chocolate being manufactured. Having achieved what he considered to be a satisfactory flavour development without conching, Mosimann investigated the possibility of achieving the other functions of conching, i.e. liquefying and moisture removal, by methods more economical than large conches.

At the same time, a manufacturer of pin mills proposed that this machine would be better than a roll refiner, so a process without conching or refiners was devised. As has been described in section 17.2, however, this type of mill is unable to operate satisfactorily with cocoa nib, owing to the high fat content. In addition, mouldy, cheeselike flavours were present in the final chocolate. By carrying out trials with the different components it was found that this undesirable effect was due to the milling of the milk powder in the presence of oxygen. The reaction appeared reversible when the milled material was stored for three or four days, but this was impractical as a continuous process was required. Future milling was therefore carried out in an oxygen-free atmosphere, which had the added advantage that it removed any explosion risk.

As the pin mill did not appear to be entirely satisfactory, tests were carried out using a jet mill. This involves using high-speed air jets to accelerate the particles against one another. Previous experience had shown that this type of mill appeared to require an excessive amount of energy, but it was thought that the energy saved by his novel conching procedure would far outweigh this. The jet mill also tended to block up with the cocoa nib, so this was treated separately and finely ground in a colloid mill to release as much fat as possible from the cells. The sugar, with or without milk powder, was pre-crushed to about 100 μm (4×10^{-3} in) before being reduced to its final size in the jet mill. Cocoa powder was also included if it was in the recipe. This mill was said to produce a good particle-size distribution, and, since the air flow and temperature were controlled, it was able to reduce the moisture content of the material.

After testing different machines to mix and liquefy the cocoa mass and milk solids, Mosiman found that the Buss Ko-Kneader gave very satisfactory results, as well as providing a continuous process. A minimum amount of cocoa butter was added at the early stages so as to give an homogeneous pumpable mass. Only in the last phases were the remaining cocoa butter and lecithin added.

The chocolate at this stage, although usable, was found not to have the flow properties and flavour of a good chocolate. It was therefore given a four-phase treatment consisting of:

  (i) 60 minutes of mechanical treatment under continuous stirring
 (ii) Treatment in a higher-shear cylindrical shearing device (the Buss Ko-Kneader did not fully mix in the fats)
(iii) Spraying in several cylindrical chambers under reduced pressure
 (iv) Treatment with ultrasound apparatus.

The full process is illustrated in Figure 17.7.

When put into use, however, the system was found to put a considerable strain on the equipment used, whilst the jet mill gave rise to particular difficulties (20). Other workers (21) found that the jet mill was expensive to operate and could only be operated with a very low fat mix. The efficiency of removal of moisture was also criticized, and although it could be improved by high-vacuum treatment, this proved to be detrimental to the flavour.

17.4.5 LSCP and BFMIRA processes and their successors

Two chocolate-making processes which have been adapted and used, at least in part, within the confectionery industry are the Lindt Sprüngli Chocolate Process (LSCP) (22) and the BFMIRA process (6). The former, described by Kleinert in 1971 (Figure 17.8) pretreats the cocoa nib under high humidity to remove unwanted volatiles and the necessity for conching. The treated nibs are then ground to form cocoa mass. Simultaneously, the sugar and milk are roughly milled before being placed in a special reactor 'with gas and steam'. The two components are then mixed and ground, firstly on a three-roll refiner, followed by two five-roll refiners in parallel. Finally, lecithin is added at a plasticizing stage. Although the idea of pretreating nibs and the sugar/milk mix in separate reactors has not been widely applied, the principle of roasting nib under high moisture is retained in the NARS process (Chapter 5), while the mixing of cocoa mass with sugar and possibly milk powder in a high-humidity reaction vessel has been developed by Carle-Montanari in the PDAT reactor vessel. The two-stage refining is also now widely publicized, although now in the form of a two-roll refiner followed by several five-roll machines (3).

The BFMIRA process (Figure 17.9) proposed that the cocoa should be roasted in the form of mass and the unwanted volatiles removed in a spray tower. Once the remaining ingredients were added, the mixture was refined and then homogenized in a high-shear stirrer. These principles are developed, amongst others, in the Petzholdt scheme of chocolate-making. In this the cocoa mass is treated and even possibly roasted in a thin-film device, whilst conching is limited to treatment by a high-shear machine known as the PIV conche (Chapter 8).

Complete chocolate-making processes have continued to be developed. The

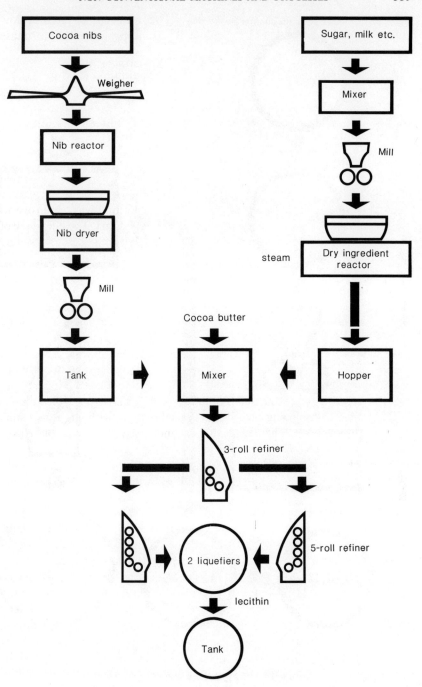

Figure 17.8 Lindt Sprüngli chocolate process.

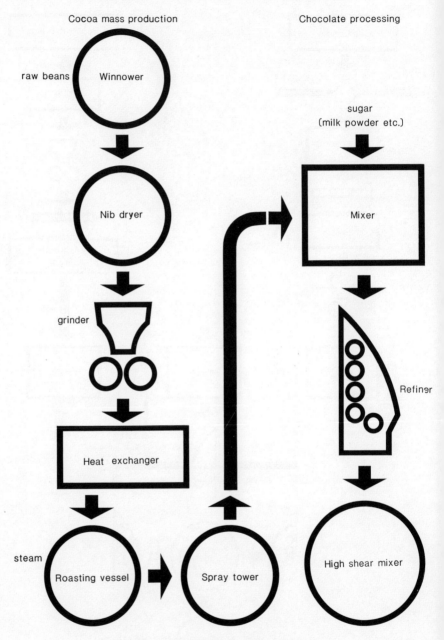

Figure 17.9 BFMIRA chocolate process.

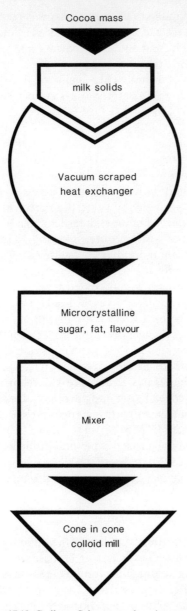

Figure 17.10 Cadbury Schweppes chocolate process.

BFMIRA principle of milling raw cocoa mass was taken up by Vakrilor and Hadshikinov (11), but instead of roasting the beans alone, sugar syrup with or without milk is mixed in, and all the ingredients spray-dried. This material is then homogenized with the remaining fat before the largest particles are removed in a 'hammer' type mill. Once again the five-roll refiner and conche have been eliminated, and a continuous system developed.

17.4.6 Cadbury process

The thin-film heat exchanger is a possible alternative to spray drying to remove moisture, and also to remove other volatiles. As was mentioned in section 17.3, it is possible to add milk powder or sugar at this stage, to give additional flavour development. Normally, however, the ingredients from such a machine are then refined and conched in the standard manner. Cadbury (23), however, have proposed an alternative procedure, involving microcrystalline sugar, that is, one with preferably not more than 2% by weight of sugar particles greater than $25 \, \mu m$ (10^{-3} in) and at least 90% by weight greater than 8 μm (3×10^{-4} in) (more fine particles would be detrimental to the flow properties of the chocolate). This fine sugar removes the need for a coventional refiner when it is added to the cocoa mass which, with or without milk powder, has been passed through a heat exchanger operating under vacuum. The latter is able to reduce the moisture level to less than 1%. This means that the only remaining procedure is to mix and liquefy the ingredients, whilst perhaps removing a few larger particles and aggregates. The process developed to do this involved adding the remaining cocoa butter and flavours to the other two components and processing the complete mixture in a cone-in-cone colloid mill. The whole chocolate-making process (Figure 17.10) is claimed to be quick and simple and less expensive in terms of process time and equipment cost. The vacuum treatment, however, is considered undesirable by many workers in that it tends to limit the range of flavour of the chocolate produced.

17.5 Chocolate tempering and coating

All chocolate temperers attempt to create the optimum number of stable fat crystals within the chocolate. Chapter 11 describes how various techniques, such as adjusting holding times and temperatures or applying pressure, have been used to change the crystal form. As more crystals are formed, there is less liquid fat present, and so the chocolate becomes thicker. A viscous chocolate will normally set up quickly and therefore require a relatively short cooling tunnel. It will, however, often cause difficulties in weight control, and excess usage may result. Conversely, thin chocolate requires extra cooling tunnels and often contains extra fat, both of which add to the cost of the product. Ideally, a chocolate should flow easily into the mould or over the sweet centre, but then set up as rapidly as possible. In order to achieve this, crystal accelerators have been developed. These take the form of a vegetable triglyceride fat in micro-fine powder form. Their crystals are in the stable β-form and are almost colourless and neutral in taste and colour. They differ from cocoa butter mainly in their high melting point (about 70 °C, 168 °F). By adding them at a concentration of 0.1–1% to the chocolate in the tempering machines at a temperature of less than 40 °C (104 °F), the crystals act as a 'seed' for the remaining fat. The presence of the crystals has been shown to more than double the setting rate of cocoa butter and a wide range of other vegetable fat samples (24).

The traditional method of making confectionery is to mould a sweet, or pour chocolate over a centre in a process known as enrobing. One alternative to this is to spray the tempered chocolate over the centre material, which may be held by a number of pointed pins. Spray nozzles have been patented that can be used with low-fat chocolates, whilst other systems have been developed for coating ice creams. The centre may cause splitting or scaling of the chocolate, and a double system was invented by Wolff whereby the centre is first sprayed with a hardened fat before it receives its spraying with chocolate.

Many other methods of coating biscuits or other stick-shaped centres have been investigated. One, patented by Meijl (25), employs two sponges soaked in chocolate. The centre is put on a holder and the chocolate deposited on it by pressing the sponge against it.

In section 17.4.3 a process was described using a series of extruders to produce chocolate. The proposed uses of this type of machine extend beyond chocolate manufacture, however, and also include methods of making certain types of sweets. The chocolate can be tempered in the usual way to ensure good gloss and finish in the final product. This is then extruded through a die to give the required shape. In a process developed by Cadbury (26), it is possible to produce a tube whose wall is composed of a net having mesh strands and intersections. If required, the centre can be filled by passing another material through a central mandrel as the net tube is formed.

17.6 Conclusion

This chapter has described a few of the more unusual processes which have developed for chocolate manufacture and use. Many more are to be found, particularly in the patent literature. Which ones are likely to be of use depend upon the circumstances of the chocolate manufacturer himself. For example, the relative importance of throughput, labour costs and quality control may make certain processes more attractive than others. It is important when deciding which techniques to employ, to understand the principles behind the processes involved in chocolate-making. It is hoped that the different authors in this book will have given the reader some knowledge of these principles and given a greater understanding of the problems and challenges of chocolate manufacturing.

References

1. Kuster, W. *Manuf. Conf.* (1984) (August) 47–55.
2. Koch, J. *Confectionery Production* (1961) (March) 233–236.
3. *The 2 Roll Refiner*, Bühler Brothers Ltd, Uzwil, Switzerland.
4. Rowntree and Company Ltd. Improvements in the Manufacture of High Class Fondant Chocolate and Similar Chocolates. UK Patent 165, 840 (1921).
5. Alikonis, J.J. Method for Making Confectioneries. US Patent 2, 459, 908 (1949).
6. Van Veen, P. and Elson, C.R. *The BFMIRA Continuous Chocolate Manufacturing Process* (BFMIRA Report). British Food Manufacturing Industries Research Association, Leatherhead, UK (1971).
7. Marshall, V.C. (ed.) *Comminution*. Institute of Chemical Engineers, London (1975).
8. Mosimann, G. 'Physical and chemical reactions in connection with Mosimann's process for automatic production'. *Proc. Conf.*, Solingen-Grafrath, April 1963.
9. Union Process International Inc. A Method of Making Chocolate and Chocolate Flavoured Compounds. UK Patent 1,568,270 (1977).
10. *Mixing For Quality With The Chocolate Production Plant SHA 5000*. Bauermeister & Co Ltd., Hamburg.
11. Vakrilov, V. and Hadshikinov, D. *Zucker und Süsswaren* **31** (1984) 84–86.
12. Turos, S. Process for Preparing Crumb Products. US Patent 4,346,121 (1982).
13. Minifie, B. *Chocolate Manuf. Conf.* (1979) (October) 43–48.
14. Cadbury Brothers Ltd. A Method of, and a Means For Manufacturing Chocolate. UK Patent 1,261,909 (1968).
15. V.E.D. Kombinat Nagema. Verfahren zur kontinuierlichen Herstellung von Schokoladenmassen. West German Patent DE 3,512,764 Al (1985).
16. Creusot-Loire. Preparation of Chocolate Paste Between Refining and Moulding Via Screw Conveyors Continuous Conching Chamber With Controlled Temperature Zones. French Patent 8,313,769 (1985).
17. *The 30-minute Chocolate Process*. Werner & Pfleiderer, Stuttgart.
18. Anon. *Confectionery Production* (1984) (August) 470–476.
19. Weiner. Device For Preparing Chocolate. European Patent Office 0 157 454 (1985).
20. Simon, E.J. *Rev. Int. Choc. (RIC)* (1969) (October) 413–421.
21. Werner & Pfleiderer. A Process and Apparatus For The Manufacture of Chocolate Mix. UK Patent 1,187,932 (1970).
22. Kleinert, J. *Rev. Int. Choc. (RIC)* (1971) (June) 158–165.
23. Cadbury Schweppes Ltd. Method For Manufacturing Chocolate. UK Patent 2 033721B (1983).
24. Kleinert, J. *CCB Review for Chocolate, Confectionery and Bakery* (1978) (December) 7–11.
25. Meijl. Manufacturing Chocolate Coated Biscuits etc. Japanese Patent JP 61056044 (1986).
26. Cadbury Schweppes Ltd and Mercer F.B. A Composite Food Product and Method of Making the Same. UK Patent 1 604 586 (1981).

18 Chocolate marketing and other aspects of the confectionery industry worldwide

C. NUTTALL

18.1 Introduction

Perhaps only 15–20% of chocolate is eaten on its own in blocks or bar form. The remainder is used to coat other food preparations such as biscuit, ice cream, or sugar confectionery centres. The true context of chocolate usage worldwide is the food industry, yet for 150 years it has been associated with sugar confectionery for marketing, administrative and legal purposes as an industry in its own right. This chapter therefore concerns itself with the confectionery industry as a whole. Chocolate and chocolate confectionery together account for about 40% by weight, and a somewhat higher proportion by value, of the confectionery industry.

Confectionery is the collective term applied to edible products usually compounded from sugar as the common ingredient. As its dictionary definition implies, such products are a melange of more than one original material which could include cocoa derivatives, milk, fats, nuts or pieces of biscuit. Products where flour is a principal ingredient along with sugar, such as biscuits and cakes, are not considered part of the confectionery industry for the purposes of this chapter. These melanges present the consumer with products which have contrasting flavours, contrasting textures such as crunchiness or chewiness, or where there is some improvement in appearance by decoration or the addition of colour.

Confectionery eating is widespread. In the more highly industrialized countries, nearly all the population have eaten it at some time, and over 90% may buy with some regularity. In Anglo-Saxon countries, about half will be bought by women and about 15% by children, who may, however, eat perhaps 25% of the total. As a result, the industry is supported by a wide-ranging system of retail outlets, backed by wholesalers and jobbers.

Confectionery products are foodstuffs, supplying nutrients to the human body. From their earliest days, however, they have seldom been eaten at regular mealtimes, and are not regarded in the same way as bread, meat, or other staples, an impression heightened by a low ratio of volume to flavour and texture sensations, and by the association of confectionery with the market for gifts.

Confectionery was the first type of 'convenience food', a product not

345

needing preparation and capable of being eaten between meals or 'on the run'. The consumption of snack foods is increasing, now accounting for over one-quarter of all consumer spending on food in the UK, to give one example. More women seek paid employment, and cook less at home. People live further from their work, and if they do not eat out, they buy a take-away snack, often of energy-rich confectionery. While this means more competition from other snack foods, the confectionery industry has benefited from this trend, and many believe that certain sections of the trade have become absorbed in the snack industry.

18.2 The organization of confectionery production

The scale of confectionery production is usually either quite small or very large. In many countries, high-priced chocolates are often made in a kitchen behind a retail shop, the chocolate covering being bought in. At the other extreme, chocolate blocks or filled chocolate bars are manufactured in advanced mass-production facilities.

Confectionery is made principally in factories in more mature industrial countries, about half the world's output coming from Western Europe and a quarter from North America. Facilities are now spreading out from this base, into South-East Asia, South America and even Saudi Arabia. The industry therefore is made up of (i) local firms, concentrating on sugar confectionery or handmade chocolates; and (ii) large firms, often parts of international food companies, operating in several countries and with a bias towards mass-produced chocolate items.

Only North America and the USSR have local populations large enough to support producers like Hershey or the Russian state monopoly. Firms like Nestlé, Mars, Cadbury or Rowntree have long since expanded outside their country of origin.

The confectionery industry is an important user of agricultural products, particularly from the Third World. While temperate areas supply glucose, milk, and dried fruits, tropical countries may depend heavily on the demands of confectionery manufacturers.

Cocoa	34% of all exports from the Ivory Coast, a principal crop in Ghana and the Dominican Republic (see Chapter 2)
Sugar	Vital to countries like Fiji, Mauritius, the West Indian islands, as well as Cuba
Hazelnuts	Turkey's second largest agricultural export
Coconut	Important to Sri Lanka and the Philippines
Liquorice	From Iraq, Iran and Afghanistan.

Taking the only figures available for production worldwide, chocolate and chocolate confectionery accounts for about 4.5 million tonnes of a total for the whole confectionery industry of about 11 million tonnes.

18.3 A short history of the industry and its products

The first surviving references to sugar-containing products were written down in ancient Persia about 2500 years ago. Middle Eastern civilizations before the Christian era developed types of sweets still sold today, like sugared almonds, Turkish delight, or *fruits confits*. Products such as these were taken westward by the Muslims and later by the Crusaders. As sugar became more plentiful in Western Europe in the sixteenth and seventeenth centuries, apothecaries began to 'sugar the pill', only to uncover a demand for confections which left out the medicine. British firms like Terry's can directly trace their descent from these early entrepreneurs. Techniques permitted only sweets which were panned or pressed.

In the early nineteenth century, the steam engine was harnessed to manufacture sweets on a bigger scale; the availability of glucose and condensed milk permitted the development of boiled sweets and toffees. Urban populations were expanding rapidly in Europe, and incomes, though still low, were expanding, allowing for the first time a network of retailers to become established in the factory towns. Rising levels of literacy encouraged advertising.

Until the Spanish invasions, the cocoa bean formed the basis of both staple and ceremonial products in Central America, and on being brought back to Europe, by 1700 it had gained popularity as a drink to rival tea and coffee. But not till the nineteenth century were methods developed in Holland, France, and Switzerland to make chocolate as we now know it. Initially, only plain (dark) chocolate was used to coat centres of nuts or fondants for the luxury trade, but British firms, particularly Cadbury, learned how to make first boxes of chocolates and then chocolate bars for a wider market. The popular filled (composition) bar emerged in the USA in about 1910.

In the present century, social circumstances such as the rapid growth in the populations of the developed world, rising consumer incomes, and increased mobility all have favoured the expansion of the confectionery industry. Throughout Europe, North America, the former British Dominions, and even in parts of the USSR, consumption increased so that by 1939 annual consumption generally exceeded 6 kg per head in these areas. In World War II, in several Allied countries the special virtues of confectionery were recognized by its inclusion in schemes of food rationing, while the British and American armies introduced confectionery products to many other countries of the world.

Because confectionery products provided an inexpensive luxury as a relief to diets which often were drab, the industry was one of the first to consider the need for mass production and mass marketing. Rowntree pioneered work study in its factories, and Cadbury were one of the first companies to use market research. In both the USA and UK, Mars developed operational research in the distributive trade. Contributions were made to particular branches of science, such as rheology. At the same time, because for various historical reasons a number of large manufacturers were of a religious bent, the industry was among the first to encourage pure food standards in production. In a notable instance of industrial co-operation, British and European firms combined to cease buying from tropical countries where cocoa was produced in conditions of near-slavery, while encouraging the setting up of a peasant-farmer alternative in West Africa. Hershey in America followed the example of British manufacturers in building model towns for its employees.

European firms were the first to export chocolate and sugar confectionery to other countries, and to set up factories overseas. Eastern Bloc countries, where development historically lagged behind the West, nevertheless permitted manufacture of confectionery, occasionally in association with Western companies.

While exporting is still important, and even within the EEC actually on the increase, as the world market grows in size and complexity local production is better able to meet the demand, albeit with the same or similar products.

18.4 Consumption of confectionery around the world

This section sets out the principal data available for making comparisons between countries. The crude data, where it exists at all, and relating either to production or to manufacturers' despatches, is derived from national statistical offices or from trade associations. Customs authorities publish details of exports and imports. 'Consumption' is either equated with despatches, or is calculated from production less exports, plus imports. There is always an element of estimation involved.

Typically, collating national returns internationally runs into difficulty because

— not all countries collect data on confectionery
— where data exists, not all countries are willing to release it
— individual industries choose different boundaries, e.g. the USA regards chewing gum as a separate industry
— categories do not agree
— periods covered by returns differ
— methods of estimating for missing or erroneous returns differ
— when comparing national trade data, one country's exports in a sector, even where the category is identical, seldom agree with those reported by the second country as imports.

International collaboration is improving, but as yet the results do not prove sufficient for use in rigorous econometric or social investigations. In these circumstances, data are presented here without much comment.

Using United Nations data as a base, and attempting to allow for its deficiencies in cover and in accuracy, Herr Pohlmann, President of the IOCCC

Table 18.1 Rough estimates of world consumption of confectionery
Sources: ISCMA/IOCC Statistical Bulletin, December 1986 U.N. Statistical Department
Year: 1984

Region	Consumption ('000 tonnes) Choco-late	Sugar confec-tionery	Total	Per head kg	Implied growth rate p.a. 1980/1975	1984/1980
Western Europe	2014	1966	3980	11.3	5	2
Soviet Bloc	887	2531	3418	8.3	4	1
North America (incl. Caribbean)	1135	974	2109	6.7	3	3
Oceania	81	76	157	6.5	−3	5
South America	209	18	227	0.7	10	2
Africa	58	259	317	0.6	5	1
Asia	186	591	777	0.3	3	2
World	4570	6415	10985	2.3	3	2

International Statistics Committee, has made estimates of consumption for each major part of the world. These are set out, with some aggregations, in Table 18.1.

In using this table, the following points should be noted:

(i) Apart from Western Europe and North America, there is no measure of accuracy of the basic data. To the present author, the figure for sugar confectionery for the Soviet bloc looks too high, and that for South America too low.

(ii) In some areas, one country dominates the total, e.g. Japan in Asia, Australia in Oceania.

The principal source of collated data is the reports of the Joint International Statistics Committee of the IOCCC. This covers the producers of Western Europe, Japan, the USA and Australia. The principal omissions are from the Soviet Bloc, Spain, and Canada, though in the full series of reports there is some data for Yugoslavia, Hungary, Israel and Brazil. The series runs from 1970. Using this data, in Table 18.2 are summarized the consumption of chocolate and sugar confectionery, with the total, for each contributing country.

The relatively high tonnages consumed in the USA, UK, West Germany and France should be noted, forming as they do a secure basis for exports.

Table 18.3 converts these tonnage figures into consumption per head, giving of course a different perspective.

Note points like the low consumption levels per head in France and Italy,

Table 18.2 Confectionery consumption data
Source: ISCMA/IOCC Statistical Bulletin, December 1986
Units: Thousands of metric tonnes
Definitions: JISC

	Choc. and choc. confectionery			Sugar confectionery			Total		
	1975	1980	1985	1975	1980	1985	1975	1980	1985
Australia	63	58	72	65	54	68	128	112	140
Austria	34	47	55	17	18	19	51	65	74
Belgium	51	62	63	38	38	47	89	100	110
Denmark	22	25	29	22	23	28	44	48	57
Finland	10	11	15	17	19	20	27	30	35
France	166	213	224	162	149	144	328	362	368
Germany (FR)	353	404	386	299	351	353	652	755	739
Ireland	14	20	25	18	18	23	32	38	48
Italy	53	53	76	116	124	115	169	177	191
Netherlands	54	70	82	80	74	74	134	144	156
Norway	22	27	32	12	16	18	34	43	50
Portugal	NA	NA	4	NA	NA	NA	NA	NA	NA
Sweden	37	44	47	38	32	36	75	76	83
Switzerland	48	54	56	16	19	19	64	73	75
UK	328	361	459	308	289	285	636	650	744
USA	797	837	1057	770	778	995	1567	1615	2052

Table 18.3 Ranking of consumption of confectionery
(IOCC/ISCMA countries, 1985)
Kg per head per year

Country	Chocolate		Sugar confectionery		Total
	kg	% of total	kg	% of total	
Ireland	7.0	53	6.3	47	13.3
UK	8.1	62	5.0	38	13.1
Germany (FR)	6.3	52	5.8	48	12.1
Norway	7.8	64	4.3	36	12.1
Switzerland	8.6	75	2.9	25	11.5
Denmark	5.7	51	5.5	49	11.2
Belgium	6.2	57	4.6	43	10.8
Netherlands	5.7	53	5.1	47	10.8
Sweden	5.7	57	4.3	43	10.0
Austria	7.3	74	2.5	25	9.8
Australia	4.6	52	4.3	48	8.9
USA	4.4	51	4.2	49	8.6
Finland	3.1	43	4.1	57	7.2
France	4.1	62	2.6	38	6.7
Italy	1.3	39	2.0	61	3.3

never properly explained: and the high proportion of chocolate eaten in Switzerland and Austria, a feature of many years' standing.

In Table 18.4–18.6, further data is provided: the top ten producing countries for chocolate and for sugar confectionery, ranked by tonnage output (Table 18.4), the top ten exporting countries (Table 18.5), and, for those countries where it is available, data for national consumption per head for the various categories used by the JISC (Table 18.6).

Table 18.4 Top ten producers of finished goods in 1985 ('000 tonnes)

	Total[†]			Chocolate			Sugar confectionery	
1	USA	3524.5	1	USA	1031.4	1	USA	909.5
2	United Kingdom	1579.9	2	United Kingdom	472.2	2	Germany (FR)	350.6
3	Germany (FR)	1114.6	3	Germany (FR)	361.0	3	United Kingdom	313.7
4	France	983.2	4	France	228.4	4	France	150.6
5	Italy	792.7	5	Netherlands	170.0[e]	5	Netherlands	110.0[p]
6	Netherlands	598.3[pe]	6	Belgium	85.6	6	Italy	109.0
7	Belgium	287.1	7	Italy	69.9	7	Australia	63.1
8**	Sweden	176.1	7	Australia	68.4	8	Belgium	53.1
9	Denmark	168.8	9	Switzerland	68.2	9	Denmark	34.8
10	Finland	135.5	10	Sweden	51.4	10	Sweden	33.2

Note: Coverage and definition of product fields are not exactly the same for all countries. Please refer to the notes on the national tables
[†]Includes biscuits and baked goods
** = 1984
[e] = Estimated
[p] = Preliminary

Table 18.5 Top ten exporters of finished goods in 1985 ('000 tonnes)

	Total[†]			Chocolate			Sugar confectionery	
1	Netherlands	293.5	1	Netherlands	114.1	1	Netherlands	62.8
2	Germany (FR)	224.1	2	United Kingdom	63.9	2	United Kingdom	62.7
3	United Kingdom	217.2	3	Germany (F.R.)	60.7	3	Germany (F.R.)	61.1
4	Belgium	160.5	4	France	47.0	4	France	31.3
5	France	155.0	5	Belgium	45.9	5	Belgium	23.8
6	Denmark	102.4	6	Switzerland	20.3	6	Italy	15.7
7	Italy	80.9	7	Italy	20.2	7	Ireland (Rep.)	13.6
8**	Sweden	44.9	8	Ireland (Rep.)	17.0	8	Denmark	13.3
9	Switzerland	33.2	9	Sweden	14.4	9	Sweden	12.5
10*	Ireland (Rep.)	30.6	10	USA	14.2[e]	10	USA	10.3

Note: Coverage and definition of product fields are not exactly the same for all countries. Please refer to the notes in the national tables
[†]Includes biscuits and baked goods
* = Chocolate + sugar only
** = 1984
[n] = Negligible amount
[e] = Estimated

Table 18.6 Consumption per head of categories of confectionery
Source: ISCMA/IOCC Statistical Bulletin December 1986
Definitions: JISC. For details consult the above publication, pp. 42–85
Year: 1985
Units: Kilogram per head per year
NA = not available

	CHOCOLATE			SUGAR CONFECTIONERY				
	Solid blocks and bars	Filled chocolate bars and tablets	Other	Boiled sugar sweets	Toffees and caramels	Gums, jellies, pastilles	Chewing gum	Other
FULL DETAILS:								
Belgium	1.50	1.75	2.91	0.70	0.84	1.11	0.48	1.42
France	2.01	0.49	1.56	0.51	0.57	0.41	0.45	0.67
Germany (FR)	2.68	1.96	1.69	1.00	0.50	0.50	0.26	3.53
UK	1.51	4.63	1.99	1.06	1.08	0.93	0.26	1.71
PARTIAL DETAILS, BOTH MAJOR CATEGORIES:								
Denmark	1.74	1.86	2.08	0.98	0.21 ⎫	1.17	NA	3.15
Italy	0.45	0.40	0.48	⎭ 1.10		0.22	0.31	0.38
DETAIL OF CHOCOLATE ONLY:								
Austria	3.12	1.07	3.06					
Portugal	0.15	0.12	0.13					
Netherlands	0.92	4.78 ⎫						
Sweden	1.28	4.87						
USA	0.97	3.45 ⎭						

18.5 Reasons for eating confectionery

In countries with high consumption levels, on average, it is probable that every person buys some confectionery every second day. In terms of the diet, confectionery can account for perhaps 5% of the average intake of calories.

Confectionery is eaten all over the world quite simply because people find pleasure in it. Through taste and texture there is probably no other product that gives so much enjoyment to so many individuals in so many countries. Chocolate and sweets bring colour and excitement into everyday lives, and throughout adult life continue to symbolize gaiety, festivity and goodwill. Confectionery is fun, and the range of products and prices is such that its enjoyment is within the reach of almost everybody.

Despite the fact that it is seldom eaten at regular mealtimes, confectionery supplies energy, and is often eaten when the body signals some need for it. The physical explanation for this is set out elsewhere; briefly, if the immediate energy resources of the body become deficient due to exercise or lack of regular food intake, the sugars in confectionery provide one of the quickest ways of restoring the balance. Because this energy is packed into a small volume, it is easily portable, and can be eaten nearly everywhere and at any time. The industry's products, although sold more for pleasure than for nutritional value, command a special place in the overall diet. Manufacturers therefore produce items like blocks of chocolate or bags of boiled sweets, typically in units of 50–150 g (2–6 oz) in weight and 1000 kilojoules in calorific value. Some items in these categories are designed for consumption at one time by one person, such as a Mars bar; some give more flexibility for consumption over an extended period of time, or for sharing, such as a Kit Kat.

A second and perhaps less obvious reason why people eat chocolate and sweets is the frequent need for one person to transmit to another by way of a physical object some element of feeling—such as love and affection, gratitude, hospitality, even remorse. In such circumstances, confectionery makes an acceptable gift. The traditional product in this category in English-speaking countries at least is the box of chocolates, or within Europe the box of *fruits confits*, each with characteristics of high product quality, where the excellence of the packaging design is important, and where the labour cost of manufacture is high, all these factors giving the impression of luxury adequate to express the degree of caring. There is a demand for these product throughout the year, but with peaks at various times in different countries. In British countries the chocolate egg is a symbol of Easter. In the USA and UK, Mothers' Day and St Valentine's Day are occasions to be marked. Almost everywhere, specially designed boxes are produced for Christmas.

Other types of products beside boxes of sweets and chocolates are also frequently bought as gifts; items principally designed to give energy and nutrition are frequently purchased as gifts for children. In the USA these form the basis of the 'treat' at Hallowe'en. Equally, because of their superiority in taste, the more expensive products are quite extensively purchased for own consumption or sharing. A few products, like chocolate mints, now even form part of a meal on special occasions.

A worldwide demand, which springs from mankind's natural liking for sweet things, is thereby satisfied by a wide range of items, at one end by artistic, hand-made products, at the other by items manufactured on highly automated plants for sale by the million each day.

18.6 The marketing of confectionery

Referring to his ambitions when he took over the firm in 1860, George Cadbury later said that they included the following:

> To manufacture a better product than competitors; to distribute it widely; and to advertise it so that people knew it was there.

With one notable omission, no better summary of the marketing function in confectionery terms has ever been stated.

No confectionery product will survive if it is not formulated to eat well. Given this, a product must be designed, or re-designed from time to time, to fulfil the function or functions required by an identified market. Continuous market research on the numbers and classes of purchasers and consumers, on the products they buy and eat, and on where and when the purchasing and eating takes place, will provide a matrix of information on which marketing decisions can be taken, from detailed changes to be made to formulation or price of an existing product to the choice of new markets and new products.

Within the firm, a new product, or an improvement to an existing product, can arise in two ways. Either the food chemist can bring forward a new combination of basic food materials, flavours, or colours, for which a potential set of consumers might subsequently be identified by product development specialists. Or these specialists can identify some market sector worthy of attention, for which the food chemist can attempt to formulate a product. To help in product development, it is standard market research practice to seek out those sectors of the buying and eating population inadequately covered by existing lines, or where competition is non-existent or weak. By a succession of product tests on potential consumers the concept and formula can be revised, culminating in a full-scale test involving every aspect of commercial selling, including price, trade terms, and advertising, on which the final decision can be taken.

The final product to be offered will have resulted from consideration of a wide variety of factors, including:

(i) The fundamental rules on taste and texture related to the chosen market
(ii) The primary function of the product: energy-giving versus high-taste/low-volume, own consumption, sharing, or gift, and preferably straddling two or more of these factors

(iii) Portion size or sizes, and the number of units to be offered in each pack size

(iv) Price per unit weight and unit price.

The whole area of product style will have been considered, covering the individualized packaging, display characteristics and advertising, each of which will have to be consistent with external requirements, such as:

(i) Relevant economic and social characteristics of the chosen market
(ii) The efficiency of packaging in transit and store
(iii) Legal requirements on formulation and labelling.

Costings will be made and related to expected selling prices, while the form, content and media to be employed for advertising will have been determined almost as early as the product design. One quite fundamental decision in this area will have been taken—whether or not to give the product or range a brand name, as opposed to accepting a generic product name or a collective house name.

Because all such factors will react with one another, all these tasks are complex and the final result one of compromises. Some optimal solution is the best that can be achieved.

The one missing element in George Cadbury's list, and one vital in the marketing of confectionery, is that of achieving and maintaining the highest practical level of product freshness at point of sale. Taste can fade over time, a product becomes stale, packaging can lose its attractiveness, and the joy of eating then departs. Freshness is a cost to the manufacturer, if only in the time spent in devising and maintaining systems to control it. Nearly all packaged confectionery is designed with freshness in mind, and items are date-coded so that distributor and consumer can check during storage and at time of purchase. Out-of-date product is usually withdrawn. Manufacturers employ special staff or agencies to estimate at regular intervals the state of product in distribution.

For a product to be sold in many countries, developers have to be aware that no two national systems of distribution are alike either in numbers of outlets per head or in types of outlets selling confectionery. In most Western countries, large self-service stores take a substantial proportion of the trade. Although not important in numbers, also common to advanced countries is the high-class shop selling highly-priced, mainly chocolate confectionery. Examples of types of outlets of local importance are:

(i) In the USA, small convenience stores, a growing number of which are attached to gasoline stations
(ii) In Australia, the milk bar
(iii) In Britain, the specialist shop retailing newspapers, periodicals, cigarettes, and greetings cards as well as chocolate and sugar confectionery.

In less well-developed countries, outlets for confectionery are usually spread out with decreasing density from the major commercial centres, with a large proportion of market stalls and itinerant vendors, as in Europe in earlier centuries.

The intermediate distribution system also can vary. For example, in Europe the wholesaler usually confines himself to breaking bulk and on-delivery to small retailers. In the USA his rough equivalent, the jobber, performs more of the marketing functions, such as encouraging display, and is moving into the market research function.

Operational research work has demonstrated that the natural level of consumer purchases of any one confectionery product depends on the number of outlets stocking the product; but each product will have at any one time a natural equilibrium of relevant shops in stock, the equilibrium changing as the effectiveness of other marketing factors like price or advertising change.

In the case of products designed for mass consumption and for energy-giving, a second strong influence on sales through retail outlets lies in the efficiency of display. Much research has been carried out in this field, and it is common ground that items should be on open display, ready to be picked up by the purchaser, that the number of display facings may be important, and that the facings should never be allowed to be empty. The sales effectiveness of display is also affected by finding the best situation for the display feature in the store—and of course, the style of the shop itself is important, its location, cleanliness, and general attractiveness. Each major manufacturer around the world will seek to influence retailers to construct and maintain displays to recommended standards.

Advertising plays an important part in the sale of confectionery. For mass production lines, television is the principal medium wherever it is available. It is expensive yet cost-effective, but prolonged campaigns cannot be mounted by medium- and small-sized firms, which must rely on other media like press or radio, or else allow for the imbalance by increasing the other elements of expenditure on marketing. Large-turnover confectionery brands are advertised more widely than most other foods, but except at the time of launch of a new product, the cost of advertising seldom amounts to more than 5% of turnover, and this is quickly balanced by the reduced costs which result from the effectiveness, both in awareness and propensity to purchase, which results from spending on advertising. For a branded product, advertising, packaging and display are all interlinked, each being planned to enhance the effects of the other two.

The last element of choice in marketing lies in the pricing of the product. The effects are perhaps better understood because in recent times many involuntary changes in price have had to be made as a result of rates of inflation, price controls and big changes in rates of sales taxes, allowing effects to be measured over an unusually wide set of changes.

For the large-turnover lines, in addition to the foundation determined by costs, 'price' has two elements: the price per unit of weight, e.g. per gram or ounce, and the price of the unit offered, e.g. 20 cents.

In some countries the unit of weight is specified by law, and in this case, as with products where the weight can vary, there are two determining factors:

(i) There are quite narrow limits of dimensions outside of which the product loses its identity, and if changes in the price per pound would take the product outside these limits at an unchanged unit price, then the unit price must be changed

(ii) The limitations of the national coinage–count lines may be bought more freely if priced at the level of a single coin, and, for children's lines, on the value of the lowest single coin.

For these types of items, in general terms, though not measurable by econometrics, more will be bought if the price per pound falls, and vice versa.

For lines which sell principally to the gift market, almost the opposite occurs. Within quite wide limits, a higher unit price may even make the product more attractive–in any case, it is usual to maintain or increase attractiveness irrespective of unit price and the limitations of the coinage system do not apply.

18.7 Nutrition and health*

The ingredients and raw materials which go to make up chocolate and sugar confectionery are also found in number of other foods. Although sold as much for pleasure as for nutrition, they make a useful contribution to the supply of all the major nutrients as well as to a wide range of vitamins and minerals required for healthy living. Weight for weight, many items compare in nutritional terms very favourably with foods consumed more regularly.

The greater informality of eating styles, and the importance of convenience foods, have already been mentioned, the latter including both sweet and savoury varieties, generally grouped into 'fast food' or more conventional types. Again, the second of these can be prepared in the home, like sandwiches or soup, or pre-prepared or ready to eat, like potato crisps, peanuts, cheese, apples, or confectionery. Six nutrients are needed to supply the three main processes which keep the body alive and healthy (Figure 18.1): of these, in the British diet, which is typical of the Western diet in general, confectionery supplies

4% by weight of all food intake
6% of all energy intake

*Data in this section is taken from *Confectionery and Nutrition*, a booklet published by the British Cocoa, Chocolate and Confectionery Alliance. The examples therefore are drawn mainly from the UK.

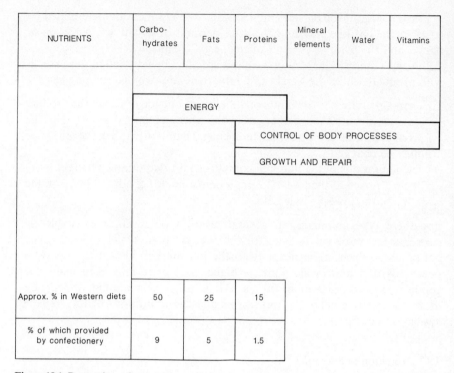

NUTRIENTS	Carbo-hydrates	Fats	Proteins	Mineral elements	Water	Vitamins
			ENERGY			
				CONTROL OF BODY PROCESSES		
				GROWTH AND REPAIR		
Approx. % in Western diets	50	25	15			
% of which provided by confectionery	9	5	1.5			

Figure 18.1 Proportion of essential nutrients provided by confectionery in a Western diet. Adapted by BCCCA from (7).

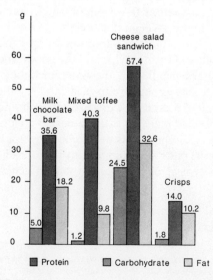

Figure 18.2 Nutrients consumed in different convenience foods (2).

9% of carbohydrate
5% of fat
$1\frac{1}{2}$% of protein
Minerals like calcium
Vitamins like riboflavin and niacin.

Confectionery is a concentrated food and adds little to water intake. Figure 18.2 shows the energy derived from two types of confectionery and two contrasting snack foods. Figure 18.3 covers the same products to show their nutritional compositions.

It is therefore wrong to think of confectionery as contributing nothing but sugar to the diet. Through confectionery, the British take in 4% of all milk and 33% of all nuts consumed. Only 14% of national daily intake of sugar comes from this source, compared with 26% dissolved in cups of tea or coffee. No single foodstuff will in itself provide a balance of nutrients, each contributing to the ideal diet by way of combination or variety. 'Bad' or 'junk' diets are just not sufficiently varied, or are nutritionally inadequate.

As nutritional and medical science has advanced in recent times, many aspects of food consumption have come under scrutiny as possibly deleterious to health. Confectionery has been no exception. In what follows, an attempt has been made to put the facts into perspective rather than to write an apologia. Most health problems have no single, simple cause. Confectionery, although not a major element in diets, has found itself being associated with obesity, food intolerance, and dental caries.

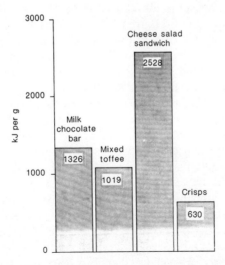

Figure 18.3 Energy derived from different convenience foods (2). Based on average portion sizes: milk chocolate bar 60 g; mixed toffee 57 g; cheese salad sandwich, bread 113 g, butter 14 g, cheese 57 g, lettuce 14 g; crisps 28 g.

N

Obesity is an undesirable condition resulting when an excessive amount of body fat is stored because energy intake exceeds energy usage. It can be defined in terms of height; for example, a person, male or female, 1.8 m in height is judged to be obese if his or her weight exceeds 80 kg. Obese persons are more likely to suffer from a wide range of health problems, from gall-bladder disease through coronary risk to hypertension. Treatment for obesity tackles both sides of the equation, by attempting to change lifestyles in respect of both diet and exercise.

It is frequently said that sugar makes one fat, and therefore eating confectionery will lead to overweight. There is no doubt that cutting down on any one food, be it confectionery or anything else, will result in weight loss, but good health requires a balance of nutrients from a variety of foods even when one is following a weight reducing diet, and there is little evidence to suggest that any one food or group of foods is more likely than any other to produce obesity.

In the special case of coronary heart disease, diet has been cited as a risk factor, indirectly in relation to obesity and directly in relation to some of its components. There is general agreement that too much fat of whatever kind in the diet is not helpful–confectionery can contain both saturated and unsaturated fat by way of milk and vegetable fats. It is also accepted that consumption of sugar should not be increased.

Many otherwise healthy people can suffer from a physical reaction to a particular food, often called an 'allergy', although the term 'food intolerance' is now preferred. The causes of some conditions, like asthma or eczema, are now understood; others, particularly those with psychological symptoms, are not solved. Milk, eggs, and nuts, all relevant ingredients of confectionery, have been associated with the first group, chocolate and strawberries with the second. Some colourings used in food preparation have also come under suspicion. Labelling regulations have been changed in several countries to inform those at risk, and formulations have sometimes been changed to eliminate the suspect substances. Using 'natural' ingredients is in line with modern consumerist thought. There is evidence that chocolate may exacerbate or provoke an attack of migraine, but there is no conclusive evidence that in normal non-migraine sufferers chocolate will actively cause a headache.

In countries with high consumption of confectionery, the main health-related concern about confectionery is the view that it is the *major* cause of tooth decay (caries). This view is widely held, but the industry has always been astonished at how little scientific evidence from studies on people can be marshalled in support. For many years, confectionery firms have been concerned that the facts should be established, so that soundly-based preventive measures can be developed which will contribute to the containment of dental decay. They have therefore been supporting independent research on the causes of this disease, with some success. In the Western world, despite increasing consumption of confectionery, the incidence of dental caries has declined perhaps because of better oral hygiene and fluoridization of water supplies.

Tooth decay results from the fermentation of carbohydrates by bacteria in the mouth, producing acid. In the plaque which surrounds the uncleaned surfaces of teeth, this acid attacks the tooth enamel which, if not checked, will demineralize, so that cavities will appear. There are few foods which do not contain some fermentable carbohydrates.

The three factors affecting the onset of the disease are the body itself and its defence mechanisms, the bacteria within the mouth, and characteristics of the diet. People vary in susceptibility, for reasons little understood. So far as the first factor is concerned, the saliva is important both in itself, as forming one of the environments of the tooth, and in its by-products, which protect the tooth against acid. The shapes and positioning of the individual teeth interact with the effects of the saliva. Tooth enamel can also be altered by a selection of mineral elements and fluoride, now the basis of many common preventive practices.

There are many bacteria capable of producing acid from a variety of fermentable carbohydrates, but most research work has concentrated on *Streptococcus mutans* and its ability to ferment sucrose. The belief that glucose and fructose are not fermentable now seems to have been abandoned. All fermentable carbohydrates are now suspect.

Carbohydrates of all kinds account for 50% of total energy intake in a normal Western diet, so that conditions are present on a large number of occasions for the formation of acid. Studies in Britain, the USA and Czechoslovakia suggest that these occasions on average number 6–9 a day for children and 3–6 for adults. Fermentable carbohydrate is eaten or drunk on 96% of these occasions. The British study shows that confectionery was eaten in about 8% and 2½% of cases in children and adults respectively.

In a ranking of foods according to their potential capacity for producing acid, caramels, chewing gum, and chocolate come near the bottom of the list, boiled sugar sweets near the top. Confectionery is not at the top of the list with respect to the average consumption of fermentable carbohydrates in snack foods. Estimates have been made as follows:

Estimated average consumption of fermentable carbohydrates in some foods which contribute to between-meals eating	Kg per head per year	% of total
Soft drinks	15.0	26
In tea and coffee	10.4	18
Confectionery	8.4	15
Biscuits	6.7	12

It is also important to consider the length of time that relevant foods remain in the mouth. It might be thought that a confectionery-eating occasion would be of long duration, but in the few experiments which have been carried out, results from the range of confectionery products tested have not differed other groups of snack foods.

Already enough is understood about tooth decay to apply preventive treatment, with dramatic results. A study in Britain has shown that between 1973 and 1983 the percentage of 5-year olds with tooth decay fell from 73% to 48%. Any attack on dental caries through diet must be slow-acting by comparison, for it would require a sharp alteration in eating patterns even where the prescription was based on understanding.

Confectionery is therefore unlikely to be the major cause of dental caries and a modification of the level of consumption would be unlikely to alter significantly the dental caries experience of most people.

18.8 Official classifications of confectionery

Confectionery producers thrive on novelty and on variety. Every group developing a new product seeks to specify something unique, and thereby defy classification. Yet for many of the requirements of the outside world, the degree of variety must be reduced by collecting together types of products which have common characteristics. The statistician wishing to analyse trends, the administrator seeking to control events, the legislator wishing to forbid some particular action, all would find their tasks impossible if faced with separate descriptions of many thousands of lines.

Official product classifications were first adopted to meet the needs of national customs authorities, and as they came first, they still exercise a strong influence on other systems. Until quite recently, each country designed a system to suit its own schedule of customs duties. Sugar and cocoa attracted duties early on, and they, together with confectionery, appeared in many national classifications. In the past forty years, the United Nations and the EEC have encouraged coordination of customs classifications, with some success.

Some of the general principles of classification are:

 (i) Distinctions made between raw materials, semi-finished and finished product
 (ii) Distinctions made between chocolate and sugar confectionery, not always logically (white chocolate treated as sugar confectionery in many systems)
(iii) Production and manufacturing differences tending to predominate, e.g. moulded and enrobed chocolate
(iv) When attempts are made at international collaboration between systems, countries are seldom willing to give up their own practices, e.g. the problem in the EEC of allowable fats in chocolate.
 (v) Computerized systems must now be devised.

Some of the principal characteristics of systems currently in use are given below. Chocolate products are given as an example.

(i) The United Nations chose a broad classification to accommodate many different national systems: chocolate and food preparations containing cocoa or chocolate in any proportion, sweetened cocoa powder, chocolate spreads. White chocolate is excluded.

(ii) For cocoa and chocolate the EEC, in its Directive L228/30, classifies formulations and product types in order to define them for legal purposes. Examples of the former are:

Parts of the cocoa bean
Cocoa fat
Chocolate, plain chocolate, vermicelli chocolate, chocolate flakes, gianduja nut chocolate, couverture, milk chocolate, high milk content chocolate, milk chocolate vermicelli, milk chocolate flakes, milk gianduja nut chocolate, couverture milk chocolate, and white chocolate.

Examples of the latter are:

Filled chocolate, where the outer part must be chocolate and at least 25% of the total weight. A chocolate, defined as a single, mouthful size, a combination of chocolate as defined above, with the chocolate parts visible and at least 25% by weight of the total piece.

(iii) CAOBISCO, the organisation of confectionery Trade Associations within the EEC, dealing as it does principally with the EEC Commission, adopts the classes of the Customs Co-operation Council (CCC) Nomenclature given below.

(iv) The Joint International Statistics Committee of the International Sugar Confectionery Manufacturers Association and the International Office of Cocoa and Chocolate (now the IOCCC) is the only body which, by meeting, comparison of treatment of specific products, and exhortation of members, has attempted a truly co-ordinated international classification. It has adopted four categories for finished chocolate products–solid chocolate, filled chocolate, pralines/bon-bons and other.

(v) National systems: each country with a trade association will define its industry in general terms as a control on membership. If it collects data about the industry from members it will adopt classifications to suit its own purposes. Only quite recently have the requirements of international statistical systems been considered in this respect. The British system is given as an example. It differentiates:

Solid milk or blended chocolate in blocks etc., with additions of discrete particles embedded in the chocolate
Solid milk or blended chocolate without additions, in blocks etc.
Filled blocks, bars and countlines with a continuous or segmented layer of biscuit or wafer

Solid plain chocolate, with or without additions
Filled blocks, bars, and countlines, other
    chocolate assortments in bite-sized pieces
'Straight' lines, i.e., collections of bite-sized pieces identical in nature
Chocolate liqueurs
Chocolate novelties, e.g. Easter eggs.

Each country with a strong Central Statistical Office will also have devised a Standard Industrial Classification, usually, but not invariably, similar to that used by the Customs authorities. In view of the number of different national systems in use, before using any figures in detail, readers should take care to examine definitions and classifications closely.

At present, the JISC data is the only collection which has been examined for consistency, and irreconcilable differences have been noted. The excellent data from France, Belgium, West Germany and Britain come from systems which are not too dissimilar. But the USA and Australia at present adopt quite different classifications reconcilable only at the broadest levels, and even then only with caution. The USSR publishes very little data.

Customs authorities currently adopt their own systems, that of the USA being the most important exception. Over 50 countries, mainly European and those historically associated with Europe, use the CCC Nomenclature. Chapter 1806 comprises 'Chocolate and other food preparations containing cocoa.' The EEC Customs and Commission further break this down into

Sweetened cocoa powder
Unfilled chocolate
Filled tablets and bars
Filled chocolates and chocolate confectionery
Other chocolate products
Sugar confectionery containing cocoa.

Note that products considered in other classifications, and by the industry generally as clearly sugar confectionery, are treated by the Customs as chocolate products even where the amounts of cocoa present are tiny.

At the time of writing (though see below) at official levels the Brussels Nomenclature is the most important of those used throughout the world. It has been in existence for so many years that long series of data, monthly and annual, are available for use in analysing international trade.

For some years now, the international Common Customs Council has been attempting to draw up a classification which, for international trade, will be adopted, it is hoped, by all major trading nations. It will use a product coding allowing ten characteristics for each class, the highest ranking codes being the broadest, and the lower ranks permitting national authorities to adopt subclassifications to meet their own national needs. The United Nations data could well be summarized from the highest-ranking code alone. The result is that drafts of this new classification are very detailed, though not dissimilar to the CCC Nomenclature. Because there is a strong possibility that major nations may begin to adopt the new system in 1988, readers requiring the use of relevant data should check the position with their local trade body or national authority.

18.9 Legal requirements specifically affecting confectionery

Legislation with regard to aspects of food production is of recent origin. In Anglo-Saxon countries non-specific pure-food laws allowed legal action to be instituted where any foodstuff could be alleged to be harmful to the eater. In Continental Europe, legislation from the outset has been more specific, particularly with regard to chocolate. In the last twenty years, bodies representing consumers' interests have lobbied successfully to widen the areas covered by statute, and the influence of the EEC Commission has been exercised in support.

For this reason, one of the most important restraints on the chocolate industry is EEC Directive L228/30. This document sets out with legal force descriptions of cocoa and chocolate products binding on all items sold in the Community, whether or not they are made there. Over one-third of all world output of chocolate is therefore subject to the rules. The text is detailed and specific, but is too lengthy to be quoted in full; examples are reproduced to give the flavour of the document.

A general definition is given for chocolate as the product obtained from cocoa nib, mass, powder and sucrose with or without added cocoa butter, having a minimum dry cocoa solids content of 35%–at least 14% of dry non-fat cocoa solids and 18% of cocoa butter, subject to omitting certain additives from the calculation. Various types of chocolate are separately defined, and that for milk chocolate (section 1.21) is given in full below.

1.21 *Milk chocolate*
the product obtained from cocoa nib, cocoa mass, cocoa powder or fat-reduced cocoa powder and sucrose, from milk or milk solids obtained by evaporation, with or without added cocoa butter, and containing, without prejudice to the definitions of milk chocolate vermicelli, gianduja nut milk chocolate and couverture milk chocolate:
—a minimum total dry cocoa solids content of 25% including at least 2.5% of dry non-fat cocoa solids;
—at least 14% of milk solids obtained by evaporation, including at least 3.5% of butter fat;
—not more than 55% of sucrose;
—at least 25% of fat;
these percentages to be calculated after the weight of the additions provided for in paragraphs 5 to 8 has been deducted.

Paragraphs 5 to 8 refer to flavourings, lecithin and phosphatides, vegetable fats, and to products where there is an outer covering other than chocolate.

The Directive also defines certain classes of finished chocolates, as mentioned in section 18.8 above. It also specifies matters concerning methods of manufacture, e.g. products permitted in the process of alkalization, citric acid, and ash content of treated matter; and sets out permissible methods of treating cocoa butter.

All EEC Directives are liable to be renegotiated from time to time, and up-to-date versions should be consulted before producing for or trading with a Community country. Member countries will have national regulations as well as those of the Community applying both to ingredients and to finished goods, not only to protect consumers but often designed as a barrier to imports.

The EEC has published lists of additives permitted in Community trade and this list has wide influence elsewhere. Regulations, wider than the edible materials specified in the Chocolate Directive, cover colouring matter, solvents, flavours, residual chemicals from agricultural operations, emulsifiers, and the like.

Products like chocolate blocks and bars are sometimes only permitted in certain sizes, and more often only at specified weights.

Modern legislation nearly always requires certain statements on the overwrapping of a confectionery product. The Chocolate Directive is no exception, but more generally a description is required of the products and its contents, the unit price and price per unit of weight, together with some indication of product freshness. The size and style of print may be specified, together with the name and address to which a purchaser can appeal in case of any fault.

Some countries have general legislation which seeks to control the content of advertising messages, particularly the accuracy of claims. In North America, attempts have been made to limit advertisements shown to children on television.

Customs and some excise duty legislation exists in most countries of the world, and can affect confectionery at various levels. Rules of the General Agreement on Tariffs and Trade apply where countries have ratified the necessary Treaties. Again, because of the complexity and frequency of changes in both legislation and product, readers are recommended to seek out the contemporary position before taking any action.

18.10 In conclusion, some trends

Probably the biggest challenge open to the confectionery industry is to begin to close the large gap which exists between consumption levels in advanced and industrializing countries. But this does not mean that opportunities are not available elsewhere.

For confectionery designed to supply appreciable quantities of quickly available energy, the old boundaries of the industry have widened, and will continue to do so. Even now, producers and distributors alike must think in terms of the total range of markets for snack foods, where confectionery competes with crisps, sandwiches, soft drinks, nuts, and the like, the market for between-meals eating. Confectionery may prove to have the advantage in ease of transportion, product size in relation to quantity of energy provided, and superior taste. The opportunities are considerable, since demographic and social forecasts suggest no reduction in the trend away from formal eating.

Another associated factor which may well intensify across the world is the degree of attention paid to 'healthy' eating. From the cereal industry has come the granola-type bar, now usually coated with chocolate. Products with a generalized 'health' appeal, whether described as 'natural', 'no artificial flavours', or 'low-calorie', become increasingly important. An industry traditionally based on sugar finds itself selling sugarless products–in the USA in 1985, one-third of all chewing gum sales were in this category. The health lobby is concerned to widen the scope of food legislation, whether on formulation or labelling.

In advanced countries the general level of consumer incomes will continue to rise, if unequally between countries and over time. This will steepen the trend, now obvious in the USA and Europe, towards the separation from the general confectionery/snack market of a sub-market for higher-priced chocolate and sugar confectionery sold through specialist outlets. The emphasis will be on the shop as a display item in its own right; the products will meet a demand for a special, delicious kind of treat, either for gratification of the purchaser alone, for sharing, or as a gift. The other goods sold will not be those traditionally found alongside confectionery, but will be other acceptable gifts such as well-designed inexpensive jewellery, even individual flowers, all giving the impression of luxury by way of the packaging.

In the parts of the world outside the regions of high incomes, the trend, shown in the statistical sections is towards higher consumption, though not necessarily increasing at the high rates of the 1970s. It is in Africa and Asia that populations are growing most quickly, and in South-East Asia where industrialization is most active. Barring tragedies of drought or war, the pattern of growth in confectionery could be expected to follow the lines of Western Europen after the first Industrial Revolution: consumption would follow the increase in consumer incomes, increasing urbanization, resulting in move retail outlets, and increasing literacy. Local manufacture is already on the move, recent announcements of new plants having recently been made in China and Saudi Arabia.

Because of the lack of good, even of published data, the progress of demand in the Soviet bloc and China is difficult to forecast. Demand has recently followed trends in the West, though by a different product route. Probably the most that can be said is that if current attempts to increase consumer goods supply are successful, confectionery will take its share.

It is possible to be optimistic that confectionery consumption worldwide will not suffer further as the understanding of dental caries increases, and preventive dental methods are more widely used. There remains an unresolved medical argument over the role of sugar in the diet, with no adequate substitute available in quantity. But for these matters, and accepting the forecasters' view that world catastrophes are unpredictable, few other factors seem to point to any downturn in confectionery manufacture as its market expands both in product types and into more countries of the world.

References

1. Nuttall, C., *Manuf. Conf.*, Jan. 1983.
2. Cocoa, Chocolate and Confectionery Alliance, London. *Marketing in the U.K. confectionery Industry*, Proceedings of the General Assembly of ISCMA/IOCC, 1966; *Confectionery in Perspective*, 2nd edn. (1982); *Confectionery and Nutrition* (1985); *An assessment of the evidence concerning confectionery and dental health in the U.K.* (1984); *Frequency of food and drink consumption* in the U.K. (1986).
3. CAOBISCO, *Statistics 1985*, Paris (1986).
4. *ISCMA/IOCC Statistical Bulletin*, Brussels, (December 1986).
5. Williams, I.A., *The Firm of Cadbury 1831–1931*, Constable (1931).
6. David, S.T., *The Statistician*, **9**, (1958).
7. Fox, B.A. and Cameron, A.G. *Food Science: a Chemical Approach*, 4th edn., University of London Press (1982).

19 Future trends

S.T. BECKETT

'According to the old adage, there is nothing new under the sun. There is, however, an infinity of possible variations on any one theme, and it is to the divergent conceptions of the chocolatier, confectioner, food chemist, packaging and mechanical experts that the industry owes its multitude of interesting product'. So wrote C. Trevor Williams (1) in his book on *Chocolate and Confectionery* over twenty years ago. This to a certain extent is still true today, and will continue to be so. However, before trying to predict future developments, it is interesting to read how this past author thought the art of chocolate-making would develop, and then to compare this with current methods.

19. Past predictions

Three areas of development were reviewed by Trevor Williams: new materials, package design, and novel processing (1).

19.1.1 *New materials*

The hydrogenation of fats was thought to open up new fields for the chocolatier with regard to texture and bloom resistance. This has been in fact surpassed by the widespread development of cocoa butter equivalents and substitutes, as partial or total replacers for cocoa butter in compounds and some chocolate markets (see Chapter 12, 13).

Some plastics and whey concentrates were regarded as possible alternative new ingredients. Although the former have yet to be developed, the latter, as whey-or lactose-derived substances are in fact incorporated in many chocolates, particularly in continental Europe.

The public's perception of the food value or harmful effects of certain products was considered a problem for the future of the industry. Over twenty years ago the need for the industry to educate the public was noted. How much more so is this true today! Natural 'healthy' products such as soya, groundnut, sunflower seed oils, pectins and yeasts were considered as possible additions. The present health-food trend is certainly following this prediction. Also included was the possibility for alternative sugars. Dextron in particular was noted. Although this has not found a major application in the confectionery

369

industry, the development of new sweeteners is a major part of present research within the sugar industry (see Chapter 3).

19.1.2 Packaging

A possible future prediction was the use of an edible moisture-proof film sprayed on confectionery to eliminate the necessity of wrapping media. This appears to be totally impractical, however, as packaging is designed to protect the product from dirt and physical damage. This necessitates the packaging material being removed to take with it the dirt. It also needs to be relatively bulky so as to withstand knocks.

The importance of the correct use of colour, packaging design, and of symbolic devices to denote different manufacturing houses, are as important today as they were in the past. The development of new machines means that the range and quality of packaging has changed rapidly. The increase in speed of the machines, and the growth in importance of large super/hypermarket outlets, has led to the rapid growth of packaged countline goods. Trevor Williams noted that in the 1960s 80% of the chocolates and confectionery produced in the USA was packaged and count goods. This trend has since continued in other markets.

19.1.3 Processing

Probably as a result of the influence of Mosimann (2), ultrasonics was considered likely to find a major role within the confectionery industry. It was thought to be able to take part in the emulsification, particle comminution and conching of chocolate. At present this role for ultrasonics appears to be unlikely.

19.2 Present position

This section looks at some of the developments which have taken place over the past twenty years and which were not reviewed above.

19.2.1 Materials

The source and quality of the basic raw material of all chocolate, cocoa, has been changing dramatically over the period. New regions such as Malaysia have expanded, while traditional West African supplies have decreased owing to external factors such as the change to an oil-based economy in Nigeria. Each location provides its individual flavour, and thus the overall flavour of some chocolate is changed as new sources and types of beans are introduced. It is interesting to note that at least four authors have stressed the importance of obtaining high-quality properly fermented and dried beans, also noting that processing is at present unable to overcome any defects. It is, therefore, of great concern to many manufacturers that changes in the source of cocoa should not lead to a deterioration in quality.

19.2.2 Processing

Here many changes have occurred. In the field of roasting, the roasting of whole beans has been replaced by nib or even cocoa mass roasting. The thin-film or batch devices developed to do the latter have also been used to reduce conching times and/or change the flavour of the chocolate. The conches themselves have tended to become bigger, while the use of the long conche has declined dramatically.

Overall there has been a movement amongst the larger manufacturers towards large-volume processing lines, which are operated as far as possible in a continuous manner. The installation of advanced computer control and instrumentation has also resulted in a vast reduction in the workforce required per ton of chocolate produced.

19.3 Possible future trends

Predicting the future is always very risky, as unforeseen circumstances can totally change the course of events. For example, an incurable disease in cocoa could destroy the industry, while the consumption of chocolate might suddenly increase manyfold. One can of course extrapolate present trends in the belief that at least some of them will continue.

It was noted above that large manufactures are installing larger and faster machinery. These firms have captured a substantial proportion of the market and their products are known internationally. At the same time small specialist firms, normally retailing their own goods, appear to be flourishing in many countries. It appears likely that this will continue, with an even greater polarization into the two types of manufacturers. The use of large machinery makes it impossible for the chocolatier to fully develop the potential of each individual type of bean. Thus although there may be a variety of 'house' flavours, it will be left to the smaller manufacturer to exploit the full range of chocolate flavours possible. It may also be that eventually the chocolate assortment box will primarily produced by this type of firm, with bars, countlines, etc. composing the chief market of the major manufacturers.

The increasing difference between the two types of manufacture is also likely to be reflected in the two types of chocolate, namely 'real' chocolate and couverture. Whilst it seems likely that legislation will become more and more strict for the former, the latter is likely to extend in its range of constituents and quality. The range of other fats and techniques for their manufacture will enable better products to be made. At the other end of the market, work on chocolate flavour will probably continue with the discovery of several hundred new contributing compounds. The probability is very low, however, of finding an economical alternative which cannot easily be distinguished from cocoa.

An increase in engineering capability has in the past been reflected by larger machines, e.g. 2.5 m (8.2 ft) roll refiners, 6-ton conches or larger, and more sophisticated machinery such as the cocoa mass treatment machines. This is likely to continue, and to lead to the development of new machinery, especially in the grinding, mass preparation and conching procedures. In addition, it should also be remembered that processes which have failed to operate satisfactorily in the past may have done so because the degree of engineering skill then available did not meet the required standard. New developments in materials and machines may mean that old ideas are worth another consideration. Perhaps, for example, the jetmill/ultrasonic system of Mosimann may be viable in some circumstances.

Research and development workers in almost every industry are frequently dispirited when the novel methods which operated satisfactorily in the laboratory or pilot scale area fail to do so under the more stringent conditions of the production line. Little progress will be made, however, unless the industry is prepared to take the financially great risk of trying very different machines and processes. The introduction of new processing technology is likely to prove of benefit both to the confectionery industry and to the many consumers of chocolate throughout the world.

References

1. Trevor Williams, C. *Chocolate and Confectionery* 3rd edn., Leonard Hill [Blackie, Glasgow and London] (1964).
2. Mosimann, G. 'Physical and chemical reactions in connection with Mosimann's process for automatic production' *Conf.*, Solingen-Grafrath, April 1963.

Glossary

Alkalizing	A treatment used during the making of cocoa powder to give particles better suspension properties when they are used in a drink (commonly known as the Dutch process)
Amorphous	Not having a distinct crystalline form.
Bloom	Fat or sugar on the surface of the sweet giving a white sheen or sometimes individual white blobs.
Cacao	Botanical name referring to the tree, pods and unfermented beans from the pods.
Chocolate liquor	Another name for cocoa mass.
Chocolate mass(e)	May refer to either cocoa mass or partially processed chocolate. In this book masse is used exclusively to mean partially processed chocolate. This was chosen to try to avoid confusion with cocoa mass.
Cocoa	Traditionally the manufactured powder used for drinks or food manufacture. At present often refers to fermented beans in bulk.
Cocoa butter	Fat expelled from the centre (kernels or nib) of cocoa beans.
Cocoa butter equivalent	Vegetable fats which are totally compatible with cocoa butter and can be mixed with it in any proportion.
Cocoa butter replacer	Vegetable fats which may be mixed with cocoa butter but only in a limited proportion.
Cocoa liquor	Another name for cocoa mass.
Cocoa mass(e)	Cocoa nib ground finely to give a liquid above 35 °C (95 °F). Cocoa mass is used in this book for all except origin liquor, as mass is used in the official EEC regulations.
Cocoa nib	Cocoa beans with the shell removed.

Cocoa powder	Cocoa nib with some of the fat removed and ground into a powder.
Conche	A machine in which the chocolate is kept under agitation, so that the flavour is developed and the chocolate becomes liquid.
	Sometimes used for machines which treat cocoa mass to remove volatile components.
Chocolatl	Drink made from crushed cocoa beans developed by the Aztecs.
Countline	An individual unit normally purchased and eaten by the consumer in informal surroundings, e.g. Mars Bar.
Couverture	Legal use, high-fat (i.e. over 31% cocoa butter), normally high-quality chocolate. In this book, these are referred to as high-fat couvetures. Common UK use as biscuit-coating chocolate, often containing other fats.
Crumb	Intermediate material in the milk chocolate making process, composed of dehydrated milk, sugar and cocoa mass.
Dietetic chocolate	'Chocolate' made for people with special dietary requirements, e.g. diabetics.
Enrober	Machine for coating sweet centre with chocolate, by pouring molten chocolate over it.
Fermentation	A process in which cocoa beans are treated such that chemical change is brought about by enzyme action. This usually involves removing the beans from the pods and placing them in covered heaps for an extended period.
Husk	The shell round the nib or kernel.
Lecithin	Class of organic compounds similar to fats but with molecules containing nitrogen and phosphorus. Used in chocolate as a surface-active agent to improve its flow properties.
Lipid	Generic term for oils, fats and waxes.
Micronizer	Device for the radiant heating of cocoa beans so as to loosen the shell.

Non-Newtonian liquid	A liquid whose viscosity varies according to the rate at which it is stirred (sheared).
Origin liquor	Cocoa mass manufactured in the country of origin of the beans.
Outer	Box containing a number of retail units.
Polymorphism	The existence of the same substance in more than two different crystalline forms.
Plastic viscosity	Relates to the amount of energy required to keep a non-Newtonian liquid moving once it has started to move (*see also* yield value).
Refiner	Roll mill, often with five rolls, used to grind solid chocolate ingredients. In some countries it also refers to machines for changing the flavour of cocoa mass. This is not used in this context in this book.
Temperer	A machine for cooling/heating chocolate to form stable fat crystals.
Winnowing	The separation of a light material from a denser one by blowing air over them. In the case of cocoa, the shell is blown away from the cocoa nib and collected separately.
Yield value	Relates to the amount of energy required to start a non-Newtonian liquid moving (see also plastic viscosity).

Useful conversion factors

Square area
1 sq. in = 6.4516 sq. cm
1 sq. cm = 0.155 sq. in
1 sq. ft = 0.0929 sq. m
1 sq. m = 10.763 sq. ft

Volume
1 cu. in = 16.387 cu.cm
1 cu. cm = 0.0610 cu. in
1 cu. ft = 0.0283 cu. m
1 cu. m = 35.314 cu. ft

Density
1 lb/cu. ft = 16.018 kg/cu. m
1 kg/cu. m = 0.0624 lb/cu. ft

Length
1 ft = 0.3048 m
1 m = 3.280 ft
1 micron (μm) = 0.0375×10^{-3} in

Pressure
1 p.s.i. = 0.0703 kg/sq. cm
1 kg/sq. cm = 14.223 p.s.i.
1 p.s.i. = 0.068 atm.
1 atm. = 14.7 p.s.i.
10 dyne/cm^2 = 1 Pa
1 p.s.i. = 0.069 Bar
1 Bar = 14.5 p.s.i.
1 p.s.i. = 6.895 kPa
1 kPa = 0.145 p.s.i.
1 kPa = 0.01 kg/sq. cm
1 kg/sq. cm = 98.1 kPa

Viscosity
1 poise = 100 mPa. s
1 poise = 0.1 Pa. s
1 mPa. s = 0.01 poise
1 centipoise = 1 mPa. s
100 centipoise = 1 poise
10 poise = 1 Pa. s

Heat transfer coefficient (thermal transmittance (U value); rate of heat exchange or conductance (C value))

1 Btu/sq. ft-h-F. = 5.678 W/sq. m-°C.
1 W/sq. m-C. = 0.176 Btu/sq. ft-h-°F
1 Btu/sq. ft-h-F. = 4.882 kcal/sq. m-h-°C
1 kcal/sq. m-h-C = 0.204 Btu/sq. ft-h-°F
1 W/sq. cm-C = 0.238 cal/sq. cm-s-°C
1 cal/sq. cm-s-C = 4.186 W/sq. cm-°C
1 W/sq. cm-C = 8.598 × 10^3 kcal/sq. m-h-°C
1 kcal/sq. m-h-C = 1.163 × 10^{-4} W/sq. cm-°C

Heat flow rate
1 Btu. h = 0.293 W
1 W = 3.1412 Btu. h
1 Btu. h = 0.2519 kcal. h
1 kcal. h = 3.9683 Btu. h

Energy per unit mass
1 Btu. 1b = 2.326 J/g
1 J/g = 0.4299 Btu. 1b
1 Btu. 1b = 0.5555 kcal/kg
1 kcal/kg = 1.800 Btu/1b

Density of heat flow rate
1 Btu/sq. ft-h = 3.1545 W/sq. m
1 W/sq. m = 0.3169 Btu/sq. ft-h
1 Btu/sq. ft-h = 2.7124 kcal/sq. m-h
1 kcal/sq. m-h = 0.3686 Btu/sq. ft-h

Thermal conductivity (K values)
1 Btu-in/sq. ft-h-fah = 0.1442 W/m-cent
1 W/m-cent = 6.9334 Btu-in/sq. ft-h-fah

Energy (work, heat)
1 Btu = 1.055 kJ
1 kJ = 0.947 Btu

Mass
1 oz. = 28.354 gram (g)
1 gm = 0.035 oz

Force
1 oz. f = 0.278 newtons (N)
1 N = 3.6 oz. f
1 1b. f = 4.448 N
1 N = 0.225 1b. f

$1 N = 0.1 kgf$
$1 kgf = 9.81 N$

Energy (work)
$1 kWh = 3.60 \times 10^6 Joule (J)$
$1 J = 2.777 \times 10^{-7} kWh$

Specific heat
$1 cal/g °C = 4186.8 J/kg K$
Spec. ht of water at $4 °C = 4.2045 kJ/kg °C$
$1 Btu/lb °F = 4.1868 \times 10^3 J/kg °C$
$1 J/kg °C = 0.238 \times 10^{-3} Btu/lb °F$

Index